Springer Tracts in Advanced Robotics

Volume 145

The Springer Tracts in Advanced Robotics (STAR) publish new developments and advances in the fields of robotics research, rapidly and informally but with a high quality. The intent is to cover all the technical contents, applications, and multidisciplinary aspects of robotics, embedded in the fields of Mechanical Engineering, Computer Science, Electrical Engineering, Mechatronics, Control, and Life Sciences, as well as the methodologies behind them. Within the scope of the series are monographs, lecture notes, selected contributions from specialized conferences and workshops, as well as selected PhD theses.

Special offer: For all clients with a print standing order we offer free access to the electronic volumes of the Series published in the current year.

Indexed by SCOPUS, DBLP, EI Compendex, zbMATH, SCImago.

All books published in the series are submitted for consideration in Web of Science.

More information about this series at https://link.springer.com/bookseries/5208

Giuseppe Averta

Human-Aware Robotics: Modeling Human Motor Skills for the Design, Planning and Control of a New Generation of Robotic Devices

Springer

Giuseppe Averta
Research Center E. Piaggio
University of Pisa
Pisa, Italy

ISSN 1610-7438 ISSN 1610-742X (electronic)
Springer Tracts in Advanced Robotics
ISBN 978-3-030-92520-8 ISBN 978-3-030-92521-5 (eBook)
https://doi.org/10.1007/978-3-030-92521-5

This Springer imprint is published by the registered company Springer Nature Switzerland AG
The registered company address is: Gewerbestrasse 11, 6330 Cham, Switzerland

To my sweet and strong Manuela, light of my life,

To my family, helm of my journey,

To the surrounding world, for being amazing.

Foreword

At the dawn of the century's third decade, robotics is reaching an elevated level of maturity and continues to benefit from the advances and innovations in its enabling technologies. These all are contributing to an unprecedented effort to bringing robots to human environment in hospitals and homes, factories, and schools; in the field for robots fighting fires, making goods and products, picking fruits, and watering the farmland, saving time and lives. Robots today hold the promise for making a considerable impact in a wide range of real-world applications from industrial manufacturing to healthcare, transportation, and exploration of the deep space and sea. Tomorrow, robots will become pervasive and touch upon many aspects of modern life.

The *Springer Tracts in Advanced Robotics (STAR)* is devoted to bringing to the research community the latest advances in the robotics field on the basis of their significance and quality. Through a wide and timely dissemination of critical research developments in robotics, our objective with this series is to promote more exchanges and collaborations among the researchers in the community and contribute to further advancements in this rapidly growing field.

The monograph by Giuseppe Averta is based on the author's doctoral thesis. The contents are focused on human-aware robotics and are organized into three parts. The complexity of human motion generation is discussed in the first part. Then, the modeling tools to describe human motor skills are used for the design of nature-inspired prostheses and the assessment of motion impairment in the second part. The third part is devoted to presenting planning and control techniques for cobots and autonomous robots.

Rich of examples developed by means of extensive experimentation on humans and robots, this volume was the co-winner of the 2021 Georges Giralt Ph.D. Award for the best doctoral thesis in Europe. A very fine addition to the STAR series!

Naples, Italy
September 2021

Bruno Siciliano
STAR Editor

Acknowledgements

The work summarized in this book is the result of almost four years of research conducted between 2016 and 2020 at the Research Center E. Piaggio and the Department of Information Engineering of the University of Pisa, and at the Soft Robotics for Human Cooperation and Rehabilitation Lab of the Italian Institute of Technology in Genova.

I would like to acknowledge, in the first place, the amazing family that lives across the two labs, that motivated, supported, and enabled this achievement.

The first and most important supporters are Prof. Antonio Bicchi, and Prof. Matteo Bianchi, to whom I am deeply grateful. The timely suggestions, the constructive criticisms, the text re-written, the ideas, the coffees, and, most importantly, the honest and friendly relationship, created a perfect humus for my research. I would like also to express my utmost gratitude to Prof. Neville Hogan, who welcomed me in his academic family when I was a fresh and motivated visiting PhD student at MIT, for teaching me a different way to do research. I hope to honor your suggestions in my career.

Thanks also to all the friends that worked with me, coauthored my manuscripts, and shared part of the journey. To Cosimo and Edoardo, for introducing me to this adventure. To the Federicas^2, Vincenzo, Franco, the Simones^2, Gemma, and many more, who represented an amazing source of inspiration, fruitful discussion, and deserved relax.

Last but not least, I would like to acknowledge my family. First, my wife Manuela, love of my life, that has been the first believer of my abilities, way more than myself. To Carmela and Luigi, for their outstanding support, and for their love, in good and bad times. To Carolina and Fiorella, for their being family, and most importantly, for their photos when I get home. And, of course, to all the others that I did not mention here, I am proud of being part of this.

Finally, let me thank the European Community and the European Research Council for their the generous funding of my work, provided by the European projects: SoMa, SoftPro, Sophia, Darko, Natural BionicS.

Pisa, Italy
October 2021

Giuseppe Averta

Contents

Chapter 1
Introduction

How do we walk? How do we manipulate objects? How do we talk, sing, play musical instruments and draw a portrait? What does it happens when—after an accident or a disease—our motor skills are modified? And how can we design and control machines able to perform tasks that most people can do effortlessly?

Understanding the capabilities of humans and other living beings to control their own body represent one of the major cross-disciplinary research topics still open in literature. This is not surprising given that movements represent *The* enabling factor for almost every action of our life. Movement—in a broad sense—is probably the most evident component that differentiate between inanimate and animate bodies, and has been central topic in scientific and philosophical discussions across the millennia.

1.1 How Do We Move? A Very Brief Historical Overview

The study of body movement is typically named *Kinesiology*, from the combination of the Greek words χίνησις (movement) and -λόγος (logos, discourse) and combine the anatomy, i.e. the study of the *structure* of the body, together with the physiology, the science of the *function* of the body.

The Athenian philosopher Plato (428–348 B.C.) was convinced that the *self-motion*, referring to the movements which are not expected by inanimate objects, was the result of the immortal soul, apparently associated to all the animals able to produce voluntary movements. His disciple, Aristotle (384–322 B.C.) is typically considered the very first ideologue of kinesiology. He described for the first time the geometry and the actions of muscles in his texts *Parts of Animals*, *Movements of Animals* and *Progression of Animals*, with specific focus on walking. Under the Roman domination, the Greek physician Galen of Pergamon (131–201 A.D.) in his

G. Averta, *Human-Aware Robotics: Modeling Human Motor Skills for the Design, Planning and Control of a New Generation of Robotic Devices*, Springer Tracts in Advanced Robotics 145, https://doi.org/10.1007/978-3-030-92521-5_1

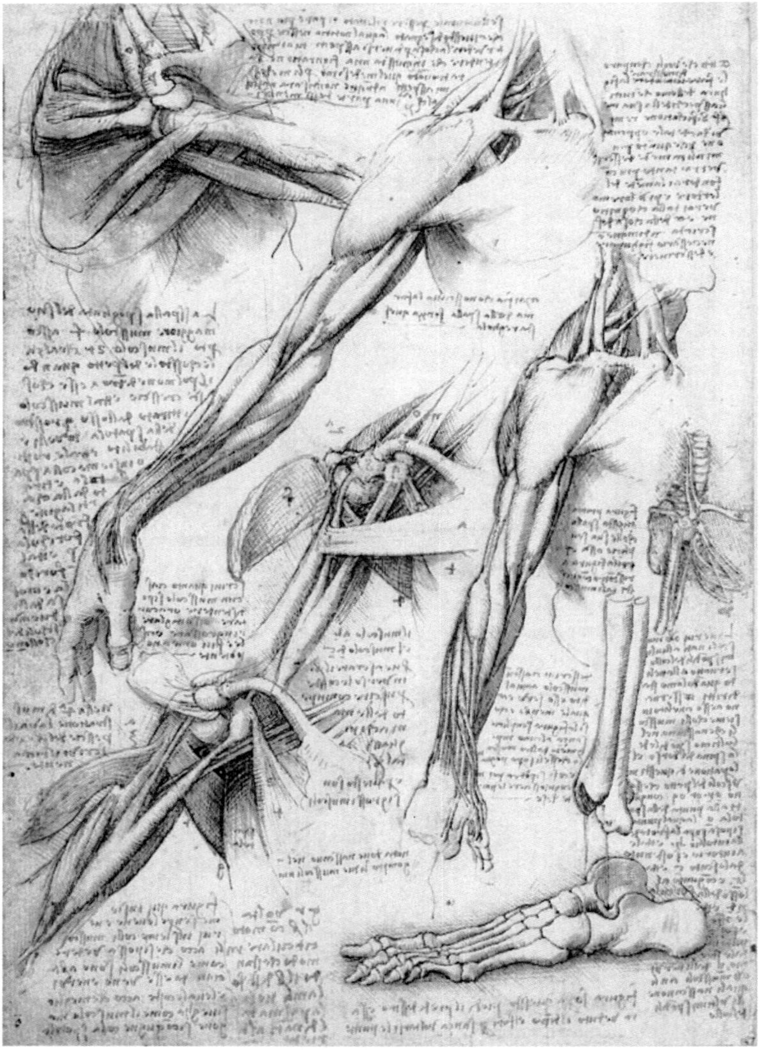

Fig. 1.1 The nature of human body movements in the studies of Leonardo Da Vinci

essay *De Motu Musculorum* described several muscles characteristics, e.g. agonist and antagonist actions, muscular tonus, motor and sensory nerves etc. Interestingly, he also proposed that the contraction of muscles resulted from the passage of *animal spirits* from the brain via the nervous system.

Kinesiology laid almost untouched for several centuries, until the Italian polymath Leonardo da Vinci (1452–1519) advanced its knowledge through his numerous studies on the structure of the human body, describing the mechanics of the body during standing, walking up and downhill, rising from a sitting position and jumping. He also explained the movement as action/interaction of multiple muscles connected to the skeleton through cords (i.e. tendons) (Fig. 1.1).

In 1665 the Italian Francesco Maria Grimaldi published (posthumous) the book *Physicomathesis de lumine, coloribus, et iride, aliisque annexis* [49], in which, for the first time, he reported on hearing sounds made by contracting muscles. The technology to study these sounds were not available yet, and only in the late '900 the work of Gerald Oster identified their origin in the vibration of single muscle fibers [88].

In *An Essay on the vital and other involuntary motions of animals*, (1751) Robert Whytt (1714–1766) discussed that movements are generated by an unconscious sentient principle [132], also named soul, which, even if anatomically inaccurate, points toward the modern definition of stretch muscular reflexes [83]. Excited by the possibility of investigating the muscular response to electrical stimulation, Guillaume Benjamin Amand Duchenne (1806–1875) started to classify the functions of individual muscles in relation to body movements, and he verified that isolated muscular action does not exist in nature [32, 33].

Around 1740, physiologists started to analyze the phenomena generated by the electrical stimulation of muscles. Among the others, it is worth mentioning Luigi Galvani (1737–1798), who in 1786 studied the effects of atmospheric electricity on dissected frog muscles. He observed that muscles contracted when touched by a scalpel, and drew the conclusion that there was *dwelling electricity which proceeded*

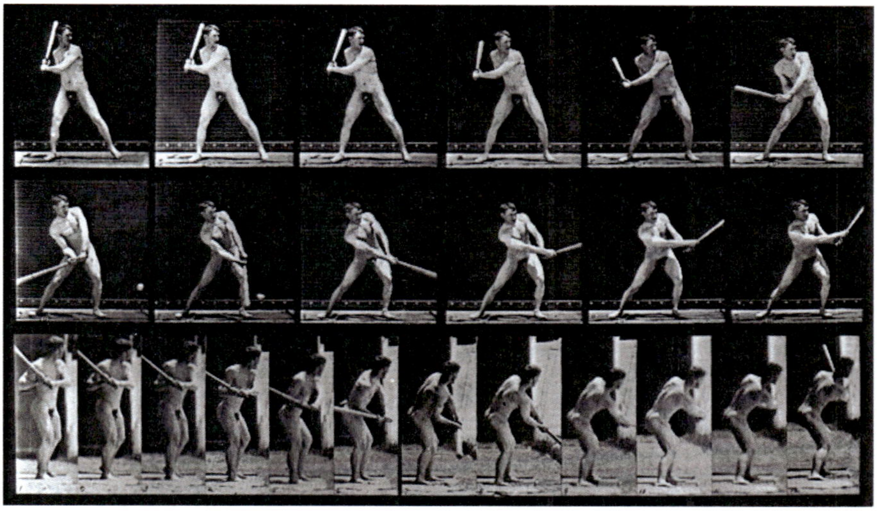

Fig. 1.2 Eadweard Muybridge, american photographer, pioneered the use of photography for the study of human and animal movement. In 1887, their works were published in 781 plates containing 20,000 pictures, in a groundbreaking collection named *Animal Locomotion: an Electro-Photographic Investigation of Connective Phases of Animal Movements*. In this figure, Plate 275, Baseball Batting

along the nerve. In his essay *De Viribus Electricitatis in Motu Musculari Commentarius* [46] he proposed that animal tissue contained an innate, vital force—that he named "animal electricity"—which activated nerve and muscle when bridged by metal conductors (Fig. 1.2).

Finally, it is also worth recalling the work of Adolf Eugen Fick (1829–1901), who provided important contributions to our knowledge of the mechanics of muscles and their movement, e.g. through the introduction of terms such as *isometric* and *isotonic*. Several other major findings were achieved during the years, however these will be not reviewed in this Introduction, leaving space for a discussion on modern theories on motor control in the next sections.

1.2 Human Motor Control

Starting from the Greek polymaths, one of the central questions on the generation of movements was (and still is) whether they are the result of an active process (by soul, will, intention) or rather is a reactive effect. In modern kinesiology[1] several methodological and analytical results, partly reviewed in this section, supported the first or the second school of thought.

1.2.1 Is Motion a Reflex or an Active Process?

In neuroscience, the study of *conditioned reflex* is one of the major topics, with—to date—over than 24000 corresponding results on PubMed. These concepts likely derive from the studies on animal behavior of the russian physiologist Ivan Pavlov (1849–1936). Such a point-of-view can essentially be summarized as the generation of movements through the combination of two main types of reflexes, i.e. *inborn* and *conditioned* [91, 92]. This idea was mainly based on his findings concerning the salivation of dogs, and the possibility to elicit such autonomic reflex—after a specific training—in response to new, conditional, stimuli. Indeed, in a famous experiment Pavlov used the sound of a bell to introduce the daily feeding to the dogs and, after a series of repetitions, the animals started to salivate when they heard the sound even if food was not provided.

Thanks to Pavlov's followers, such idea of conditioned and inborn reflexes had a great impact also for the description of animal motor behavior.

Roughly in the same period, the English neurophysiologist Sir Charles Sherrington showed that muscular contractions are followed by relaxation, and that different reflexes represent a complex network in which the brain and the spinal nerves are deputed to process incoming nerve impulses and produce response—in terms of nerve impulses—toward muscles and organs. Importantly, in this network, reflexes cannot

[1] With this term, I will refer, hereinafter, to the research conducted from the beginning of '900.

be considered in isolation but in the context of neural regulations in which these reflexes are integrated. In fact, in 1906 Sir Charles Sherrington published the book *"The Integrative Action of the Nervous System"*, changing the subsequent course of neurophysiology. In this fundamental textbook, he defined the reflex arc as a whole chain of structures—receptor, conductor and effector—, stating that the unit reaction in nervous integration is the reflex. He also qualified the notion of a simple reflex as *"probably a purely abstract conception, because all parts of the nervous system are connected together"* and, finally, proposed that *"the main secret of nervous co-ordination lies evidently in the compounding of reflexes"* [17]. However, he did not claim that movements are generated as a mosaic of reflexes but these were rather produced by modulation of parameters of reflexes [113].

The idea of movement generated only passively through reflexes was strongly opposed by Graham Brown (1882–1965) and Nikolai Bernstein (1896–1966), who viewed animals as active rather than reactive systems, and stated that no reflex combination could explain normal animal behaviour [13], see also [68, 78] for reviews. Indeed, as Bernstein properly pointed out, *"to obtain simple reflexes, it is not necessary to break an integral movement into separate pieces, but it is necessary to put all the possible intact movements for the man in a row, in order of their increasing complexity; and then, at that end of the line, where we will report the least complex, we will also find all the reflexes* [13]. For these reasons, he argued in favor of the so-called *engrams*, i.e. motor programs stored in memory and recalled when a coordinated movement was required. Graham Brown, instead, discussed about locomotion in animals without reflexes and started a line of research which will lead to the definition of central pattern generators.

Note that the view of Bernstein does not exclude the existence and the importance of reflexes for the control of movement, but rather, in Bernstein' vision, the degrees of freedom abundancy is managed through the principle of sensory corrections, which take place according to the basic structural formula of the reflex ring. Therefore, coordination is not some special accuracy or subtlety of effector nerve impulses, but rather a particular group of physiological mechanisms, which create a continuous cyclic interaction between the reception and efferent process [13].

Indeed, a generally accepted idea from the classical work of the Soviet neurophysiologist is that the Central Nervous System is unable, by construction, to predict the exact pattern of mechanical movement, due to the variable external loads and states of the neurons involved in movement generation [11]. In a complex kinematic chain, the forces that determine the actual movement in each link may be represented in the form of a geometric sum of three components: external forces, reactive forces and internal (muscle) forces. External forces (i.e., gravity, resistance of the external environment, etc.), arise by definition outside the body. Reactive forces, instead, appear by definition within the kinematic chain itself during its movement, since each force generated by a specific link will necessarily generate a system of reactive forces that are transferred to the other links. External and reactive forces combine to generate movement, together with internal (muscular) forces, resulting in a coherent dynamic picture, complex and unpredictable. Among these, only the first category of forces, although only to a certain extent, is under the control of the organism. Indeed, the internal (muscle) forces are not totally controlled by the Central Nervous System, since the muscular efforts (tensions) developed are functions of two variables—the central commands and the current muscular length—and the latter variable is not completely dependent from the Central Nervous System. This results in a not univocal correspondence between nervous excitement, developed muscular tension and resulting movement. The way for addressing and overcoming these issues is directly suggested by the dual conditioning of muscular tensions. Given a specific central command, muscles tension depends on current muscle length and, hence, the actual force is controlled via sensory corrective signals [13]. As a consequence, the final peripheral movement patterns could not be encoded by neural signals only, but rather emerge with shared participation of signals from peripheral receptors and the direct mechanical effects introduced by the environment [66].

It is also worth mentioning that Bernstein did not intend engrams as direct patterns of motor torques, but rather as time functions of abstract variables encoding salient features of movements (such as their topology). This is in slight contrast with the theory of Generalized Motor Programs—proposed as a derivation of Bernstein's Motor Programs [106]—where patterns of mechanical variables (i.e. muscle forces or joint torques) were assumed encoded in the CNS.

All these theories find a natural evolution in the equilibrium-point hypothesis, proposed by Anatol Feldman [38, 39]. The core idea takes inspiration from (i) the definition of modulated reflexes parameters proposed by Sherrington and (ii) the engrams of Bernstein intended as patterns of control variables related to changes in the reference body configuration and stored in memory as the basis for voluntary movements.

In a nutshell, the equilibrium-point hypothesis proposes that motion is generated by the nervous system[2] through a transition of consequent *Equilibrium points* along a specific trajectory (Fig. 1.3). The equilibrium is intended between internal (opposing muscles) and external forces. Control signals provided by the Central Nervous System are used to modulate the threshold length of each muscle, i.e. the length at which each muscle begins to be activated. The role of the high-level control is then to define a reference configuration of the whole body—in terms of muscles thresholds, also called λ values—to which the muscle-skeletal system is attracted as per connection via elastic components. Such control is counterbalanced by internal (i.e. the interaction effect between muscles) and external forces, due to the effect of the external environment.[3]

[2] The Nervous System is intended in his central and peripheral components.

[3] Note the cross-disciplinary similarities with the Impedance Control [59].

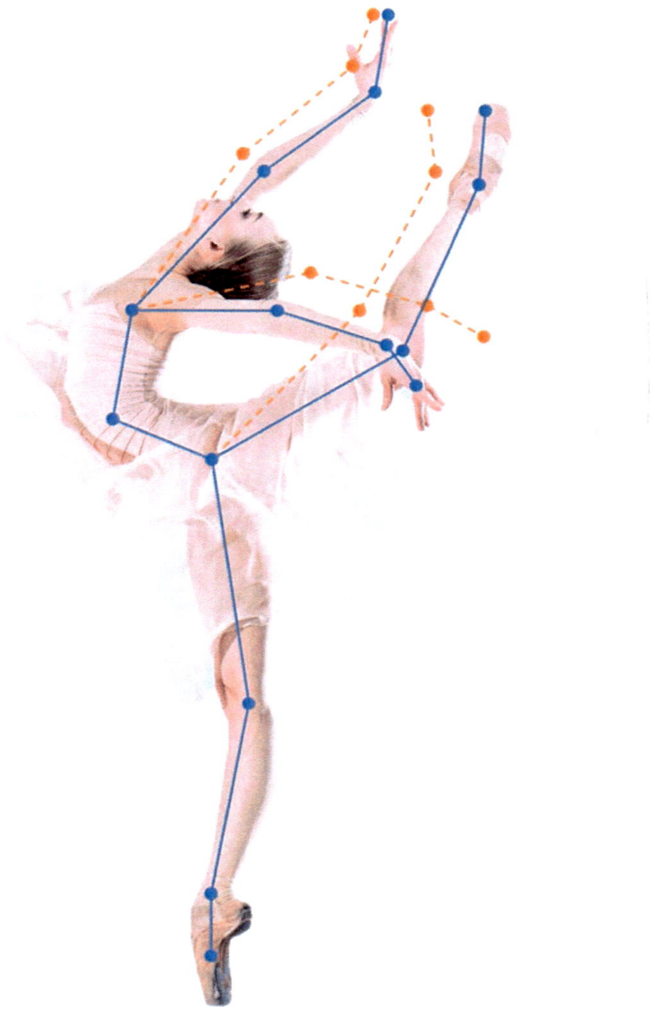

Fig. 1.3 A dancer. According to the equilibrium-point hypothesis, the Nervous System specifies a set of λ-parameters to generate forces in equilibrium with the environment. This results in a reference configuration (the orange one) to which the body (the blue one) is attracted, reaching an equilibrium with external forces

1.3 The Problem of Dimensionality Reduction

So far in this Introduction, it seems clear that the problem of understanding the solutions that Nature implemented to manage the enormous complexity of its own body and motion is far to be solved. Nevertheless, it is manifest that the main conceptual question regards the number of *controllable* and *non-controllable* Degrees of Freedom. In other terms, there are—in general—multiple ways for an animal body to perform a specific task, and this represent the cause and the solution to the marvelous variability of movements. Bernstein, the first theorist of the *Degrees of Freedom Problem*, affirmed that:

> It is clear that the basic difficulties for co-ordination consist precisely in the extreme abundance of degrees of freedom, with which the [nervous] centre is not at first in a position to deal.

It is nowadays broadly accepted that the nervous system does not control directly each available degree of freedom, but rather multiplies a very reduced number of control signals through a series of hierarchical redundant structures, with the ultimate goal to produce coordinated actions [66, 85, 102]. A very common approach in literature to explain such behavior is through the concept of synergies. The word *synergy* comes from the Greek συνεργία (synergia), meaning *working together*, and indicates a series of different mechanisms in human motor control that produce commonalities in structures recruitment, either at the kinematic, muscular and neural level.

Latash and colleagues [67] defined synergies as a neural organization resulting in task specific covariation of elemental variables, with the ultimate goal of generating stable and effective performance variables. This theory, also known as Uncontrolled Manifold [67, 107], suggests that a distinction can be made between the so-called *bad variance*, namely, variability of the elemental variables that would affect the precision in the task, and *good variance*, that is, variability of elemental variables that would not affect performance. If the variability of the elemental variables implements a good variance, then the system of elemental variables must not be controlled by hierarchically higher controller, leaving the unit elements not individually controlled but rather free to vary (Uncontrolled Manifold). The three main mechanisms identified in [67] for implementing the synergistic control are: (i) the *elimination* of specific kinematic Degrees of Freedom (e.g. using muscles spanning several kinematic joints [124]); (ii) the optimization of specific cost functions (e.g. end effector jerk) to get coordinated movements [42, 75]; (iii) the functional linkage between unit elements [28, 101, 103, 104, 116, 123]. The implication of the latter point at the level of muscle activation led to the introduction of the concept of m-modes, i.e. groups of muscles controlled together [63], which act as single functional units.

The study of optimality principles in sensory-motor control represents a central discussion topic in literature and, as discussed in [118], *"Their appeal lies in the ability to transform a parsimonious performance criterion into elaborate predictions regarding the behavior of a given system."* The problem has been addressed at multiple level of analysis, such as movement, exchanged forces, joint torques, muscles activation etc. [1, 22, 25, 29, 37, 42, 50, 52–58, 60, 61, 64, 69, 74, 82, 84, 89, 90, 95, 99, 110–112, 117, 119, 120, 125, 128]. Each of these contributions would deserve a detailed discussion, which however is omitted here to preserve the readability of this Introduction. Nonetheless, since in the following of this book some of the concepts firstly proposed in these papers have been used, it is significant to briefly mention here [58], where the author propose a novel mathematical model to approximate observed voluntary movements in primates which lays its foundations on a basis criterion of "natural" movements as trajectories that minimize the jerk (third derivative of the position). Under a theoretical point of view, these methods usually rely on the fact that the "final version" of the sensory-motor system is a product of the evolution, hence a result of a long optimization process that continues to improve behavioral performance. Practically, it is (almost) always possible to define a cost function whose minimization leads to the prediction of an observed behavior (or at least to its average representation).

Regarding the description of functional linkages between units, a huge effort has been carried out at the muscular level. Indeed, the concept of synergy-based dimensionality reduction has produced numerous relevant results, leading to the identification of muscle synergies as the building blocks for natural motion generation [28]. The general approach is to record electromyographic (EMG) signals from a large set of muscles during the execution of specific tasks. Matrix factorization techniques are then used to identify patterns of muscles co-activations, associated to independent synergies, which are then related to relevant action variables [121]. In [28], the authors affirmed that these components are actively enrolled, following task-specific strategies, although some muscle activation patterns are preserved across tasks. Notwithstanding the very reduced number of synergies observed in these tasks, a high variability of the muscle activity profile can be generated by combining muscle synergies, modulating their amplitude and temporal structure [2]. The existence of muscle synergies in humans was confirmed by several other studies, e.g. during static hand postures [19, 131], for active force generation with one or more fingers [76, 96, 127], and at the upper limb level during fast reaching movements [27] and at the lower limb level for walking [62]. For an overview on these topics the interested reader may refer to [26].

A strong synergistic characterization has been observed also at the kinematic level, with particular focus for the hand movements [4, 31, 77, 103, 104, 114, 129]. These studies agreed on the description of movements through a low-dimensional ordered basis of covariation patterns between joints. More specifically, given a dataset of human movements during specific activities (e.g. reach to grasp), it is possible to demonstrate—via dimensionality reduction techniques—that most of the variability in human postures can be explained through a combination of principal components (or covariation patterns of joints). What is noticeable is that the number of components required to approximate the dataset with a suitable precision ($> 95\%$ of the total dataset variance) is typically much lower than the number of Degrees of Freedom. However, these approaches typically focused on the statics of the phenomenon and did not considered their temporal evolution. To bridge this gap and extend the synergistic description also to the temporal domain, in this book (Part I) the problem has been tackled considering the functional variation of movements at joint level (Chap. 2) and the variability of joint covariation patterns during the task execution (Chap. 3). Additional details are reported in the following of this Introduction and in Part I.

1.4 The Implication for Robotics and the Advancement of Technologies

The idea of simplifying motion complexity is extremely interesting not only for understanding human/animal motor control, but also for its implication in the development of technologies. Indeed, in the last fifty years, robotics highly benefited from the findings discussed in the previous section, and proposed nature-inspired design, planning and control for a new generation of mechatronic devices. One of the first applications of dimensionality reduction in robotics was presented in [23], as an approach to reduce the dimensionality of the search space for a grasp planning algorithm, thus resulting in a strong improvement of algorithm performances (in terms of time required) [24] and in the reduction of the control variables [40]. The observation of human movement was also used to map human motion to anthropomorphic robot motion via non-linear, constrained optimization methods (see e.g. [70–73]). An informed coupling between joint movements was also profitably exploited for the design of under-actuated hands. The first notable mechanical implementation of postural grasp synergies is discussed in [16], where two eigen-postures were physically encoded via pulleys within a 17-DOFs 5-fingered robotic hand. However, it has been argued that such purely geometric approach does not necessarily account for the final shape of the hand while grasping a real object. For this reason, a dynamic extension of the synergistic model was proposed introducing the mechanical compliance of the hand, also leveraging on the Equilibrium Point Hypothesis [38], with the so called *soft synergy paradigm*. More specifically, as discussed in [15], the physical hand is intended to be attracted by an elastic field towards a reference posture configuration, identified by the geometric synergy coordinates. At the same time, the forces arising from contact with the external environment, mediated by the impedance of the hand–object mechanical system, repels the physical hand from the reference hand, generating an equilibrium between the reference attraction and the external forces. As a consequence, the soft-synergy model also results in a synergistic shaping of contact forces with objects, as discussed in [44].

The paradigm of soft-synergies has inspired a series of studies on the problem of robotic grasp analysis with soft synergistic hands, i.e. robotic under-actuated hands that move along the joint covariation patters commonly associated to postural synergies and include compliant elements to enable the grasp adaptation w.r.t. external objects [20, 103]. For this reason, in [98] the authors proposed a quasi-static model of grasping with soft-synergies and discussed contact force and object motion controllability properties, while in [97] they studied the manipulability of synergistic hands via the concept of manipulability ellipsoids, with the ultimate goal of identifying the directions that maximize the manipulation capability and efficiency. Finally, a general framework for grasp analysis of under-actuated compliant hands was then introduced in [45].

It is worth considering that the mechanical implementations of the soft-synergies paradigm is not trivial. This problem was tackled through the translation of soft synergies into a series of corresponding *adaptive* synergies, i.e. a physical implementation of joint covariation patters through mechanical coupling and joint stiffness. This resulted in the Pisa/IIT SoftHand [20, 21], an anthropomorphic robotic hand with 19 Degrees of Freedom and one degree of actuation. The design ensures that the hand free movement follows the joint covariation pattern correspondent to the first grasp synergy presented in [103]. Fingers are actuated via a single tendon and endowed with elastic elements. The particular design is a good approximation of the theorized soft-synergistic actuation strategy, with a single commanded posture that can be regarded as an equilibrium configuration to which the whole hand is attracted [38].

The embedded simplexity[4] of under-actuated devices represent one of the key enabling factors for the development of new prosthetic devices. Indeed, the clinical and industrial state-of-the-art in controlling active prostheses is based on muscle electrical signals that generate control inputs for active prostheses by processing EMG signals. The identification of more than one independent signals is not a trivial problem and machine learning methods need to be used to increase the number of available commands [36, 51, 80, 86], even if so far results are still unsatisfactory for the upper limb [87, 130] and completely absent for the lower limb. For this reason, having a very reduced number of degrees of actuation and, hence, control inputs is mandatory to design usable devices for prosthetic and rehabilitation applications.

1.5 Open Questions and Proposed Solutions

As discussed in the previous sections of this Introduction, it is nowadays broadly accepted that a synergistic coordination is detectable in human free movements at different levels—although it is still debated whether such behavior is completely driven by brain control or rather is a consequence of the bio-mechanics of the body [12, 65, 66, 118, 122]—and robotics highly benefited from neuroscientific studies for the design of bio-inspired robotic devices [102]. In the field, however, several problems are still open. More specifically, there is a strong need for the development of novel, more simple, robotic devices to be used as a substitution of body parts (i.e. robotic prostheses)—targeting not only the hand but also the rest of the upper limb. This is fundamental to reduce the cost and increase robustness and usability of these tools [94].

Besides the limbs substitution, also the effective evaluation of motion impairment for disabled people (e.g. after a stroke or muscular spasticity) represent a key problem which still needs a suitable solution. Indeed, there exist multiple different standard clinical upper limb assessments in stroke research, among which the Fugl–Meyer Assessment of the Upper Extremity [43] is widely used to describe voluntary

---

[4] The term simplexity stands for a complementary relationship between simplicity and complexity. Nature's control of animal body, and by extension the synergistic control of mechatronic devices, can be intended—at least in principle—as a very simple implementation of a very complex solution.

movement control [105]. Despite the satisfactory measurement properties [48, 109], however, most of these clinical assessments exhibit floor and ceiling effects [115] and heavily rely on subjective observer scoring on broad ordinal scales.

Finally, moving toward autonomous robotic applications and co-botics [93], there are several works in literature that have highlighted the importance of human-likeness to ensure a safe and effective Human-Robot Interaction [10, 41]. This aspect is particularly relevant for the control of devices that interact with users for rehabilitation purposes. Indeed, human-inspired movement profiles may be preferred for the active control of exoskeletons, while having a robot that moves embedding the same underlying characteristics of the user was proven to increase the predictability and acceptability of the device [10, 100]. This plays a key role for the usability and the effectiveness of assistive and/or companion robots/cobots [34, 35, 81]. Several approaches have been presented to solve this problem, typically relying on the optimization of arbitrary cost functions or using machine learning approaches. However, optimal control-based methods usually come with hypotheses on motion generation that may limit the variability of the planned movement and, sometimes also lacks of experimental support [79], while, on the other hand, learning methods typically requires a large dataset whose dimensionality dramatically increases with task complexity. A direct exploitation of human observations for robotic motion generation has not been applied yet. This approach would come with several advantages, since human-likeness would be intrinsically guaranteed, without underlying hypothesis. In addition, it could be also considered to include Deep Learning -based system as an high-level intelligence in a more general framework of distributed intelligence [30].

I do believe that all these questions are characterized by one underlying common pattern, related to the strong relationship with the human counterpart. For this reason, I propose in this book to face these problems starting from human observations. To do this, it is mandatory to start from the systematical modeling of human behavior to cover some aspects of motor control that have been not addressed in literature so far.

In particular, one fascinating problem still open is related to the modifications of human behavior when our limbs interact with the external environment, and specifically on how this affects the synergistic organization of motion execution. Focusing on hands, I have studied hand synergies during the so called Environmental Constraint Exploitation, i.e. actions which aim to manipulate or grasp an object, and involve the use of a constraint as central element of the strategy. To bridge this gap, part of the work discussed in this book focuses on the description of synergistic hand motion during the grasp of objects difficult to be grasped, i.e. resulting in the implementation of strategies that heavily rely on the interaction with the environment.

It is also worth recalling that, notwithstanding the fundamental role played by the whole upper limb to move and orientate the hand, low attention has been devoted to the arm motion description. As demonstrated for hands, potential outcomes of these analysis may represent a significant impact not only for the design of new robotic devices, but also for the development of new prosthesis and to inspire the planning and control of anthropomorphic manipulators. For this reason, a significant contribution of this book is dedicated to the description of human motor control with specific focus on upper limb kinematics during Activities of Daily Living. Given the extraordinary complexity and time-dependence of arm movements, one fundamental aspect of the analysis proposed in this book is related to the temporal domain. For this reason, approaches of functional analysis are used to describe the variability of single and multi joint movements along the task execution. Subsequently, also the problem of dimensionality reduction has been addressed while preserving the time-related information, with a new technique I developed, named Repeated Principal Component Analysis.

These analyses have opened fascinating perspectives for the development of technologies. Indeed, part of this book shows how the functional description of human arm movement could be used as a ground truth to quantify—in an objective fashion—the level of impairment of arm motion following a stroke. At the same time, the low dimensional synergistic description of upper limb movements could be used to inspire the design of under-actuated devices. In this book, such an idea has been exploited for the design of a soft robotic wrist which has three Degrees of Freedom and one single motor, thus enabling and favoring translations toward prosthetic applications. Finally, the problem of Human-Likeliness of motion generation for autonomous robots and cobots has been addressed leveraging on the functional description of arm movements. More specifically, I developed an algorithm with the final goal to generate Human-Like joint patterns, which are then mapped on a generic anthropomorphic manipulator through a control-based mapping strategy. To account also for movement dynamics, I also considered the problem of planning motion impedance during the interaction with the environment.

An overview on how these topics are addressed along this book is reported in the following section.

1.6 Contents of the Monograph

Figure 1.4 graphically summarizes the general methodological approach behind this work. This book is organized in three Parts. The first one concerns the mathematical modeling human natural movements, with specific focus on the hand and upper limb. In Chap. 2 I propose a functional characterization of human upper limb movements during activities of daily living [7]. Chapter 3 deals with the same problem under a perspective of kinematic dimensionality reduction [9]. Finally, Chap. 4 focuses on the hierarchical, synergistic, control of the hand, with the twofold perspective of kinematic and kinetic domain [6, 31].

Lessons learned in Sect. 1.1 are exploited in the following of the book. More specifically, Part II deals with the application for prosthetics and rehabilitation. Chapter 5 propose the design of an under-actuated soft robotic wrist, whose design exploit the synergistic description proposed in the first part of this book [18]. In Chap. 6 I propose a new approach—based on the functional characterization of arm movements—to quantify the level of impairment following a disabling pathology

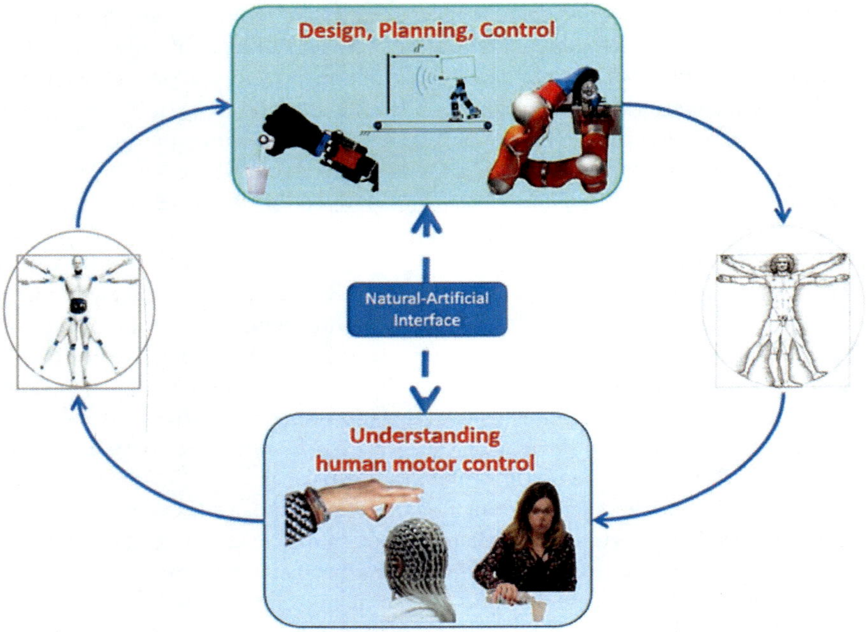

Fig. 1.4 General approach followed in this research. The human motor control is studied under several points of view and methodological guidelines are devised for the design, planning and control of bio-inspired robotic devices

(e.g. stroke) and assess the rate of capabilities improvement during rehabilitation [108]. In Chap. 7, I present a new mechatronic device that can be used to evaluate the severity of muscular spasticity at the elbow level relying on the estimation of the tonic Stretch Reflex threshold [3].

The last Part presents the applications of simple principles devised from human observation for the planning and control of robotic manipulators. Chapter 8 proposes a strategy to generate human-like movements, with the possibility to generalize to an arbitrary anthropomorphic kinematic structure [5, 8]. In Chap. 9, I deal with motion dynamics and introduce the planning of Cartesian impedance behavior for the case of physical interaction. Then, Chap. 10 proposes the exploitation of human-inspired behaviors to enhance autonomous robotic reactive grasping skills [14]. Chapter 11, builds up on the previous one and deals with the learning of the most suitable approaching direction to grasp objects using a soft robotic hand and an anthropomorphic manipulator [30]. Chapter 12 proposes a framework based on Deep Neural Networks able to predict—from the on-line reading of inertial sensors placed on hand fingers—when an object grasped by a soft robotic hand is sliding and, hence, the grasp is failing [126]. Chapter 13 discusses a study on how dexterity augments with a hierarchical enrollment of synergistic actuations, with a specific focus on the kinetic domain [4] and, finally, Chap. 14 shows an application of functional analysis to the design of a sub-optimal planner for autonomous bipedal robots in walking tasks [47].

Each Chapter is written to be as much self-contained as possible, so that it can be read either as part of the book or as a stand-alone. Furthermore, other chapters are referred in the text when needed, to ensure an easier navigation through the manuscript.

1.6.1 Novelties

In this book I present a number of contributions to the State of the Art in multiple research fields, moving from the neuroscience and the study of human motor control, discussing applications for the design of prosthetic devices and the evaluation of motor impairments, and finally proposing novel approaches to introduce human skills in anthropomorphic manipulators.

All these results are well rooted in the multi-variate statistical analysis, extended to the functional case, which I have applied, for the first time, to extensively unveil the mechanisms of upper limb motion control, and for force generation. More specifically, functional analysis has been applied to characterize the temporal evolution of movements in both physiological and pathological conditions (which has opened to a new to quantitatively and effectively evaluate the motion impairment leveraging on the observation of the motion itself), while a novel implementation of Principal Component Analysis (named r-PCA) is used to study the co-ordination of multi-joint movements of the arm. These studies shade light on a topic still not well described by the research community, and provide multiple significant opportunities in neuroscience and technology development.

On the other side, this book also provide novel evidence to the very well studied problem of hand hierarchical control, still tackling the problem from a completely new perspective, i.e. analysing whether and how such synergistic control of human hands is modified when an interaction with the environment is purposefully exploited to enable difficult grasps.

These results have been formalized in a mathematical language that can have a significant impact in prosthetics, e.g. for the design of a novel solution for an under-actuated wrist.

The results of the functional investigation have been applied to devise new techniques for improving the rendering of Human Likeliness in robots' motion. Indeed, the findings I discussed in the first part of the book enabled the development of a planning algorithm based on motion generators (i.e. functional Principal Components) extracted from movements observed from real subjects and, hence, endowed with desired characteristics of Human Likeliness. The problem of planning robots' motion has been tackled not only at the kinematic level, but also with a focus on motion dynamics and considering the real control-enabled implementation of these approaches.

Finally, this book also contributes with a new framework to generate complex movements for robots during the interaction with the environment. Coherently with all the work I did, the philosophy is to learn from humans. Then, looking at humans, I proposed a touch-based reactive framework, able to automatically adapt in feed-forward the relative hand-object position. This approach solves the problem of imperfect localization of objects in the 3D space, which is common for vision-based systems, and enabled the possibility to propose, for the first time in the research community, an integrated framework able to assign, given an arbitrary unstructured scene, a preferred (human-inspired) approaching strategy to maximize the probability of grasp success with soft hands (with a dedicated methodology for grasp failure prediction).

References

1. Anderson, F.C., Pandy, M.G.: Dynamic optimization of human walking. J. Biomech. Eng. **123**(5), 381–390 (2001)
2. d Avella, A., Tresch, M.C.: Modularity in the motor system: decomposition of muscle patterns as combinations of time-varying synergies. Adv. Neural Inf. Process. Syst. **1**, 141–148 (2002)
3. Averta, G., Abbinante, M., Orsini, P., Felici, F., Lippi, P., Bicchi, A., Catalano, M.G., Bianchi, M.: A novel mechatronic system for evaluating elbow muscular spasticity relying on tonic stretch reflex threshold estimation. In: Proceedings of the Engineering in Medicine and Biology Society, 2020. EMBC'20. 42th Annual International Conference of the IEEE (2020)
4. Averta, G., Angelini, F., Bonilla, M., Bianchi, M., Bicchi, A.: Incrementality and hierarchies in the enrollment of multiple synergies for grasp planning. IEEE Robot. Autom. Lett. **3**(3), 2686–2693 (2018)
5. Averta, G., Caporale, D., Della Santina, C., Bicchi, A., Bianchi, M.: A technical framework for human-like motion generation with autonomous anthropomorphic redundant manipulators. In: 2020 IEEE International Conference on In Robotics and Automation (ICRA), Paris. IEEE (2020)
6. Averta, G., Della Santina, C., Battaglia, E., Ciotti, S., Arapi, V., Fani, S., Bianchi, M.: From humans to robots: the role of cutaneous impairment in human environmental constraint exploitation to inform the design of robotic hands. In: 2017 9th International Congress on Ultra Modern Telecommunications and Control Systems and Workshops (ICUMT), pp. 179–184. IEEE (2017)
7. Averta, G., Della Santina, C., Battaglia, E., Felici, F., Bianchi, M., Bicchi, A.: Unvealing the principal modes of human upper limb movements through functional analysis. Front. Robot. AI **4**, 37 (2017)
8. Averta, G., Della Santina, C., Valenza, G., Bianchi, M., Bicchi, A.: Exploiting upper-limb functional synergies for human-like motion generation of anthropomorphic robots. J. Neuro Eng. Rehabil.
9. Averta, G., Valenza, G., Catrambone, V., Barontini, F., Scilingo, E.P., Bicchi, A., Bianchi, M.: On the time-invariance properties of upper limb synergies. IEEE Trans. Neural Syst. Rehabil. Eng. **27**(7), 1397–1406 (2019)

10. Bartneck, C., Kulić, D., Croft, E., Zoghbi, S.: Measurement instruments for the anthropomorphism, animacy, likeability, perceived intelligence, and perceived safety of robots. Int. J. Soc. Robot. **1**(1), 71–81 (2009)
11. Bemstein, N.: The problem of interrelation between coordination and localization. Arch. Biol. Sci. **38**, 1–35 (1935)
12. Berger, D.J., Gentner, R., Edmunds, T., Pai, D.K., d'Avella, A.: Differences in adaptation rates after virtual surgeries provide direct evidence for modularity. J. Neurosci. **33**(30), 12384–12394 (2013)
13. Bernstein, N.A.: O postrojenii dvizhenij [on the construction of movements]. Medgiz, Moscow (1947)
14. Bianchi, M., Averta, G., Battaglia, E., Rosales, C., Bonilla, M., Tondo, A., Poggiani, M., Santaera, G., Ciotti, S., Catalano, M.G., et al.: Touch-based grasp primitives for soft hands: applications to human-to-robot handover tasks and beyond. In: 2018 IEEE International Conference on Robotics and Automation (ICRA), pp. 7794–7801. IEEE (2018)
15. Bicchi, A., Gabiccini, M., Santello, M.: Modelling natural and artificial hands with synergies. Philos. Trans. R. Soc. B: Biol. Sci. **366**(1581), 3153–3161 (2011)
16. Brown, C.Y., Asada, H.H.: Inter-finger coordination and postural synergies in robot hands via mechanical implementation of principal components analysis. In: IEEE/RSJ International Conference on Intelligent Robots and Systems, 2007. IROS 2007, pp. 2877–2882. IEEE (2007). https://doi.org/10.1109/iros.2007.4399547
17. Burke, R.E.: Sir charles sherrington's the integrative action of the nervous system: a centenary appreciation. Brain **130**(4), 887–894 (2007)
18. Casini, S., Tincani, V., Averta, G., Poggiani, M., Della Santina, C., Battaglia, E., Catalano, M.G., Bianchi, M., Grioli, G., Bicchi, A.: Design of an under-actuated wrist based on adaptive synergies. In: 2017 IEEE International Conference on Robotics and Automation (ICRA), pp. 6679–6686. IEEE (2017)
19. Castellini, C., van der Smagt, P.: Evidence of muscle synergies during human grasping. Biol. Cybern. **107**(2), 233–245 (2013)
20. Catalano, M.G., Grioli, G., Farnioli, E., Serio, A., Piazza, C., Bicchi, A.: Adaptive synergies for the design and control of the Pisa-IIT softhand. Int. J. Robot. Res. **33**, 768–782 (2014)
21. Catalano, M.G., Grioli, G., Serio, A., Farnioli, E., Piazza, C., Bicchi, A.: Adaptive synergies for a humanoid robot hand. In: 2012 12th IEEE-RAS International Conference on Humanoid Robots (Humanoids 2012), pp. 7–14. IEEE (2012)
22. Chow, C., Jacobson, D.: Studies of human locomotion via optimal programming. Math. Biosci. **10**(3–4), 239–306 (1971)
23. Ciocarlie, M., Goldfeder, C., Allen, P.K.: Dimensionality reduction for hand-independent dexterous robotic grasping (2007)
24. Ciocarlie, M.T., Allen, P.K.: Hand posture subspaces for dexterous robotic grasping. Int. J. Robot. Res. **28**(7), 851–867 (2009)
25. Collins, J.: The redundant nature of locomotor optimization laws. J. Biomech. **28**(3), 251–267 (1995)
26. d'Avella, A., Giese, M., Ivanenko, Y.P., Schack, T., Flash, T.: Modularity in motor control: from muscle synergies to cognitive action representation. Front. Comput. Neurosci. **9**, 126 (2015)
27. d'Avella, A., Portone, A., Fernandez, L., Lacquaniti, F.: Control of fast-reaching movements by muscle synergy combinations. J. Neurosci. **26**(30), 7791–7810 (2006)
28. d'Avella, A., Saltiel, P., Bizzi, E.: Combinations of muscle synergies in the construction of a natural motor behavior. Nat. Neurosci. **6**(3), 300 (2003)
29. Davy, D., Audu, M.: A dynamic optimization technique for predicting muscle forces in the swing phase of gait. J. Biomech. **20**(2), 187–201 (1987)
30. Della Santina, C., Arapi, V., Averta, G., Damiani, F., Fiore, G., Settimi, A., Catalano, M.G., Bacciu, D., Bicchi, A., Bianchi, M.: Learning from humans how to grasp: a data-driven architecture for autonomous grasping with anthropomorphic soft hands. IEEE Robot. Autom. Lett. **4**(2), 1533–1540 (2019)

31. Della Santina, C., Bianchi, M., Averta, G., Ciotti, S., Arapi, V., Fani, S., Battaglia, E., Catalano, M.G., Santello, M., Bicchi, A.: Postural hand synergies during environmental constraint exploitation. Front. Neurorobot. **11**, 41 (2017)
32. Duchenne, G.: Physiology of motion. Translated by E.B. Kaplan (1959)
33. Duchenne, G.B.: Physiology of motion demonstrated by means of electrical stimulation and clinical observation and applied to the study of paralysis and deformities. WB Saunders (1959)
34. Duffy, B.R.: Anthropomorphism and the social robot. Robot. Auton. Syst. **42**(3), 177–190 (2003)
35. Edsinger, A., Kemp, C.C.: Two arms are better than one: a behavior based control system for assistive bimanual manipulation. In: Recent progress in robotics: viable robotic service to human, pp. 345–355. Springer (2007)
36. Englehart, K., Hudgins, B., Parker, P.A., Stevenson, M.: Classification of the myoelectric signal using time-frequency based representations. Med. Eng. Phys. **21**(6–7), 431–438 (1999)
37. Fagg, A.H., Shah, A., Barto, A.G.: A computational model of muscle recruitment for wrist movements. J. Neurophysiol. **88**(6), 3348–3358 (2002)
38. Feldman, A.G.: Functional tuning of the nervous system with control of movement or maintenance of a steady posture-ii. controllable parameters of the muscle. Biofizika **11**, 565–578 (1966)
39. Feldman, A.G.: Once more on the equilibrium-point hypothesis (λ model) for motor control. J. Mot. Behav. **18**(1), 17–54 (1986)
40. Ficuciello, F., Palli, G., Melchiorri, C., Siciliano, B.: Experimental evaluation of postural synergies during reach to grasp with the ub hand iv. In: 2011 IEEE/RSJ International Conference on Intelligent Robots and Systems, pp. 1775–1780. IEEE (2011)
41. Fink, J.: Anthropomorphism and human likeness in the design of robots and human-robot interaction. In: International Conference on Social Robotics, pp. 199–208. Springer (2012)
42. Flash, T., Hogan, N.: The coordination of arm movements: an experimentally confirmed mathematical model. J. Neurosci. **5**(7), 1688–1703 (1985)
43. Fugl-Meyer, A.R., Jääskö, L., Leyman, I., Olsson, S., Steglind, S.: The post-stroke hemiplegic patient. 1. a method for evaluation of physical performance. Scand. J. Rehabil. Med. **7**(1), 13–31 (1975)
44. Gabiccini, M., Bicchi, A., Prattichizzo, D., Malvezzi, M.: On the role of hand synergies in the optimal choice of grasping forces. Auton. Robot. **31**(2–3), 235–252 (2011)
45. Gabiccini, M., Farnioli, E., Bicchi, A.: Grasp analysis tools for synergistic underactuated robotic hands. Int. J. Robot. Res. **32**(13), 1553–1576 (2013)
46. Galvani, L., Aldini, G.: Aloysii Galvani. De viribus electricitatis in motu musculari commentarius cum Ioannis Aldini Dissertatione et notis. Accesserunt Epistolæ ad animalis electricitatis theoriam pertinentes. Apud Societatem Typographicam (1792)
47. Gasparri, G.M., Manara, S., Caporale, D., Averta, G., Bonilla, M., Marino, H., Catalano, M., Grioli, G., Bianchi, M., Bicchi, A., et al.: Efficient walking gait generation via principal component representation of optimal trajectories: application to a planar biped robot with elastic joints. IEEE Robot. Autom. Lett. **3**(3), 2299–2306 (2018)
48. Gladstone, D.J., Danells, C.J., Black, S.E.: The Fugl-Meyer assessment of motor recovery after stroke: a critical review of its measurement properties. Neurorehabil. Neural Repair **16**(3), 232–240 (2002)
49. Grimaldi, F.M.: Physico-mathesis de lumine, coloribus et iride. Victorii Benatii (1663)
50. Guigon, E., Baraduc, P., Desmurget, M.: Constant effort computation as a determinant of motor behavior. Adv. Comput. Mot. Control **2** (2003)
51. Hahne, J., Rehbaum, H., Biessmann, F., Meinecke, F.C., Müller, K.R., Jiang, N., Farina, D., Parra, L.C.: Simultaneous and proportional control of 2D wrist movements with myoelectric signals. In: 2012 IEEE International Workshop on Machine Learning for Signal Processing, pp. 1–6. IEEE (2012)
52. Hamilton, A.F.D.C., Wolpert, D.M.: Controlling the statistics of action: obstacle avoidance. J. Neurophysiol. **87**(5), 2434–2440 (2002)

53. Harris, C.M., Wolpert, D.M.: Signal-dependent noise determines motor planning. Nature **394**(6695), 780 (1998)
54. Harwood, M.R., Mezey, L.E., Harris, C.M.: The spectral main sequence of human saccades. J. Neurosci. **19**(20), 9098–9106 (1999)
55. Hatze, H., Buys, J.D.: Energy-optimal controls in the mammalian neuromuscular system. Biol. Cybern. **27**(1), 9–20 (1977)
56. Hoff, B., Arbib, M.A.: Models of trajectory formation and temporal interaction of reach and grasp. J. Mot. Behav. **25**(3), 175–192 (1993)
57. Hoff, B.R.: A computational description of the organization of human reaching and prehension (1992)
58. Hogan, N.: An organizing principle for a class of voluntary movements. J. Neurosci. **4**(11), 2745–2754 (1984)
59. Hogan, N.: Impedance control-an approach to manipulation. i-theory. ii-implementation. iii-applications. ASME Trans. J. Dyn. Syst. Meas. Control B **107**, 1–24 (1985)
60. Ijspeert, A.J.: A connectionist central pattern generator for the aquatic and terrestrial gaits of a simulated salamander. Biol. Cybern. **84**(5), 331–348 (2001)
61. Ivanchenko, V., Jacobs, R.A.: A developmental approach aids motor learning. Neural Comput. **15**(9), 2051–2065 (2003)
62. Ivanenko, Y.P., Poppele, R.E., Lacquaniti, F.: Five basic muscle activation patterns account for muscle activity during human locomotion. J. Physiol. **556**(1), 267–282 (2004)
63. Krishnamoorthy, V., Goodman, S., Zatsiorsky, V., Latash, M.L.: Muscle synergies during shifts of the center of pressure by standing persons: identification of muscle modes. Biol. Cybern. **89**(2), 152–161 (2003)
64. Kuo, A.D.: An optimal control model for analyzing human postural balance. IEEE Trans. Biomed. Eng. **42**(1), 87–101 (1995)
65. Kutch, J.J., Valero-Cuevas, F.J.: Challenges and new approaches to proving the existence of muscle synergies of neural origin. PLoS Comput. Biol. **8**(5) (2012)
66. Latash, M.: Evolution of motor control: from reflexes and motor programs to the equilibrium-point hypothesis. J. Hum. Kinet. **19**, 3–24 (2008)
67. Latash, M.L., Scholz, J.P., Schöner, G.: Toward a new theory of motor synergies. Mot. Control **11**(3), 276–308 (2007). https://doi.org/10.1123/mcj.11.3.276
68. Latash, M.L., Zatsiorsky, V.M.: 6—reflexes. In: Latash, M.L., Zatsiorsky, V.M. (eds.) Biomechanics and Motor Control, pp. 99 – 120. Academic Press, San Diego (2016). https://doi.org/10.1016/B978-0-12-800384-8.00006-5, http://www.sciencedirect.com/science/article/pii/B9780128003848000065
69. Li, W., Todorov, E., Pan, X.: Hierarchical optimal control of redundant biomechanical systems. In: The 26th Annual International Conference of the IEEE Engineering in Medicine and Biology Society, vol. 2, pp. 4618–4621. IEEE (2004)
70. Liarokapis, M.V., Artemiadis, P., Bechlioulis, C., Kyriakopoulos, K.: Directions, methods and metrics for mapping human to robot motion with functional anthropomorphism: a review. School of Mechanical Engineering, National Technical University of Athens, Technical Report (2013)
71. Liarokapis, M.V., Artemiadis, P.K., Kyriakopoulos, K.J.: Task discrimination from myoelectric activity: a learning scheme for emg-based interfaces. In: 2013 IEEE 13th International Conference on Rehabilitation Robotics (ICORR), pp. 1–6. IEEE (2013)
72. Liarokapis, M.V., Artemiadis, P.K., Kyriakopoulos, K.J., Manolakos, E.S.: A learning scheme for reach to grasp movements: on EMG-based interfaces using task specific motion decoding models. IEEE J. Biomed. Health Inform. **17**(5), 915–921 (2013)
73. Liarokapis, M.V., Dollar, A.M., Kyriakopoulos, K.J.: Humanlike, task-specific reaching and grasping with redundant arms and low-complexity hands. In: 2015 International Conference on Advanced Robotics (ICAR), pp. 490–497. IEEE (2015)
74. Loeb, G., Levine, W., He, J.: Understanding sensorimotor feedback through optimal control. In: Cold Spring Harbor symposia on quantitative biology, vol. 55, pp. 791–803. Cold Spring Harbor Laboratory Press (1990)

75. Loeb, G.E.: Optimal isn't good enough. Biol. Cybern. **106**(11–12), 757–765 (2012)
76. Maier, M., Hepp-Reymond, M.: EMG activation patterns during force production in precision grip. ii. muscular synergies in the spatial and temporal domain. Exp. Brain Res. **103**(1), 123–136 (1995)
77. Mason, C.R., Gomez, J.E., Ebner, T.J.: Hand synergies during reach-to-grasp. J. Neurophysiol. **86**(6), 2896–2910 (2001)
78. Meijer, O., Latash, M.: Bernstein versus pavlovianism: an interpretation. Progress in motor control, vol. 2: structure-function relations in voluntary movements, pp. 229–250 (2002)
79. Miossec, S., Kheddar, A.: Human motion in cooperative tasks: moving object case study. In: IEEE International Conference on Robotics and Biomimetics, 2008. ROBIO 2008. pp. 1509–1514. IEEE (2009)
80. Muceli, S., Farina, D.: Simultaneous and proportional estimation of hand kinematics from EMG during mirrored movements at multiple degrees-of-freedom. IEEE Trans. Neural Syst. Rehabil. Eng. **20**(3), 371–378 (2011)
81. Mukai, T., Hirano, S., Nakashima, H., Sakaida, Y., Guo, S.: Realization and safety measures of patient transfer by nursing-care assistant robot riba with tactile sensors. Journal of Robotics and Mechatronics **23**(3), 360–369 (2011)
82. Nakano, E., Imamizu, H., Osu, R., Uno, Y., Gomi, H., Yoshioka, T., Kawato, M.: Quantitative examinations of internal representations for arm trajectory planning: minimum commanded torque change model. J. Neurophysiol. **81**(5), 2140–2155 (1999)
83. Nashner, L.: Adapting reflexes controlling the human posture. Exp. Brain Res. **26**(1), 59–72 (1976)
84. Nelson, W.L.: Physical principles for economies of skilled movements. Biol. Cybern. **46**(2), 135–147 (1983)
85. Newell, K.M., Vaillancourt, D.E.: Dimensional change in motor learning. Hum. Mov. Sci. **20**(4–5), 695–715 (2001)
86. Ortiz-Catalan, M., Håkansson, B., Brånemark, R.: Real-time and simultaneous control of artificial limbs based on pattern recognition algorithms. IEEE Trans. Neural Syst. Rehabil. Eng. **22**(4), 756–764 (2014)
87. Ortiz-Catalan, M., Rouhani, F., Brånemark, R., Håkansson, B.: Offline accuracy: a potentially misleading metric in myoelectric pattern recognition for prosthetic control. In: 2015 37th Annual International Conference of the IEEE Engineering in Medicine and Biology Society (EMBC), pp. 1140–1143. IEEE (2015)
88. Oster, G.: Muscle sounds. Sci. Am. **250**(3), 108–115 (1984). http://www.jstor.org/stable/24969327
89. Pandy, M., Garner, B., Anderson, F.: Optimal control of non-ballistic muscular movements: a constraint-based performance criterion for rising from a chair. J. Biomech. Eng. **117**(1), 15–26 (1995)
90. Pandy, M.G., Zajac, F.E., Sim, E., Levine, W.S.: An optimal control model for maximum-height human jumping. J. Biomech. **23**(12), 1185–1198 (1990)
91. Pavlov, I.P., Gantt, W.: Lectures on conditioned reflexes: twenty-five years of objective study of the higher nervous activity (behaviour) of animals (1928)
92. Pavlov, I.P., Gantt, W.H.: Conditioned reflexes and psychiatry, vol. 2. International publishers New York (1941)
93. Peshkin, M., Colgate, J.E.: Cobots. Ind. Robot.: Int. J. **26**(5), 335–341 (1999)
94. Piazza, C., Catalano, M.G., Godfrey, S.B., Rossi, M., Grioli, G., Bianchi, M., Zhao, K., Bicchi, A.: The softhand pro-h: a hybrid body-controlled, electrically powered hand prosthesis for daily living and working. IEEE Robot. Autom. Mag. **24**(4), 87–101 (2017)
95. Popovic, D., Stein, R.B., Oguztoreli, M.N., Lebiedowska, M., Jonic, S.: Optimal control of walking with functional electrical stimulation: a computer simulation study. IEEE Trans. Rehabil. Eng. **7**(1), 69–79 (1999)
96. Poston, B., Danna-Dos Santos, A., Jesunathadas, M., Hamm, T.M., Santello, M.: Force-independent distribution of correlated neural inputs to hand muscles during three-digit grasping. Am. J. Physiol.-Hear. Circ. Physiol. (2010)

97. Prattichizzo, D., Malvezzi, M., Gabiccini, M., Bicchi, A.: On the manipulability ellipsoids of underactuated robotic hands with compliance. Robot. Auton. Syst. **60**(3), 337–346 (2012)
98. Prattichizzo, D., Malvezzi, M., Gabiccini, M., Bicchi, A.: On motion and force controllability of precision grasps with hands actuated by soft synergies. IEEE Trans. Rob. **29**(6), 1440–1456 (2013)
99. Rasmussen, J., Damsgaard, M., Voigt, M.: Muscle recruitment by the min/max criterion–a comparative numerical study. J. Biomech. **34**(3), 409–415 (2001)
100. Riek, L.D., Rabinowitch, T.C., Chakrabarti, B., Robinson, P.: How anthropomorphism affects empathy toward robots. In: Proceedings of the 4th ACM/IEEE International Conference on Human Robot Interaction, pp. 245–246. ACM (2009)
101. Saltiel, P., Wyler-Duda, K., D'Avella, A., Tresch, M.C., Bizzi, E.: Muscle synergies encoded within the spinal cord: evidence from focal intraspinal NMDA iontophoresis in the frog. J. Neurophysiol. **85**(2), 605–619 (2001)
102. Santello, M., Bianchi, M., Gabiccini, M., Ricciardi, E., Salvietti, G., Prattichizzo, D., Ernst, M., Moscatelli, A., Jörntell, H., Kappers, A.M., et al.: Hand synergies: integration of robotics and neuroscience for understanding the control of biological and artificial hands. Phys. Life Rev. (2016)
103. Santello, M., Flanders, M., Soechting, J.F.: Postural hand synergies for tool use. J. Neurosci. **18**(23), 10105–10115 (1998)
104. Santello, M., Flanders, M., Soechting, J.F.: Patterns of hand motion during grasping and the influence of sensory guidance. J. Neurosci. **22**(4), 1426–1435 (2002)
105. Santisteban, L., Térémetz, M., Bleton, J.P., Baron, J.C., Maier, M.A., Lindberg, P.G.: Upper limb outcome measures used in stroke rehabilitation studies: a systematic literature review. PLoS ONE **11**(5), e0154792 (2016)
106. Schmidt, R.A.: A schema theory of discrete motor skill learning. Psychol. Rev. **82**(4), 225 (1975)
107. Scholz, J.P., Schöner, G.: The uncontrolled manifold concept: identifying control variables for a functional task. Exp. Brain Res. **126**(3), 289–306 (1999). https://doi.org/10.1007/s002210050738
108. Schwarz, A., Averta, G., Veerbeek, J.M., Luft, A., Held, J.P., Valenza, G., Bicchi, A., Bianchi, M.: A functional analysis-based approach to quantify upper limb impairment level in chronic stroke patients: a pilot study. In: Engineering in Medicine and Biology Society, 2019. EMBC'19. 41th Annual International Conference of the IEEE. IEEE (2019)
109. See, J., Dodakian, L., Chou, C., Chan, V., McKenzie, A., Reinkensmeyer, D.J., Cramer, S.C.: A standardized approach to the Fugl-Meyer assessment and its implications for clinical trials. Neurorehabil. Neural Repair **27**(8), 732–741 (2013)
110. Shimansky, Y.P.: Spinal motor control system incorporates an internal model of limb dynamics. Biol. Cybern. **83**(4), 379–389 (2000)
111. Shimansky, Y.P., Kang, T., He, J.: A novel model of motor learning capable of developing an optimal movement control law online from scratch. Biol. Cybern. **90**(2), 133–145 (2004)
112. Smeets, J.B., Brenner, E.: A new view on grasping. Mot. Control **3**(3), 237–271 (1999)
113. Stuart, D., Pierce, P., Callister, R., Brichta, A., McDonagh, J.: Sir Charles Sherrington: humanist, mentor, and movement neuroscientist. In: Classics in Movement Science, pp. 317–374 (2001)
114. Thakur, P.H., Bastian, A.J., Hsiao, S.S.: Multidigit movement synergies of the human hand in an unconstrained haptic exploration task. J. Neurosci. **28**(6), 1271–1281 (2008)
115. Thrane, G., Sunnerhagen, K.S., Persson, H.C., Opheim, A., Alt Murphy, M.: Kinematic upper extremity performance in people with near or fully recovered sensorimotor function after stroke. Physiother. Theory Pract. 1–11 (2018)
116. Ting, L.H., Macpherson, J.M.: A limited set of muscle synergies for force control during a postural task. J. Neurophysiol. (2005)
117. Todorov, E.: Cosine tuning minimizes motor errors. Neural Comput. **14**(6), 1233–1260 (2002)
118. Todorov, E.: Optimality principles in sensorimotor control. Nat. Neurosci. **7**(9), 907 (2004)

119. Todorov, E., Jordan, M.I.: Smoothness maximization along a predefined path accurately predicts the speed profiles of complex arm movements. J. Neurophysiol. **80**(2), 696–714 (1998)
120. Todorov, E., Jordan, M.I.: Optimal feedback control as a theory of motor coordination. Nat. Neurosci. **5**(11), 1226 (2002)
121. Tresch, M.C., Cheung, V.C., d'Avella, A.: Matrix factorization algorithms for the identification of muscle synergies: evaluation on simulated and experimental data sets. J. Neurophysiol. (2006)
122. Tresch, M.C., Jarc, A.: The case for and against muscle synergies. Curr. Opin. Neurobiol. **19**(6), 601–607 (2009)
123. Tresch, M.C., Saltiel, P., Bizzi, E.: The construction of movement by the spinal cord. Nat. Neurosci. **2**(2), 162 (1999)
124. Tuller, B., Turvey, M.: The bernstein perspective: ll. the concept of muscle linkage or coordinative. In: Human Motor Behavior: An Introduction, p. 253 (1982)
125. Uno, Y., Kawato, M., Suzuki, R.: Formation and control of optimal trajectory in human multijoint arm movement. Biol. Cybern. **61**(2), 89–101 (1989)
126. Arapi, V., Zhang, Y., Averta, G., Catalano, M.G., Rus, D., Della Santina, C., Bianchi, M.: To grasp or not to grasp: an end-to-end deep-learning approach for predicting grasping failures in soft hands. In: IEEE International Conference on Soft Robotics, Yale University, USA. IEEE (2020)
127. Valero-Cuevas, F.J.: Predictive modulation of muscle coordination pattern magnitude scales fingertip force magnitude over the voluntary range. J. Neurophysiol. **83**(3), 1469–1479 (2000)
128. Valero-Cuevas, F.J., Zajac, F.E., Burgar, C.G.: Large index-fingertip forces are produced by subject-independent patterns of muscle excitation. J. Biomech. **31**(8), 693–703 (1998)
129. Vinjamuri, R., Sun, M., Chang, C.C., Lee, H.N., Sclabassi, R.J., Mao, Z.H.: Temporal postural synergies of the hand in rapid grasping tasks. IEEE Trans. Inf Technol. Biomed. **14**(4), 986–994 (2010)
130. Vujaklija, I., Roche, A.D., Hasenoehrl, T., Sturma, A., Amsuess, S., Farina, D., Aszmann, O.C.: Translating research on myoelectric control into clinics–are the performance assessment methods adequate? Front. Neurorobot. **11**, 7 (2017)
131. Weiss, E.J., Flanders, M.: Muscular and postural synergies of the human hand. J. Neurophysiol. **92**(1), 523–535 (2004). https://doi.org/10.1152/jn.01265.2003
132. Whytt, R.: The Works of Robert Whytt. Becket, T., De Hondt, P.A. (eds.). London (1768)

Part I
Taming the Complexity of Human Motion Generation

Chapter 2
Understanding the Principal Modes of Natural Movements in Temporal Domain

The rich variety of human upper limb movements requires an extraordinary coordination of different joints according to specific spatio-temporal patterns. However, unveiling these motor schemes is a challenging task. Principal components have been often used for analogous purposes, but such an approach relies on hypothesis of temporal uncorrelation of upper limb poses in time. To overcome these limitations, in this work, we leverage on functional principal component analysis (fPCA). We carried out experiments with 7 subjects performing a set of most significant human actions, selected considering state-of-the-art grasp taxonomies and human kinematic workspace. fPCA results show that human upper limb trajectories can be reconstructed by a linear combination of few principal time-dependent functions, with a first component alone explaining around 60/70% of the observed behaviors. This allows to infer that in daily living activities humans reduce the complexity of movement by modulating their motions through a reduced set of few principal patterns. Finally, we discuss how this approach could be profitably applied in robotics and bioengineering, opening fascinating perspectives to advance the state of the art of artificial systems, as it was the case of hand synergies.

2.1 Introduction

Human hands represent an extraordinary tool to explore and interact with the external environment. Not surprisingly, a lot of studies have been devoted to model how the nervous system can cope with the complexity of hand sensory-motor architecture [23, 30, 45, 48, 50]. These studies have led to the definition of the so called synergies, broadly intended as covariation patterns that can be represented at different levels [44]. More specifically, at the level of motor units, neural activation shows a synergistic control in the time and/or frequency domain [42]. At the muscle level,

G. Averta, *Human-Aware Robotics: Modeling Human Motor Skills for the Design, Planning and Control of a New Generation of Robotic Devices*, Springer Tracts in Advanced Robotics 145, https://doi.org/10.1007/978-3-030-92521-5_2

different works explored patterns of muscle activity whose timing and/or amplitude modulation enables the generation of different movements, see [16] for a review. Synergies have also been identified and defined as covariation patterns of joint angles, e.g., hand postural synergies [30, 43, 45], or covariation patterns among digit forces (for a review see [51]). However, to correctly understand human manipulation, in addition to hand analysis the role of whole upper limb movements should be also taken into account. Indeed, the whole upper limb motions are devoted to guide and optimize position and orientation of the hand w.r.t. external targets.

For these reasons, in addition to many works devoted to analyze hand behavior, it is also possible to find studies modeling human upper limb motor workspace, either from a kinematic point of view, or from a muscular or neural point of view. In [26] the authors studied the kinematic movements of upper limb during selected tasks in order to compare stroke patients and normal subjects. In [10] the authors developed a quantitative method to assess upper limb motor deficits in children with cerebral palsy using three-dimensional motion analysis during the Reach and Grasp Cycle. Other papers studied muscular synergies in upper limb activities, as in [3], where the authors introduced a model based on combinations of muscle time-varying synergies, and in [17], where authors recorded electromyographic activity from shoulder and arm muscles during point-to-point movements. As for hand synergies, whose robotic applications are reviewed in [44], synergies have also been applied to movement generation for virtual arms [22] as well as myocontrol of a multi-DoF planar robotic arm using muscle synergies [29]. However, none of the previous studies considered the dynamic aspects of human upper limb motion, i.e. that different temporal evolutions and shapes of upper limb joints trajectories would result in different final hand poses.

To solve this problem, in this Chapter we used functional Principal Component Analysis (fPCA) to study upper limb motions. fPCA is a statistical method for investigating dominant modes of variation of functional data in time and has been widely used in one dimensional or multi dimensional time series analysis in chemistry, weather phenomena, medicine [2, 15, 24]. The interested reader in functional data analysis could refer to [35–37]. In human movements studies, this method has been used to explore the presence of variations in repetitions of a specific task, e.g. in [40] an analysis of knee joint kinematics in the vertical jump was performed. In [12], fPCA was used to analyze a bio-mechanical data set examining the mechanisms of chronic Achilles tendon injury and the functional effects of orthoses by comparing injured and healthy subjects. In [15] fPCA was used in conjunction with PCA for the analysis of grasping motion. Of note, in [39], a similar approach, namely Temporal Decomposition, was used to decompose temporal variations observed in joint torques and muscle patterns, through PCA or NNMF algorithms respectively.

In this Chapter, the choice to use fPCA as main data analysis tool is motivated by the fact that it allows to include some important features of the signal, such as shape and time dependence. To achieve this goal, we propose an experimental setup for studying upper limb movements, based on a Motion Capture (MoCap) system (*Phase Space*®). Using this tool, we carried out a series of experiments with human considering a comprehensive dataset of daily living activities (ADLs) and grasping/manipulation actions. These actions were selected relying on the study of grasping taxonomies [14, 20], and considerations on human upper limb movement workspace [1, 28, 33]. Our analysis has led to the reduction of complexity of upper limb trajectories by describing these as linear combinations of few principal functions (or modes). Implications for robotics are also discussed. The content of this chapter is adapted from [4].

2.2 Experimental Protocol and Setup

2.2.1 A Set of Daily Living Tasks

In order to develop a comprehensive study of human upper limb movements, one of the key features for the generation of a valid dataset is the definition of a set of meaningful actions [30, 45, 49, 50]. For this reason, we selected a set of movements driven by the study of grasping taxonomies [14, 20], and the analysis of human upper limb movement workspace [1, 28, 33]. The output of this selection resulted in a set of 30 different actions, divided in intransitive, transitive and tool-mediated actions to avoid bias due to the affordances of the objects used for the grasp investigation. Indeed, as Cubelli in [13] suggested starting from apraxia investigation and Handjaras confirmed in [25] with cortical imaging, different movements are generated by different cortical activations, because require different motor schema, based on the type of interaction with the environment. These movements can be classified in three classes, according to the presence or absence of an object and, if the object is present, on the approach with it: intransitive class, which collects actions that does not need the use of an object; transitive class, which collects actions that introduces the use of an object; tool mediated class, which collects actions where an object is used to interact with another one. Tasks are meant to be executed three times with dominant hand, the subject seating on a chair, with the objects placed on a frontal table at a fixed distance. At the end of the task the subject returned to the starting point. The complete list of actions is reported in Table 2.1.

2.2.2 An Experimental Setup for Data Acquisition

I focused on kinematic recordings, which was achieved using a commercial system for 3D motion tracking with active markers (*Phase Space*®). Ten stereo-cameras working 480 Hz tracked 3D position of markers, which were fastened to supports rigidly attached to upper limb links. In this manner, 20 markers were accommodated on the upper limb so that the distance between elements of each support was fixed. Supports were suitably designed for these experiments and printed in ABS (see Fig. 2.1a). The acquisition was implemented through a custom application developed

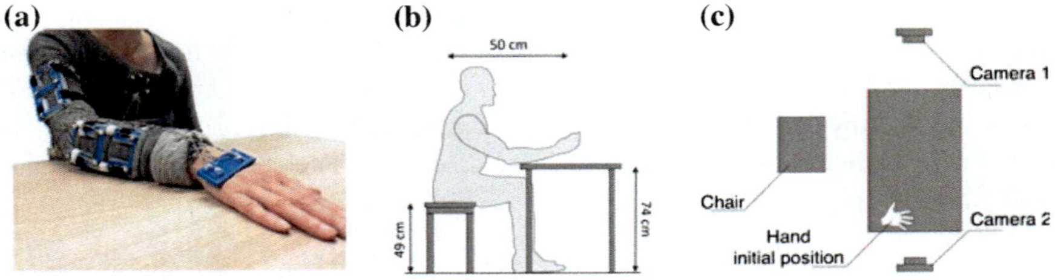

Fig. 2.1 These figures show the experimental setup. In **a** markers accommodation along the upper limb. In **b** and **c**, we report a scheme of the experimental setup. Subjects were comfortably sit in front of the table. In the starting position, subject's hand was located at the right side of the table. Two cameras were also included to record the scene

in C++, employing Boost libraries [47] to enable the synchronization between Phase Space data and other sensing modalities, such as force/torque sensors and EEG, and the Phase Space OWL library to get the optical tracking system data.

Seven adult right-handed subjects (5 male and 2 female, aged between 20 and 30), performed the experiment. Each task was repeated three times in order to increase the robustness of collected data. The experimenter gave the starting signal to subjects. In the instructions, the experimenter emphasized that the whole movement should be performed in a natural fashion. No instruction was provided on the velocity of the action, leaving the subject free to implement the movement with a preferred speed. The object order was randomized for every subject. Each subject performed the whole experiment in a single day. No subject knew the purpose of the study, and had no history of neuromuscular disorders. Each participant signed an informed consent to participate in the experiment, and the experimental protocol was approved by the Institutional Review Board of University of Pisa, in accordance to the declaration of Helsinki. The complete experimental setup is reported in Fig. 2.1.

Moreover, we used two cameras (Logitech hd 1080p) to record the scene of the experiments in order to visually compare the real and the reconstructed movement.

2.3 Motion Identification

2.3.1 Modeling of Upper Limb Kinematics

An accurate description of human upper limb is challenging due to the high complexity of the kinematic structure, e.g. for axis location and direction, which are usually time varying. In order to explore the system complexity, the interested reader can refer to [27, 31]. In this work, we used a trade-off between complexity and accuracy. This allows to get an acceptable computational time, still maintaining a good level of explanation of physical behaviour. Taking inspiration from [6], we adopted a model with 7 degrees of freedom (DoFs), and 3 invariable shape links. Joints angles

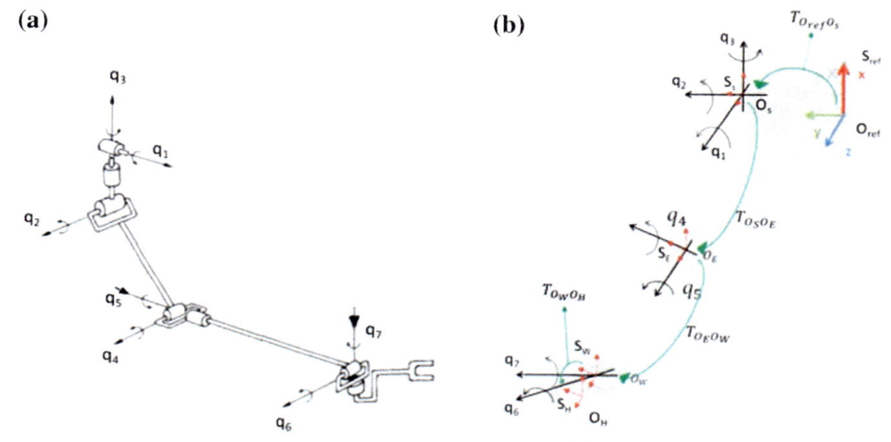

Fig. 2.2 System parametrization. In **a** we show the kinematic model used in this chapter. The figure is adapted from [6]. In **b** we report the kinematic parametrization. The labels $q_1, \ldots, q_7$ refers to joints of the model. S_{ref} refers to the reference system centered in O_{ref}, fixed to the epigastrium; S_S refers to the reference system centered in O_S, Center of Rotation (CoR) of shoulder joints, fixed to the arm; S_E refers to the reference system centered in O_E, CoR of elbow joints, fixed to the forearm; S_W refers to the reference system centered in O_W, CoR of wrist joints, fixed to the hand; S_H refers to the reference system centered in O_H, fixed to the hand. The label $T_{O_{ref}O_S}$ refers to the rigid transform between S_{ref} and S_S; $T_{O_SO_E}$ refers to the rigid transform between S_S and S_E; $T_{O_EO_W}$ refers to the rigid transform between S_E and S_W; $T_{O_WO_H}$ refers to the rigid transform between S_W and S_H

are defined as $q_1, \ldots, q_7$: q_1 is associated to the shoulder abduction-adduction; q_2 is associated to the shoulder flexion-extension; q_3 is associated to the shoulder external-internal rotation; q_4 is associated to the elbow flexion-extension; q_5 is associated to the elbow pronation-supination; q_6 is associated to the wrist abduction-adduction; q_7 is associated to the wrist flexion-extension. In Fig. 2.2a a scheme of the model is reported.

2.3.2 Model Parameters

In order to describe the forward kinematics of the arm, 5 different reference systems was defined: S_{ref}, centered in O_{ref}, fixed to the epigastrium; S_S, centered in O_S, Center of Rotation (CoR) of shoulder joints, fixed to the arm; S_E, centered in O_E, CoR of elbow joints, fixed to the forearm; S_W, centered in O_W, CoR of wrist joints, fixed to the hand; S_H, centered in O_H, fixed to the hand. The rigid transform between S_{ref} and S_S is $T_{O_{ref}O_S}$; the rigid transform between S_S and S_E is $T_{O_SO_E}$; the rigid transform between S_E and S_W is $T_{O_EO_W}$; the rigid transform between S_W and S_H is

$T_{O_W O_H}$. The defined reference systems are shown in red in Fig. 2.2b. Green arrows indicate rigid transforms from a reference system and the next one in the kinematic chain. To parameterize the *i-th* segment we use the *Product of Exponentials* (POE) formula [9]:

$$g_{O_{ref} O_j}(\theta) = \left[\prod_{k=1}^{j} e^{\hat{\xi}_k \theta_k} \right] g_{O_{ref} O_j}(0), \tag{2.1}$$

where $\hat{\xi}_k$ are the twists of the joints defining the kinematic chain, $\theta = [\theta_1, \ldots, \theta_k, \ldots, \theta_j]^T$ are the exponential coordinates of the $2nd$ kind for a local representation of SE(3) (Special Euclidean group, 4×4 rototranslation matrices) for the j-th link, and $g_{O_{ref} O_j}(0)$ is the initial configuration. For further details, the interested reader can refer to [23].

2.3.3 Markers Modeling

Links movements were tracked by fastening optical active markers to upper limb links. Markers positioning is inspired by [8]. In order to improve tracking performance, a redundant configuration of marker was used, in particular 4 markers fixed to the chest, 6 markers fixed to the lateral arm, 6 markers fixed to the dorsal forearm, 4 markers fixed to the hand dorsum. A picture showing marker distribution is reported in Fig. 2.1a. The position of each marker can be calculated as rigid transform w.r.t. the center of the corresponding support. The supports kinematic can be described as a rigid transform from the link reference system to the support reference system, as depicted in Fig. 2.3.

Reference systems of links and marker supports are chosen to be all parallel when all the joint angles are equal to zero. The zero-configuration is by-design with the arm completely extended in front of the subject, with the wrist flexion/extension axis aligned with the vertical direction, and with the forearm in a neutral position (thumb up). In this particular case, the model is completely parametrized using 14

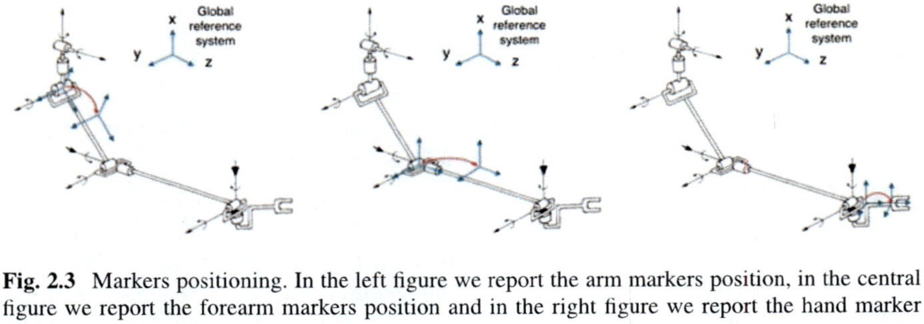

Fig. 2.3 Markers positioning. In the left figure we report the arm markers position, in the central figure we report the forearm markers position and in the right figure we report the hand marker position

coefficients (different for each subject) collected in a vector p_G: Bones length (arm and forearm, 2 parameters); rigid transform from epigastrium to the shoulder CoR (3 parameters); rigid transform from shoulder CoR to the center of arm marker support (3 parameters); rigid transform from elbow CoR to the center of forearm marker support (3 parameters); rigid transform from wrist CoR to the center of hand marker support (3 parameters).

Note that this model is not able to differentiate between rotations of the marker's support w.r.t. the link (which is an undesired effect) and actual pronation/supination (for the forearm) or shoulder internal/external rotation (for the arm). This is an assumption required for marker-based motion tracking. For this reason, it is assumed hereinafter that all the relative motions between markers are associated with actual joint rotations. This assumption is supported by the minimization of possible sliding between markers and arm/forearm links, achieved by covering the supports at the arm and forearm level with an high-friction silicon (see Fig. 2.1-A), thus improving the grip between markers and subject's skin.

The parameter vector p_G was calibrated for each subject. Given p_G, the upper limb pose is described by 7 joints angles $[q_1, \ldots, q_7]^T$ collected in a vector x.

2.3.4 Model Calibration and Angles Estimation

As previously mentioned, the parameters of the kinematic model must be adapted for the specific subject that performs the experiments. The optimal parameters were obtained by solving a constrained least-squares minimization problem:

$$(x^*, p_G^*) = arg \min_{x_k \in D_x, p_G \in D_p} \frac{1}{2} \sum_{k=1}^{N_p} r_k^T r_k$$

The residual function r_k is calculated as $r_k(x_k, p_G) := y_k - f(x_k, p_G)$, where: y_k is the marker position vector measured with PhaseSpace; x_k is the vector of estimated joint angles; p_G is the vector of model kinematic parameters; D_x is the upper limb joints range of motion; D_p is the variation around a preliminary estimation of parameters performed with manual measurements; $f(x_k, p_G)$ is the estimated positions vector of markers using the forward kinematics. The vector of measures y_k and the vectors of estimations $f(x_k, p_G)$ can be obtained by concatenating the measures of marker positions and estimations at different time frames. To obtain an effective calibration output, the selected frames for the calibration procedure must consider different poses of the kinematic chain. For the experiments reported in this chapter, we had r_k normalized w.r.t. the dimension of y_k equal to 15.30 ± 16.25 mm. Taking inspiration from [23], the calibrated model was then used to identify the joints angles using an Extended Kalman Filter (EKF). Indeed, the model can be considered as an uncertain noisy process where at time frame k the joints angle vector x_k is the state of the process, y_k is the markers position vector, w_k and v_k are process and observation zero mean gaussian noises, with covariance Q_k and R_k respectively, and $f(x_k)$ is the forward kinematics. The system can be described using the following equations:

$$\begin{cases} x_k = x_{k-1} + w_k \\ y_k = f(x_k) + v_k \end{cases} \tag{2.2}$$

Given the state at time frame $k-1$, the state at time k was obtained using a 2-steps procedure: *prediction* of the future state $\hat{x}_{k|k-1} = \hat{x}_{k-1}$; *update* of the state estimated in the first step by calculating $\hat{x}_{k|k} = \hat{x}_{k|k-1} + K_k\tilde{r}_k$. The correction amount of the state prediction is the product between the residual values vector $\tilde{r}_k = y_k - f(\hat{x}_{k|k-1})$ and the Kalman Gain K_k. This gain is calculated as product between the covariance matrix estimation of the predicted state $P_{k|k-1}$, the jacobian matrix, i.e. $H_k = \frac{\partial(f(x))}{\partial(x)}$, and the inverse matrix of the residual covariance.

2.3.5 *Experimental Results*

The performance of the estimation tool for time frame k can be evaluated by calculating the Mean Squared Error (MSE) R_k as

$$R_k = \frac{1}{N_{markers}}||(y_k - f(\hat{x}_{k|k}))||$$

where $N_{markers}$ is the number of markers, y_k is the marker positions vector and $f(\hat{x}_{k|k})$ is the vector of markers estimated positions using the joint angles calculated with the EKF. Tipical values of R_k are ≈1 cm (Fig. 2.4), with a mean error for hand and forearm markers of ≈0.5 cm and a mean error for arm markers of ≈1.8 cm. Identified angles was used to reconstruct the movements, which were represented using a virtual model of upper limb suitably implemented in MatLab for these reconstructions. The reconstructions was used for a visual check w.r.t. the real movement recorded with cameras. This allows to verify the human-likeliness of the reconstructed movement.

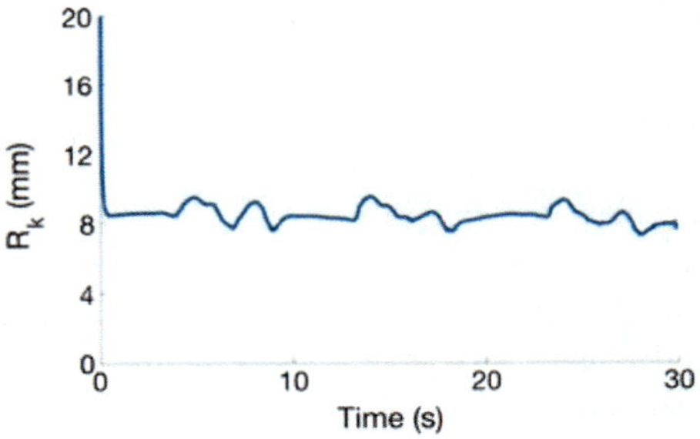

Fig. 2.4 Mean Squared Error obtained in the estimation procedure in a sample. Initial error value is 57.1 mm, related to the filter initialization

2.4 Data Analysis

The goal of this chapter is the study of functional motor synergies of upper limb. This is accomplished using functional PCA, a statistical method that allows to study the differences in shapes between functions. In order to avoid the inclusion in this analysis of undesired features due to misalignments in time or in velocity of the samples, we performed the following pre-processing techniques: segmentation, to divide the repetition of each task, time warping, to synchronize in time all the elements of the dataset.

2.4.1 *Segmentation*

For each task the three repetitions have been segmented using the following procedure (see Fig. 2.5):

1. select the data elements of the third DoF (q_3);
2. find the first two peaks p_1, p_2 of the signal;
3. evaluate the mean slope s_1, s_2 of the signal in a section close to each peak;
4. calculate the segmentation point as $t_1 = \frac{p_1 s_1 + p_2 s_2}{s_1 + s_2}$;
5. repeat points 2–4 using the second and the third peak and obtain t_2.

q_3 data (i.e. shoulder flexion-extension) was used for segmentation because it almost always contains three distinct peaks. If the peaks were not detectable, another DoF with detectable peaks was used instead. Note that the segmentation is performed using the same couple of segmentation points for all the 7 DoFs.

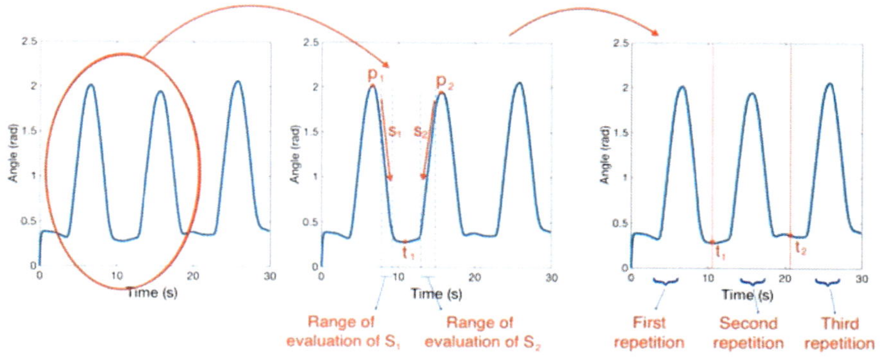

Fig. 2.5 Segmentation procedure. In the left figure we report a sample of joint evolution in time. In the central figure we show two peaks of the signal p_1, p_2 and the mean slope of the signal s_1, s_2 evaluated in two ranges close to each peak. The segmentation point t_1 is evaluated as $t_1 = \frac{p_1 s_1 + p_2 s_2}{s_1 + s_2}$. The same procedure is repeated using the second and the third peak of the signal and the output of the segmentation procedure is reported in the right figure

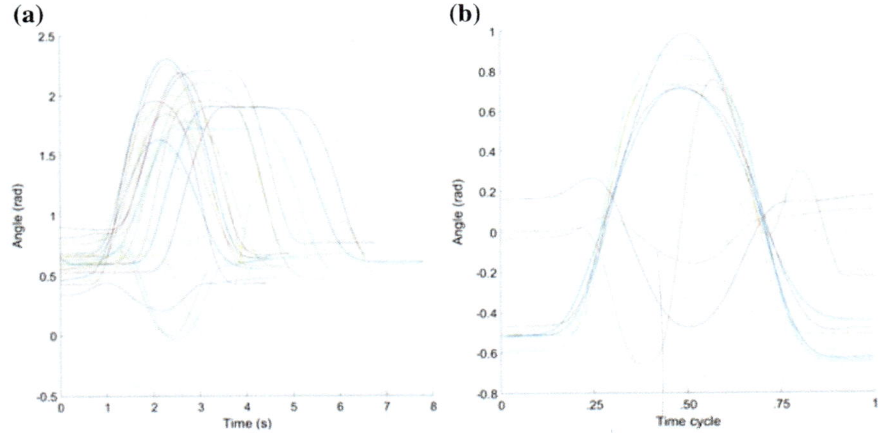

Fig. 2.6 Segmentation and time warping. **a** Sample of segmented data reported to show the differences in starting-time and velocity of the action. **b** Sample data are warped in time and the three repetitions are replaced with the mean signal. Note that the time after warping is normalized, hence it is counted as a percentage of the time cycle

Considering different subjects and tasks, differences between shapes are evident (see Fig. 2.6a). The three repetitions of each task should be replaced by the corresponding mean vector to increase robustness. This replacement can be performed only after signal synchronization, achieved using a time-warping procedure.

2.4.2 Time Warping

The synchronization between two signals allows to increase the affinity by conforming starting-time and speed of the action. This can be achieved by finding the optimal time-shift and time-stretch of one signal w.r.t. the other one. This problem is known in literature as *Dynamic Time Warping* (DTW) and widely explored in sound engineering and pattern recognition [7, 32, 34, 41]. In this chapter, DTW is needed to allow the mediation of the three repetitions, to avoid misalignment and to compare different tasks and subjects data. As a consequence, the following assumptions were done: monotonicity, to preserve data integrity, and linear distortion of time. Given two time series, v_1 and v_2, the affinity between the two signals is increased by the solution of the following least-squares minimization problem:

$$(S, T) = arg \min_{S>0,T} (||v_1(t) - v_2(St - T)||)$$

where S is a scaling factor for the velocity of signal v_2, and T is the amount of shifting in time applied to v_2. The dataset elements were time warped w.r.t. a reference time

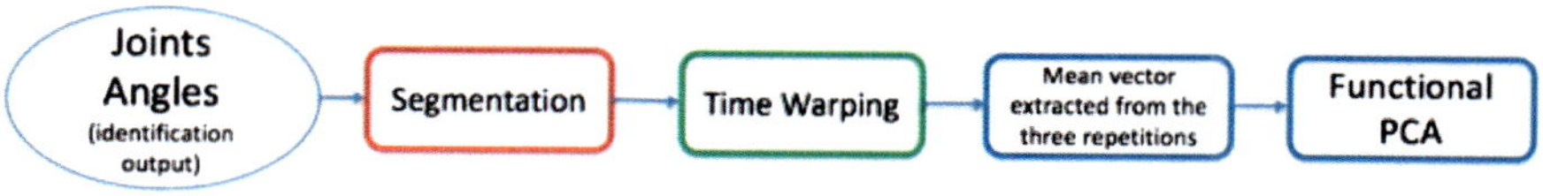

Fig. 2.7 Scheme of data analysis

series, selected in the set as the element whose length is the mean value w.r.t the length of all dataset elements. For each element, S and T are calculated by performing DTW on DoFs used for segmentation, then all the components are time-warped using the optimum set of parameters. The time warped vectors have the same number of frames (number of elements). Once the time warping was performed on all the dataset elements, the three repetitions for each task can be replaced by the corresponding mean vector. A sample output of this procedure is reported in Fig. 2.6a. In Fig. 2.7 a scheme of the data analysis procedure is reported.

2.4.3 Principal Component Analysis

To explain fPCA it is useful to start from classic Principal Component Analysis (PCA). Principal component analysis (PCA) is a statistical procedure that uses an orthogonal transformation to convert a set of observations of possibly correlated variables into a set of values of linearly uncorrelated variables called principal components. This transformation is defined in such a way that the first principal component (PC) has the largest possible variance (that is, it accounts for the largest part of the variability in the data). The other components explain an amount of variance in decreasing order, with the constraint that each principal component is orthogonal to the previous ones. Hence, the resulting vectors represents an orthogonal basis set. Principal components are calculated as eigenvectors of the covariance matrix of data $V = N^{-1}X^TX$, where X is the matrix that collects the samples x_i. The variance explained by each PC is calculated as normalization of the corresponding eigenvalue. Given the first eigenvector ξ_1, the principal component score $f_{i1} = \xi_1' x_i$ maximizes $\sum_i f_{i1}^2$ subject to $||\xi_1|| = 1$; the second eigenvector ξ_2 maximizes $\sum_i f_{i2}^2$ subject to $||\xi_2|| = 1$ and $\xi_2'\xi_1 = 0$, and so on.

2.4.4 Functional Principal Component Analysis

Functional PCA can be described as a functional extension of PCA. Given a set of functions $x_i(t)$, the first functional principal component $\xi_1(t)$ is the function for which the principal component score $f_{i1} = \int \xi_1(t)x_i(t)dt$ maximizes $\sum_i f_{i1}^2$ subject to $\int \xi_1^2(t)dt = ||\xi_1|| = 1$; the second functional principal component $\xi_2(t)$ maximizes

$\sum_i f_{i2}^2$ subject to $||\xi_2|| = 1$ and $\int \xi_2(t)\xi_1(t)dt = 0$, and so on. In practice, this is done implementing the following steps:

1. Consider a set of functions x_i and extract the mean signal $\bar{x}$ as $\bar{x_j} = \frac{1}{N}\sum_{i=1}^{N} x_{ij}$;
2. Remove the mean calculated in step 1 from each data element by $\tilde{x_i} = x_i - \bar{x}$;
3. Define a basis function. The basis must contain a number of functions large enough to consider all possible modes of variations of data. Usually basis elements are exponential functions, splines, Fourier basis [35–37];
4. Given the basis functions $b_1, \ldots, b_K$, each data element can be described as combination of basis elements $\tilde{x_i} = \sum_{k=1}^{K} \theta_k b_k$;
5. Then each function is described by a vector of coefficients $\Theta = (\theta_1, \ldots, \theta_K)'$;
6. A problem of eingenvectors/eigenvalues can now be solved on the covariance matrix. Note that, in case the basis function $b_1, \ldots, b_N$ are not orthogonal (as for the case of the work presented in this Chapter), it is necessary to compute the matrix W of the scalar products between the basis elements as $W = \int \Phi(t)\Phi^T(t)\mathrm{d}t$. Please refer to [36] (Chap 8), and to the appendix of this chapter, for further details.
7. Each PC is, then, transformed into the corresponding functional Principal Component (fPC) using the basis elements as $\xi_k = c_1 b_1 + \cdots + c_K b_K$, where b_k are the basis functions and c_k the corresponding coefficient that map the ξ_k functions as combination of the basis elements, i.e. the eigenvectors calculate in the previous step (see appendix);
8. Each fPC explains a certain percentage of variance. The variance explained by a fPC is quantified normalizing (w.r.t. the sum of the eigenvalues) the corresponding eigenvalue (ρ in the appendix).

2.4.5 *Movement Reconstruction and Performance Analysis*

I used fPCA on this dataset after the post-processing phase reported in previous sections. 15 5th order spline basis elements were used, taking inspiration for the polynomial description in [21]. Each basis function is defined by piecewise polynomial functions. The places where the pieces of the spline intersect are known as knots. Each piece has the following form

$$s_k(t) = \sum_{i=1}^{5} a_{ik}(t - t_k)^i$$

where t_k is the kth knot. The 15 basis functions are depicted in Fig. 2.8.

The fPCs can be used to reconstruct the data sample by adding M fPCs weighted by coefficients c_i, i.e. $x_{rec} = \bar{x} + c_1\xi_1 + \cdots + c_i\xi_i + \cdots + c_M\xi_M$ with $M \leq N$.

Observing the normalized eigenvalues associated to the fPCs, it is possible to infer that the first fPC by itself account for 60–70% of the variation w.r.t. the mean function, as reported in Fig. 2.9, with a mean value between the DoFs of 65.2%, a

Fig. 2.8 15 splines are used as basis functions for the fPCA implementation

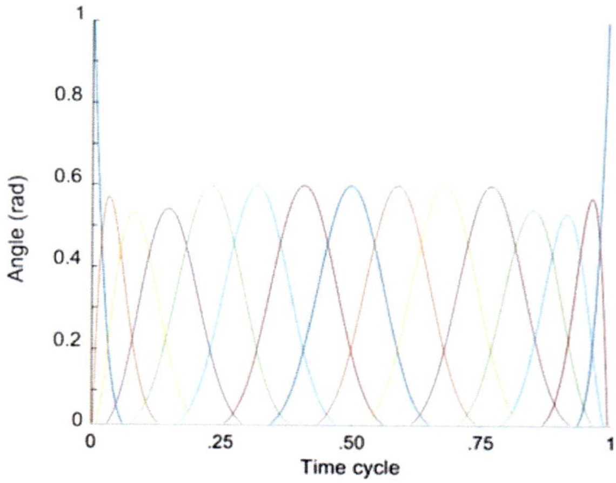

minimum value of 54.4% and a maximum of 76.9%. It is significant to notice that reconstruction with just one (the first) fPC provides a very good approximation of the real behavior, and that the explained variance of the first three fPCs altogether is higher than 84% for all DoFs. In Fig. 2.10 we show how the main principal functions can shape the reconstruction of the joints trajectories. Individual basis function do not need to represent meaningful movements. What is needed is that a combination of basis elements (plus an offset) could reproduce any original trajectory of the joints of the dataset.

The reconstruction performance is showed in Fig. 2.11a, in which a reconstruction using 1, 2 and 3 fPCs is reported. In order to quantify the reconstruction performance, an index of reconstruction error can be evaluated as

$$E_{RMS} = \sqrt{\frac{1}{N_{DoF}} \sum_{i=1}^{N_{DoF}} \left(\sqrt{\frac{1}{N_{frames}} \sum_{j=1}^{N_{frames}} (x - x_{rec})^2} \right)^2}$$

where x is the real function and x_{rec} is the reconstructed function.

In Fig. 2.11b is reported a plot of the normalized error, calculated as $E_{RMS}/max(E_{RMS})$, for different number of fPCs used. Initial point refers to the case where only mean function is used for reconstruction and the value of E_{RMS} is 0.6 rad. The reconstruction using one fPC has an E_{RMS} value lower than 0.2 rad, adding other fPCs, the reconstruction error decay, i.e. using three fPCs the E_{RMS} value is around 0.1 rad. Furthermore, the whole reconstructed movement for the upper limb (considering all DoFs) was displayed using a visualization tool developed in MatLab, showing an high level of anthropomorphism and realism. We can conclude that the kinematic complexity of upper limb trajectories can be simplified and easily described using the mean function and few principal functional modes.

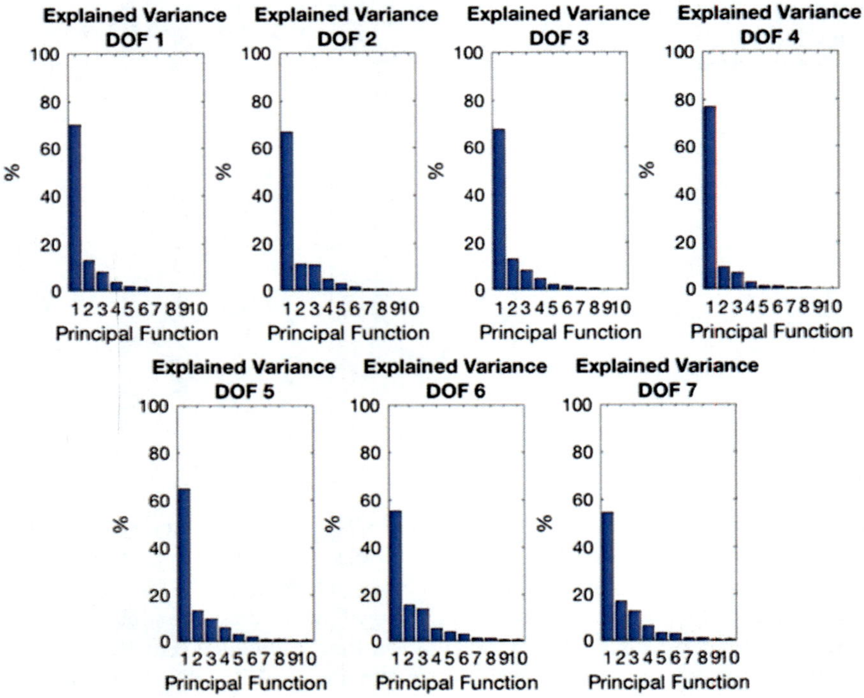

Fig. 2.9 Explained variance for different DoFs and for each fPC

2.5 Conclusions and Implications for Robotics and Bioengineering

In this chapter we have shown that the complexity of upper limb movements in activities of daily living can be described using a reduced number of functional principal components. To achieve this goal we developed an experimental setup, which is based on kinematic recordings but also allows to include additional sensing modalities. Kinematic data are based on a 7 DoFs model and are quantified through a calibration-identification procedure. Collected data was used to characterize upper limb movements through functional analysis. The findings of this chapter can be used to pave the path towards a more accurate characterization of human upper limb principal modes, opening fascinating scenarios in rehabilitation, e.g. for automatic recognition of physiological and pathological movements. (e.g. stroke affected subjects) through machine learning.

At the same time, the here reported results and future investigations could also offer a valuable inspiration for the design and control of robotic manipulators. First, recognizing that few principal modes describe most of kinematic variability could provide insights for a more effective planning and control of robotic manipulators.

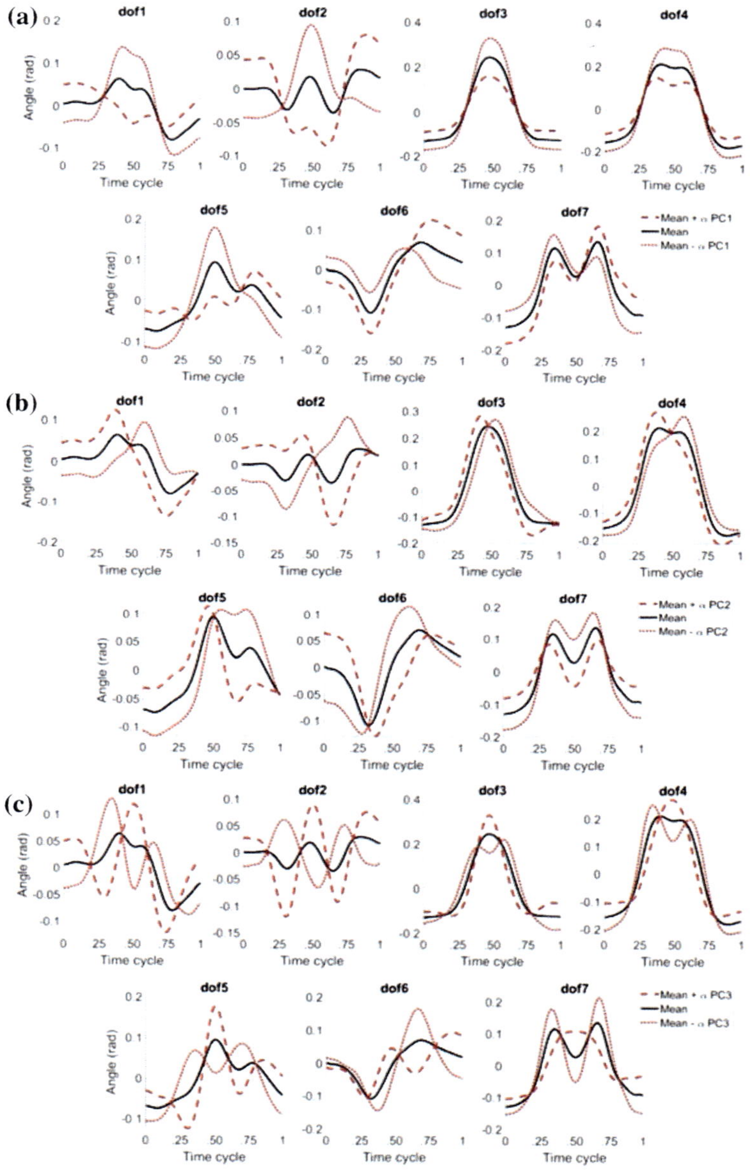

Fig. 2.10 In **a**, mean function (in black) and the same mean function with the contribution of the first principal function, weighted with a coefficient α equal to one (with positive sign in red dashed line, with negative sign in red dotted line); in **b**, mean function (in black) and the same mean function with the contribution of the second principal function with a coefficient α equal to one (same legend of panel (**a**)); in **c** mean function (in black) and the same mean function with the contribution of the third principal function with a coefficient α equal to one (same legend of panel (**a**))

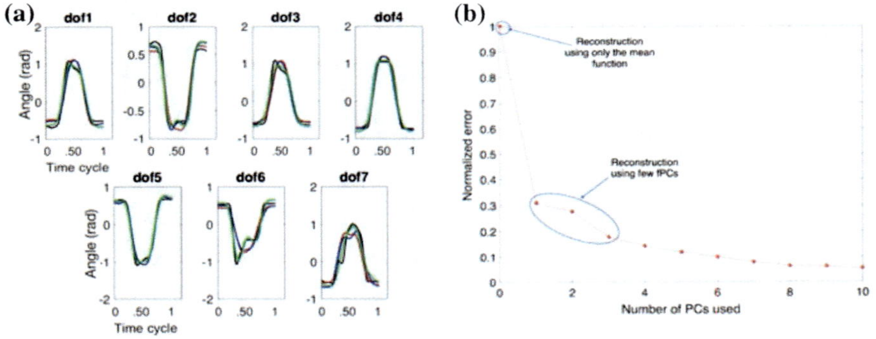

Fig. 2.11 In **a** we show an example of trajectory reconstruction. The black line is the real data. The red line is the reconstructed data using the mean values and the first principal component. The blue line is the reconstructed data using the mean values and the first two principal components. The green line is the reconstructed data using the mean values and the first three principal components. In **b** we report the normalized reconstruction error (RMS). The initial point refers to the error when only the mean function is used for reconstruction. The other points refers to the error when one or more fPCs are used for the reconstruction

For the planning phase, using input trajectories as combinations of the main functional components, which explain most of the kinematic variability, could represent a succesful initial guess to control the movement of the robot—eventually combined with a feedback correction. This combination of feedforward and feedback components could be successfully employed also with soft robotic manipulators, i.e. robots designed to embody safe and natural behaviors relying on compliant physical structures purposefully used to achieve desirable and sometimes variable impedance characteristics. In these cases, standard methods of robotic control can effectively fight against or even completely cancel the physical dynamics of the system, replacing them with a desired model—which defeats the purpose of introducing physical compliance. To overcome this limitation in [46] an anticipative model of human motor control was proposed, which used a feedforward action combined with low-gain feedback, with the goal of obtaining human-like behavior through iterative learning. Results presented in this chapter could be used to define the feedforward component for the control of soft robots. Second, using human-like primitives for controlling robotic systems could improve the effectiveness and safety of Human—Robot Interaction (HRI). Indeed, several studies identified anthropomorphism as one of the key enabling factor for successful, acceptable, predictable and safe HRI in many fields, such as human robot co-working and rehabilitative/assistive robotics [5, 18, 19, 38]. Furthermore, the here reported experimental and analytical framework could be used to identify principal actuation schemes for under-actuated robotic devices. As an example, in [11], we used the identification procedure and the kinematic model reported in this chapter to estimate the contribution of wrist joints in the most common poses for grasping. We performed PCA on the estimated joints of the wrist pre-grasp poses and we found that the flexo-extension DoF plays a dominant role.

We used these results to calibrate an under-actuated wrist system, which is also adaptable and allows to implement different under-actuation schemes, demonstrating its effectiveness to accomplish grasping and manipulation tasks. Future works will aim at using functional data to allow a dynamic implementation of principal kinematic modes of human upper-limb in robotic systems. Finally, the integration of other sensing modalities, such as Electro-encephalographic recordings, could be used to study neural correlates of human upper limb motions, thus possibly inspiring the development of effective Brain-Machine Interfaces for assistive robotics.

Appendix: Computational Methods for Functional PCA

In classic multivariate analysis, the principal components correspond to the eigenvectors of the covariance matrix of data, i.e. they satisfy the following equation:

$$V\xi = \rho\xi, \tag{2.3}$$

where $V = N^{-1}X^TX$ is the covariance matrix of the data arranged in the matrix X after removing the mean, and N is the number of samples. ξ is the eigenvector of V and ρ the corresponding eigenvalue. In the functional approach, the Eq. 2.3 can be generalized with the following:

$$\int v(s,t)\xi(t)\mathrm{d}t = \rho\xi(s) \tag{2.4}$$

where $v(s, t)$ is the covariance function, defined as

$$v(s,t) = N^{-1}\sum_{i=1}^{N} x_i(s)x_i(t), \tag{2.5}$$

$\xi(t)$ is the eigenfunction and ρ the eigenvalue.
The problem of finding $\xi(t)$ and ρ can be tackled following different approaches. On the following we briefly report the approach used in my work, while we suggest the interested reader to refer [35] for further details.

As mentioned before, the computational approach used here to calculate the functional PCs relies on a set $b_1, \ldots, b_K$ of basis function. The idea is to express each element of the dataset as a combination of basis functions:

$$x_i = \sum_{k=1}^{K} \theta_k b_k \tag{2.6}$$

In matrix form, Eq. 2.6 can be expressed as

$$X = \Phi(t)\Theta \tag{2.7}$$

where $\Phi(t)$ is a matrix which collects the basis functions $b(t)$ and $\Theta \in \mathbb{R}^{NxK}$ the matrix of the coefficients.

Also each eigenfunction can be defined as a combination of basis element, i.e.

$$\xi_i = \sum_{k=1}^{K} c_k b_k. \tag{2.8}$$

Given Eqs. 2.7, the 2.5 can be rewritten as

$$v(s, t) = N^{-1}\Phi^T(s)\Theta^T\Theta\Phi(t). \tag{2.9}$$

By substituting Eqs. 2.8 and 2.9 in 2.4 we get:

$$N^{-1}\Phi^T(s)\Theta^T\Theta \int \Phi(t)\Phi^T(t)\mathrm{d}t \quad c = \rho\Phi^T(s)c. \tag{2.10}$$

We define now the matrix W of the scalar products between the basis elements as

$$W = \int \Phi(t)\Phi^T(t)\mathrm{d}t \tag{2.11}$$

and rewrite Eq. 2.10 in the form:

$$N^{-1}\Phi^T(s)\Theta^T\Theta W c = \rho\Phi^T(s)c. \tag{2.12}$$

Since this equation must hold for all s, this implies the purely matrix equation:

$$N^{-1}\Theta^T\Theta W c = \rho c. \tag{2.13}$$

As we mentioned before, the eigenfunctions are orthonormal, then the following equalities hold: $c_j^T W c_j = 1 \;\; \forall \;\; j$ and $c_j^T W c_i = 0 \;\; \forall \;\; j \neq i$.

We now define $u = W^{\frac{1}{2}}c$ and, by substituting in 2.13, we get the following symmetric eigenvalue problem:

$$N^{-1}W^{\frac{1}{2}}\Theta^T\Theta W^{\frac{1}{2}}u = \rho u. \tag{2.14}$$

The problem in Eq. 2.14 can be solved numerically as for the classic PCA problem. Then, the coefficients of the eigenfunctions c can be calculated as $c = W^{-\frac{1}{2}}u$.

Note that if the basis functions b_k are selected as orthogonal, then $W = I$. Also note that the matrix $W^{\frac{1}{2}}$ can be calculated through Cholesky decomposition.

Tables

See Table 2.1.

Table 2.1 Protocol actions

#	#Cutkosky	Class	Description
1		Intransitive	Ok gesture (lifting hand from the table)
2		Intransitive	Thumb down (lifting hand from the table)
3		Intransitive	Exultation (extending the arm up in the air and keeping it in with closed fist)
4		Intransitive	Hitchhiking (extending the arm along the frontal plane, laterally, parallel to the floor, with extended elbow, closed fist, extended thumb)
5		Intransitive	Block out sun from own face (with open hand, touching the face with the palm and covering the eyes)
6		Intransitive	Greet (with open hand, moving wrist) (3 times)
7		Intransitive	Military salute (with lifted elbow)
8		Intransitive	Stop gesture (extending the arm along the sagittal plane, parallel to the floor, with extended elbow, open palm)
9		Intransitive	Pointing (with index finger) of something straight ahead (with outstretched arm)
10		Intransitive	Silence gesture (bringing the index finger, with the remainder of the hand closed, on the lips)
11	2	Transitive	Reach and grasp a small suitcase (placed along own frontal plane) from the handle, lift it and place it on the floor (close to own chair, along own sagittal plane)
12	3	Transitive	Reach and grasp a glass, drink for 3 s (stop signal by the examiner) and place it in the initial position
13	4	Transitive	Reach and grasp a phone receiver (placed along own sagittal plane), carry it to own ear for 3 s (stop signal by the examiner) and place it in the initial position
14	6	Transitive	Reach and grasp a book (placed overhead on a shelf), put in on the table and open it (from right side to left side)
15	8	Transitive	Reach and grasp a small cup from the handle (2 fingers + thumb), drink for 3 s (stop signal by the examiner) and place it in the initial position
16	11	Transitive	Reach and grasp an apple, mimic biting and put it in the initial position
17	12 13	Transitive	Reach and grasp a hat (placed on the right side of the table) from its top and place it on own head
18	12	Transitive	Reach and grasp a cup from its top, lift it and put it on the left side of the table
19	15	Transitive	Receive a tray from someone (straight ahead, with open hand) and put it in the middle of the table
20	16	Transitive	Reach and grasp a key in a lock (vertical axis), extract it from the lock and put it on the left side of the table
21	1	Tool Mediated	Reach and grasp a bottle, pour water into a glass and put the bottle in the initial position
22	2 3 4	Tool Mediated	Reach and grasp a tennis racket (placed along own frontal plane) and play a forehand (the subject is still seated)
23	5	Tool Mediated	Reach and grasp a toothbrush, brush teeth (horizontal axis, one time on left side one time on right side) and put the toothbrush inside a cylindrical holder (placed on the right side of the table)
24	6	Tool Mediated	Reach and grasp a laptop and open the laptop (without changing its position) (4 fingers + thumb)

(continued)

Table 2.1 (continued)

#	#Cutkosky	Class	Description
25	7 8 9	Tool Mediated	Reach and grasp a pen (placed on the right side of the table) and draw a vertical line on the table (from the top to the bottom)
26	7	Tool Mediated	Reach and grasp a pencil (placed along own frontal plane) (3 fingers + thumb) and put it inside a squared pencil holder (placed on the left side of the table)
27	9	Tool Mediated	Reach and grasp a tea bag in a cup (1 finger + thumb), remove it from the cup and place it on the table on the right side of the table
28	10	Tool Mediated	Reach and grasp a doorknob (disk shape), turn it clockwise and counterclockwise and open the door
29	13	Tool Mediated	Reach and grasp a tennis ball (with fingertips) and place it in a basket placed on the floor (close to own chair)
30	14	Tool Mediated	Reach and grasp a cap (2 fingers + thumb) of a bottle (held by left hand), unscrew it and place it overhead on a shelf

References

1. Abdel-Malek, K., Yang, J., Brand, R., Tanbour, E.: Towards understanding the workspace of human limbs. Ergonomics **47**(13), 1386–1405 (2004)
2. Aguilera, A.M., Ocaña, F.A., Valderrama, M.J.: Forecasting time series by functional PCA: discussion of several weighted approaches. Comput. Stat. **14**(3), 443–467 (1999)
3. d Avella, A., Tresch, M.C.: Modularity in the motor system: decomposition of muscle patterns as combinations of time-varying synergies. Adv. Neural Inf. Process. Syst. **1**, 141–148 (2002)
4. Averta, G., Della Santina, C., Battaglia, E., Felici, F., Bianchi, M., Bicchi, A.: Unvealing the principal modes of human upper limb movements through functional analysis. Front. Robot. AI **4**, 37 (2017)
5. Bartneck, C., Kulić, D., Croft, E., Zoghbi, S.: Measurement instruments for the anthropomorphism, animacy, likeability, perceived intelligence, and perceived safety of robots. Int. J. Soc. Robot. **1**(1), 71–81 (2009)
6. Benati, M., Gaglio, S., Morasso, P., Tagliasco, V., Zaccaria, R.: Anthropomorphic robotics. Biol. Cybern. **38**(3), 125–140 (1980)
7. Berndt, D.J., Clifford, J.: Using dynamic time warping to find patterns in time series. In: KDD Workshop, vol. 10, pp. 359–370. Seattle, WA (1994)
8. Biryukova, E., Roby-Brami, A., Frolov, A., Mokhtari, M.: Kinematics of human arm reconstructed from spatial tracking system recordings. J. Biomech. **33**(8), 985–995 (2000)
9. Brockett, R.W.: Robotic manipulators and the product of exponentials formula. In: Mathematical Theory of Networks and Systems, pp. 120–129. Springer (1984)
10. Butler, E.E., Ladd, A.L., Louie, S.A., LaMont, L.E., Wong, W., Rose, J.: Three-dimensional kinematics of the upper limb during a reach and grasp cycle for children. Gait posture **32**(1), 72–77 (2010)
11. Casini, S., Tincani, V., Averta, G., Poggiani, M., Santina, C.D., Battaglia, E., Catalano, M.G., Bianchi, M., Grioli, G., Bicchi, A.: Design of an under-actuated wrist based on adaptive synergies. In: IEEE International Conference of Robotics and Automation, ICRA2017 (In Press)
12. Coffey, N., Harrison, A.J., Donoghue, O., Hayes, K.: Common functional principal components analysis: a new approach to analyzing human movement data. Hum. Mov. Sci. **30**(6), 1144–1166 (2011)
13. Cubelli, R., Marchetti, C., Boscolo, G., Della Sala, S.: Cognition in action: testing a model of limb apraxia. Brain Cogn. **44**(2), 144–165 (2000)
14. Cutkosky, M.R.: On grasp choice, grasp models, and the design of hands for manufacturing tasks. IEEE Trans. Robot. Autom. **5**(3), 269–279 (1989)

15. Dai, W., Sun, Y., Qian, X.: Functional analysis of grasping motion. In: 2013 IEEE/RSJ International Conference on Intelligent Robots and Systems, pp. 3507–3513. IEEE (2013)
16. D'avella, A., Lacquaniti, F.: Control of reaching movements by muscle synergy combinations. Front. Comput. Neurosci. **7**, 42 (2013)
17. d'Avella, A., Portone, A., Fernandez, L., Lacquaniti, F.: Control of fast-reaching movements by muscle synergy combinations. J. Neurosci. **26**(30), 7791–7810 (2006)
18. Dragan, A., Srinivasa, S.: Integrating human observer inferences into robot motion planning. Auton. Robot. **37**(4), 351–368 (2014)
19. Duffy, B.R.: Anthropomorphism and the social robot. Robot. Auton. Syst. **42**(3), 177–190 (2003)
20. Feix, T., Romero, J., Schmiedmayer, H.B., Dollar, A.M., Kragic, D.: The grasp taxonomy of human grasp types. IEEE Trans. Hum.-Mach. Syst. **46**(1), 66–77 (2016)
21. Flash, T., Hogan, N.: The coordination of arm movements: an experimentally confirmed mathematical model. J. Neurosci. **5**(7), 1688–1703 (1985)
22. Fu, K.C.D., Nakamura, Y., Yamamoto, T., Ishiguro, H.: Analysis of motor synergies utilization for optimal movement generation for a human-like robotic arm. Int. J. Autom. Comput. **10**(6), 515–524 (2013)
23. Gabiccini, M., Stillfried, G., Marino, H., Bianchi, M.: A data-driven kinematic model of the human hand with soft-tissue artifact compensation mechanism for grasp synergy analysis. In: 2013 IEEE/RSJ International Conference on Intelligent Robots and Systems, pp. 3738–3745. IEEE (2013)
24. Gokulakrishnan, P., Lawrence, A., McLellan, P., Grandmaison, E.: A functional-PCA approach for analyzing and reducing complex chemical mechanisms. Comput. Chem. Eng. **30**(6), 1093–1101 (2006)
25. Handjaras, G., Bernardi, G., Benuzzi, F., Nichelli, P.F., Pietrini, P., Ricciardi, E.: A topographical organization for action representation in the human brain. Hum. Brain Mapp. **36**(10), 3832–3844 (2015)
26. Heidari, O., Roylance, J.O., Perez-Gracia, A., Kendall, E.: Quantification of upper-body synergies: a case comparison for stroke and non-stroke victims. In: ASME 2016 International Design Engineering Technical Conferences and Computers and Information in Engineering Conference, pp. V05AT07A032–V05AT07A032. American Society of Mechanical Engineers (2016)
27. Holzbaur, K.R., Murray, W.M., Delp, S.L.: A model of the upper extremity for simulating musculoskeletal surgery and analyzing neuromuscular control. Ann. Biomed. Eng. **33**(6), 829–840 (2005)
28. Lenarcic, J., Umek, A.: Simple model of human arm reachable workspace. IEEE Trans. Syst. Man Cybern. **24**(8), 1239–1246 (1994)
29. Lunardini, F., Casellato, C., d'Avella, A., Sanger, T., Pedrocchi, A.: Robustness and reliability of synergy-based myocontrol of a multiple degree of freedom robotic arm (2015)
30. Mason, C.R., Gomez, J.E., Ebner, T.J.: Hand synergies during reach-to-grasp. J. Neurophysiol. **86**(6), 2896–2910 (2001)
31. Maurel, W., Thalmann, D.: Human shoulder modeling including scapulo-thoracic constraint and joint sinus cones. Comput. Graph. **24**(2), 203–218 (2000)
32. Müller, M.: Dynamic time warping. Information Retrieval for Music and Motion, pp. 69–84 (2007)
33. Perry, J.C., Rosen, J., Burns, S.: Upper-limb powered exoskeleton design. IEEE/ASME Trans. Mechatron. **12**(4), 408 (2007)
34. Rabiner, L.R., Rosenberg, A.E., Levinson, S.E.: Considerations in dynamic time warping algorithms for discrete word recognition. J. Acoust. Soc. Am. **63**(S1), S79–S79 (1978)
35. Ramsay, J.O.: Functional data analysis. Wiley Online Library (2006)
36. Ramsay, J.O., Hooker, G., Graves, S.: Functional data analysis with R and MATLAB. Springer Science & Business Media (2009)
37. Ramsay, J.O., Silverman, B.W.: Applied functional data analysis: methods and case studies, vol. 77. Citeseer (2002)

38. Riek, L.D., Rabinowitch, T.C., Chakrabarti, B., Robinson, P.: How anthropomorphism affects empathy toward robots. In: Proceedings of the 4th ACM/IEEE International Conference on Human Robot Interaction, pp. 245–246. ACM (2009)
39. Russo, M., D'Andola, M., Portone, A., Lacquaniti, F., d'Avella, A.: Dimensionality of joint torques and muscle patterns for reaching. Front. Comput. Neurosci. **8**, 24 (2014)
40. Ryan, W., Harrison, A., Hayes, K.: Functional data analysis of knee joint kinematics in the vertical jump. Sport. Biomech. **5**(1), 121–138 (2006)
41. Salvador, S., Chan, P.: Toward accurate dynamic time warping in linear time and space. Intell. Data Anal. **11**(5), 561–580 (2007)
42. Santello, M.: Synergistic control of hand muscles through common neural input. In: The Human Hand as an Inspiration for Robot Hand Development, pp. 23–48. Springer (2014)
43. Santello, M., Baud-Bovy, G., Jörntell, H.: Neural bases of hand synergies. Front. Comput. Neurosci. **7**, 23 (2013)
44. Santello, M., Bianchi, M., Gabiccini, M., Ricciardi, E., Salvietti, G., Prattichizzo, D., Ernst, M., Moscatelli, A., Jörntell, H., Kappers, A.M., et al.: Hand synergies: integration of robotics and neuroscience for understanding the control of biological and artificial hands. Phys. Life Rev. (2016)
45. Santello, M., Flanders, M., Soechting, J.F.: Postural hand synergies for tool use. J. Neurosci. **18**(23), 10105–10115 (1998)
46. Santina, C.D., Bianchi, M., Grioli, G., Angelini, F., Catalano, M.G., Garabini, M., Bicchi, A.: Controlling soft robots: Balancing feedback and feedforward elements. IEEE Robot. Autom. Mag. **PP**(99), 1–1 (2017). https://doi.org/10.1109/MRA.2016.2636360
47. Schäling, B.: The boost C++ libraries. Boris Schäling (2011)
48. Thakur, P.H., Bastian, A.J., Hsiao, S.S.: Multidigit movement synergies of the human hand in an unconstrained haptic exploration task. J. Neurosci. **28**(6), 1271–1281 (2008)
49. Todorov, E., Ghahramani, Z.: Analysis of the synergies underlying complex hand manipulation. In: Engineering in Medicine and Biology Society, 2004. IEMBS'04. 26th Annual International Conference of the IEEE, vol. 2, pp. 4637–4640. IEEE (2004)
50. Vinjamuri, R., Sun, M., Chang, C.C., Lee, H.N., Sclabassi, R.J., Mao, Z.H.: Temporal postural synergies of the hand in rapid grasping tasks. IEEE Trans. Inf. Technol. Biomed. **14**(4), 986–994 (2010)
51. Zatsiorsky, V.M., Latash, M.L.: Prehension synergies. Exerc. Sport Sci. Rev. **32**(2), 75 (2004)

Chapter 3
Quantifying the Time-Invariance Properties of Upper Limb Synergies

In this chapter, we present a novel approach to dynamically describe human upper limb trajectories, addressing the question on whether and to which extent synergistic multi-joint behavior is observed and preserved over time evolution and across subjects. To this goal, we performed experiments to collect human upper limb joint angle trajectories and organized them in a dataset of daily living tasks. We then characterized the upper limb poses at each time frame through a technique that we named repeated-principal component analysis (R-PCA). We found that, although there is no strong evidence on the predominance of one principal component (PC) over the others, the subspace identified by the first three PCs takes into account most of the motion variability. We evaluated the stability of these results over time, showing that during the reaching phase, there is a strong consistency of these findings across participants. In other words, our results suggest that there is a time-invariant low-dimensional approximation of upper limb kinematics, which can be used to define a suitable reduced dimensionality control space for upper limb robotic devices in motion phases.

3.1 Introduction

It is nowadays broadly recognized that the human nervous system is able to control the marvelous complexity of the muscular-skeletal apparatus, enabling a seamless and effective execution of different motor actions. Understanding the processes that underline this behavior represents a very active and fascinating topic in neuroscience, where a lot of attention has been devoted to the study on the so-called *synergies*, with a special attention devoted to the hand (see next section for a discussion on the related work). Synergies are defined as principal control patterns that the brain exploit to

G. Averta, *Human-Aware Robotics: Modeling Human Motor Skills for the Design, Planning and Control of a New Generation of Robotic Devices*, Springer Tracts in Advanced Robotics 145, https://doi.org/10.1007/978-3-030-92521-5_3

handle the abundance of degrees of freedom (DoFs), to generate meaningful motions [5]. The study of this generalized simplification approach has shed light on the neural mechanisms of kinesiology and also opened promising perspectives in technologies development, such as prosthetics, rehabilitation and assistive robotics [7, 11, 23, 31, 33], for the design and control of human-inspired artificial devices with a reduced number of inputs [39].

Looking at the human upper limb, the study of the synergistic control, and its characteristics over time, has received very limited attention so far, compared with analogous studies on human hand postural synergies. The latter usually employ dimensionality reduction to identify principal joint covariation patterns, e.g. through Principal Component Analysis (PCA). State-of-the art approaches to extend the applicability of PCA with temporal kinematic data usually rely on the augmentation of the dimensionality of the dataset, expanding the number of samples along the temporal coordinate [19, 52]. A similar problem was tackled in [29], where authors, leveraging on the Spatiotemporal Decomposition discussed in [37], used PCA-based algorithms to decompose spatio-temporal kinematic patterns observed in subjects during overarm throwing movements.

This chapter contributes to the state of the art by proposing a new method, named Repeated PCA (R-PCA), that applies PCA on upper limb joint angles, at each time frame of the motion trajectories, to identify joint covariation patters in time. More specifically, after a pre-processing phase based on Dynamic Time Warping, principal components are calculated for each temporal frame of motion execution. In this manner, we are able to study the temporal evolution of postural synergies while preserving the funding hypotheses of PCA. To this goal, we built a dataset of motor tasks by asking 33 right-handed able bodied participants (17 Female, age 26.56 ± 2.77) to perform 30 daily living actions with their arm, belonging to three different classes [10]. Then, we evaluated the in-time stability of these components by quantifying the angular deviation between the subspaces identified by a reduced number of PCs at each frame. Surprisingly, despite the non-negligible variability of the single component in time, the norm of the 3D angles between the subspaces identified by the first three PCs remains lower than 20 degrees for all the motion phases, increasing only during the contact-manipulation one. These findings suggest that human upper limb movements can be generated through a time-invariant model with a good approximation during motion phases. Such an approximation degrades when the hand is in contact with the environment or is adapting to perform a specific task. To verify these results, we performed a two-stage validation: (i) a cross-validation, where we repeated 10-times our analysis on a random subset of 20 subjects; (ii) validation on a synthetic dataset that we obtained by fitting a time invariant statistical model on real data. Results show that our description appears robust across subjects and coherent with the analytic outcomes obtained with the synthetic data. The latter point allows to exclude possible effects due to the pre-processing algorithms in use. Moreover, the results on the synthetic dataset further characterize the proposed measure as a quantification of the time-invariance properties of the analyzed phenomenon during the motion phases. The content of this chapter is adapted from [4].

3.2 Related Work

The investigation of the synergistic organization in the motor domain has been studied from different points of view and at various levels (including muscular and kinematic), with a specific focus on the hand [39]. The main outcomes of all these studies is that the number of DoFs directly controlled by the human nervous system is lower than the one resulting from a pure mechanical count [32]. Latash and colleagues [26] defined synergies as a neural organization resulting in task specific covariation of elemental variables, with the ultimate goal of generating stable and effective movements. This theory, also known as Uncontrolled Manifold [26, 43], suggests that unit elements are not individually controlled but rather are free to vary. This appears to be exploited by the Central Nervous System (CNS) to maximize motion performances and minimize the error in pose control. The three main mechanisms identified in [26] to reduce the number of independent DoFs are: (i) the *elimination* of certain kinematic DoFs (e.g. using muscles spanning several kinematic joints [50]); (ii) the optimization of specific cost functions (e.g. the end effector jerk) to get coordinated movements [18, 27]; (iii) the covariation between functional elements [14, 38, 40, 41, 47, 49]. The implication of the latter point at the level of muscle activation led to the introduction of the concept of m-modes, i.e. groups of muscles that are elicited by a common activator [25], therefore acting as single functional units. At the muscular level, the concept of synergy-based dimensionality reduction has received considerable attention and produced numerous relevant results, leading to the identification of muscle synergies as the building blocks for natural motion generation [14]. The general approach is to record ElectroMyoGraphic (EMG) signals from a wide set of muscles during the execution of specific motor tasks. Matrix factorization techniques are then used to identify patterns of muscles co-activations, associated to independent synergies, which are then related to relevant action variables [48]. In [14], the authors demonstrated that these components are actively enrolled, following task-specific strategies, although some muscle activation patterns are preserved across tasks. Notwithstanding the very reduced number of synergies observed in these tasks, a high variability of the muscle activity profiles can be generated by modulating the amplitude and temporal enrollment of muscular synergies [1]. The existence of muscle synergies in humans was confirmed by several other studies, e.g. during static hand postures [8, 53], for active force generation with one or more fingers [28, 34, 51], and at the upper limb level during fast reaching movements [12]. Similar outcomes were observed at the lower limb level, e.g. in [21] the authors demonstrated that the EMG signals from 32 muscles in humans during walking are organized in patterns of temporal activation of five independent components. Alterations in this synergistic description were observed in case of motion impairment, e.g. as a consequence of stroke events. In [36], analyzing chronic stroke survivors performing a task of 3-D force matching, the authors observed the recruitment of altered shoulder muscle synergies, which was strongly associated to abnormal task performance. In [44] the authors proposed a index of impairment level based on the difference, between the affected and non-affected upper limb, in terms of movement complexity, calculated as the cumulative of variance explained by functional principal components. In [16] the authors reported on training-related changes in synergies while using a robotic therapy device to train and analyze impaired arm movements.

At the kinematic level, the synergistic description was mainly studied looking at the hand. Besides the well known papers of Santello and colleagues on hand postural synergies during grasp of imagined and real objects [40, 41], several other studies on this topic can be found in literature (see e.g. [2, 15, 30, 46, 52]). In these papers, dimensionality reduction techniques, namely PCA, Non Negative Matrix Factorization, etc., were applied to datasets of hand poses (joint values) recorded during specific tasks. The major findings of these studies demonstrate that a reduced number of correlation patterns between hand joints, e.g. described in terms of PCs, explains most of the poses variability. These results were then used to develop under-actuated robotic hands that implemented these principal co-variation schemes, e.g. directly embedded in their mechanical design see e.g. [9]. However, to the best of authors' knowledge, the rest of upper limb has not received the same level of attention, despite the crucial role that it plays in the orientation of the hand for task execution. The attention was mainly devoted to a comparative analysis of physiological and pathological conditions, such as in [20], where upper limb movements were studied during selected tasks to compare stroke patients and able-bodied participants. Similarly, in [35] it was observed that subjects with hemiparesis exploit the redundancy of their motor control to stabilize performance variables, in a way similar to healthy subjects—although with different patterns of joint couplings. In [6] the authors developed a method to quantitatively assess upper limb motor deficits in children with cerebral palsy, using three-dimensional motion analysis on data gathered during reach and grasp cyclic movements. Other papers studied muscular synergistic activation in upper limb, as in [13], where electromyographic activity was recorded from shoulder and arm muscles during point-to-point movements. However, the relation between the synergistic description at muscle level and its impact on the kinematics is far from being understood and still under investigation [37].

3.3 Experimental Setup

The experimental campaign carried out to enable the analyses of this chapter is shared with the Chap. 2, to which the interested reader is referred for technical details. In the following, we will report the major concepts needed to make this chapter standalone.

3.3.1 Setup and Experiments

Data were collected from a cohort of 33 subjects (17 Female, age 26.56 ± 2.77, all right-handed). Participants were asked to perform a set of 30 different daily-living tasks (see Tab. 2.1), while a Phase Space system (sampling frequency 100Hz) was

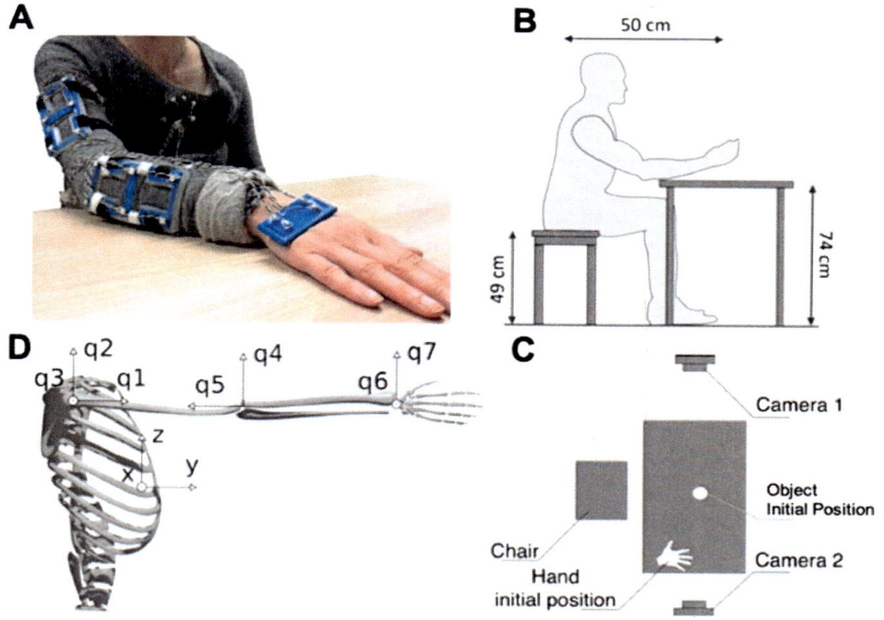

Fig. 3.1 In figure A we show marker location on upper limb and chest, in figure B and C we report a scheme of the experimental setup. Two RGB cameras are included to record the scene. Finally, in figure D, we show the system parametrization. The labels $q_1, \ldots, q_7$ refer to the joint angles of the model. The global reference frame is fixed to the chest (epigastrium) and defined by the axes x, y and z

used to track the 3D position of 20 active optical markers fastened to the upper limb and the chest - see Fig. 3.1.

Leveraging on a 7 DoFs kinematic model of the upper limb - three DoFs for the shoulder, one for the elbow Flexion-Extension, one for the forearm Pronation-Supination and two for the wrist Abduction-Adduction and Flexion-Extension, see also Fig. - joint angular values were estimated through a two-phases procedure. First, the kinematic model was calibrated through an optimization procedure. This step is used to estimate a set of subject specific kinematic parameters (e.g. bones length etc.). Then, an Extended Kalman Filter is used to estimate the joint angular values given the 3D positions of markers.

3.4 Data Analysis

To answer the question on whether and to which extent a low-dimensional representation of upper limb trajectories and inter-joint couplings exists, one approach could be to use a dimensionality reduction technique, such as Principal Component Analysis (PCA). PCA has been extensively used in the state of the art of motion analysis (see section Related Work) to identify principal directions of data variance, which may be intended as the principal DoF coupling scheme, which explain most of pose variability (see e.g. [15, 40, 41]). As reported in the Introduction and Related Work sections, several studies have focused on hand investigation, while the rest of upper limb has often been discarded. In addition, time-dependence of joint trajectories is scarcely considered when using standard PCA [19]. Indeed, the latter point must deal with the underlying observation that kinematic description at time step i intrinsically depends on the previous poses of the upper limb. As discussed in [22], classic PCA assumes that the elements of the datasets are i.i.d. This essentially excludes the usage of temporally-correlated samples. To overcome this limitation, yet preserving the temporal information, we introduced a new methodological application of PCA that we have named Repeated-PCA (R-PCA). R-PCA allows to compute time-varying PCs while a descriptive approach - reported later - enables to evaluate the robustness of PC-based kinematic description over time, through a frame-by-frame result comparison. To correctly implement this approach, it is important to have kinematic recordings synchronized in time. In other terms, taken one sample as reference, all the acquisitions should be modified in such a way to match the same start and stop frame. This is typically solved by applying Dynamic Time Warping (DTW) [3] to the dataset consisting of upper limb joint evolutions recorded during the experimental tasks of the protocol. DTW is a technique that *re-shape* temporal profiles so as they are synchronized, i.e. they have the same duration and timing. Before DTW, a moving average filter (window width 10 samples, sampling 100 Hz) was applied to each element of the dataset.

3.4.1 Dynamic Time Warping

To enable a comparison between different samples, collected data were pre-processed via Dynamic Time Warping, an approach used to normalize the temporal domain in such a way to synchronize different movements. Technical details on how this was performed in this work are provided in Chapter 2. The same method is used here, and hence is omitted.

DTW was applied to the whole dataset of joint trajectories, having as reference signal (with respect to which the optimization is carried out) one random elements of the dataset. The set of parameters identified with such procedure was then used to warp all the joint signals coherently. After this step, all the elements have the same number of elements (i.e. same time-frames) and are represented in a matrix $\in R^{N_t m}$, where N_t is the number of time frames considered and m is the number of DoFs.

3.4.2 Repeated Principal Component Analysis

As previously stated, PCA represents one of the most common approaches to identify principal patterns of variation within a dataset. PCA computes the orthogonal transformation that converts a set of correlated observations into a set of linearly uncorrelated variables called Principal Components (PCs). In other terms, let us consider X to be the matrix $\in R^{N\text{x}m}$ that collects N observations from a stochastic process with m degrees of freedom. Principal Components (PCs) can be identified as the eigenvectors S_i of the data covariance matrix $\Sigma = XX^T - \mu\mu^T$, where $\mu = \text{mean}(X)$. When applied to the description of kinematic configurations, these vectors are typically named Postural Synergies [40]. The amount of data variability that each PC can explain is calculated as the normalized corresponding eigenvalue. In other terms, the first synergy is the vector S_1 that maximizes the function $\sum_i f_{i1}^2$ subject to $||S_1|| = 1$, where $f_{i1} = S_1^T S_i$ is the first principal component score; the second eigenvector S_2, then, maximizes $\sum_i f_{i2}^2$ subject to $||S_2|| = 1$ and $S_2^T S_1 = 0$, and so on. Note that this problem is well posed only if $N >> m$.

It is worth noticing that all the elements of the dataset are intended as single independent observations of the same stochastic process. The ith row of matrix X is a vector of m variables. This severely limits the application of PCA for the investigation of human movements, given the intrinsic time-correlation of the motion samples (see Introduction). For this reason, a lot of studies on PCA for motion analysis focus on single frames of the movement, typically the last one, i.e. the final grasp pose (see e.g. [30, 40, 41]), discarding the temporal evolution of the signals.

In this chapter, we propose an innovative approach to include the temporal information into PCA, while preserving the dimensionality of the problem. This algorithm, which is described in the following, is named Repeated Principal Component Analysis (R-PCA).

Let us consider X_i to be the matrix $\in \mathbb{R}^{N\text{x}m}$ that collects the upper limb poses of all the N elements of the dataset for the specific time frame i. In this case, X_i collects a set of i.i.d. samples in R^m and, then, it can be characterized using the classic PCA implementation described before. The outcomes of this analysis provides a definition of the principal joint couplings that emerge for each time frame i. Consequently, repeating this analysis while varying i between 1 and N_t allows to describe the temporal evolution of the synergistic behavior.

It is, in general, plausible that the synergistic coupling between joints is not constant in time. To evaluate the variation of PCs during a task execution, it is possible to quantify the difference between the subspaces defined by time-varying PCs and those identified by a corresponding set of synergies used as reference. Without any loss of generality, here we used as reference PCs the ones calculated considering the average poses (over time) during motion, over the whole dataset. Hereinafter, we will refer to these PCs as *global static synergies*. Note that the mean pose is used only as a single-frame representation of the movement. In particular, in case of two one-dimensional manifolds, defined by the vectors v_1 and v_2, the minimum angular distance between them is quantified as $\theta_{1,2} = \cos^{-1}(v_1^T v_2)$. This is the typical case in which we want to compare two single synergies. In case of two or more dimensions, the distance between subspaces is provided through a number of angles equal to the cardinality of the vector set considered. In this chapter we rely on the definition of principal angles between subspaces given in [24]. Given the vector of the principal angles, the distance between two subspaces is finally calculated as norm of this vector.

3.5 Results

In Fig. 3.2 we report the variance explained in time by the first three PCs. While for hand movements several studies identified one principal component that explains a considerable part of dataset variability, our results on upper limb seem suggesting that there is no strong dominance of one inter-joint coupling among the others, but rather we can identify three main PCs that, together, explain between 70 and 85% of the total variance ($73.13 \pm 4.64\%$). The variance explained by each PC is almost stable over time, respectively $36.07 \pm 3.24\%$, $24.37 \pm 1.79\%$, $16.68 \pm 2.44\%$. These values indicate that the Euclidean space defined by the first three components represents a low dimensional control manifold that accounts for most of the upper limb movement variability.

What is also noticeable - as expected - is that the variance explained by the first three components is lower during the interaction phase, i.e. when subjects are in contact with the objects or are performing specific gestures. This observation supports the idea that the complexity of motion control increases when subjects are interacting with the environment.

Looking at individual PCs computed through R-PCA, we can observe that these tend to be not stable in time, but rather present variations during the motion execution (see Fig. 3.3) which could be interpreted as a dynamic modulation along trajectories. Therefore, to systematically evaluate synergies variability in time, we used the definition of distance between subspaces introduced in 3.4.2. In particular, the stability of each single synergy is quantified as the time-varying distance with respect to the corresponding PC defined on the average dataset (*global static synergy*). The term "correspondin" refers hereinafter to the cardinality, i.e. the first time-varying kinematic synergy is compared with the first global static synergy and so on. This definition can be easily extended to multi-dimensional manifolds: the distance between two k-dimensional subspaces is calculated as the norm of k principal angles, as defined in [24].

In Fig. 3.4 we report the angular distance between each time-varying kinematic synergy (only the first three are reported here for the sake of space), and the corresponding global static synergy. As expected, synergies present a remarkable variation in time—not directly correlated with the action phases. However, interestingly, when we consider the subspace defined by the first three PCs together, the norm of the 3D distance w.r.t. the corresponding global static synergy remains below 20 degrees for the whole movement phases (see Fig. 3.5), while it increases during the con-

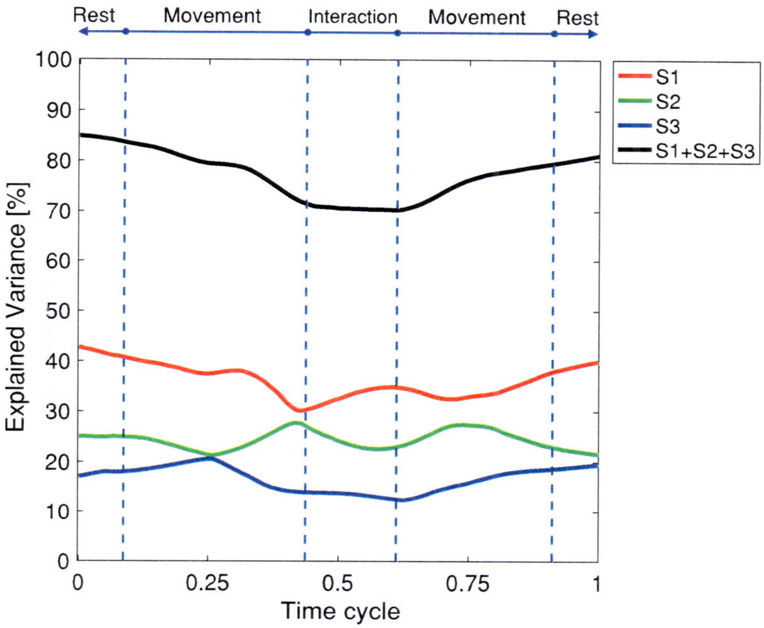

Fig. 3.2 Percentage of variance explained by the first three Principal Components in time. Red plot is associated to S_1, green plot to S_2 and blue plot to S_3. The sum is finally reported in black

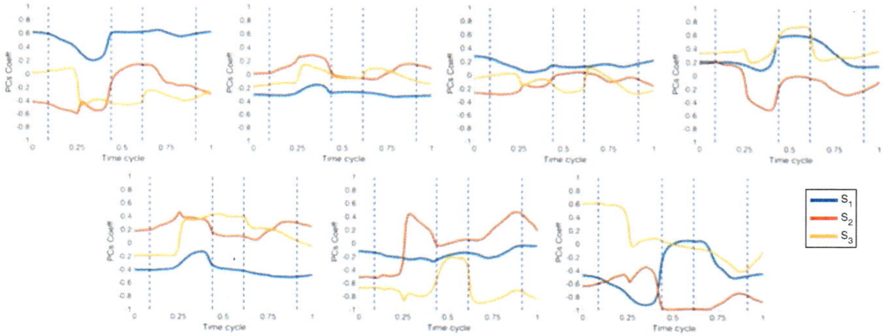

Fig. 3.3 Coefficients of the first three PCs in Time. From top-left to bottom-right we report the values for joints q_1-q_7 according to the labeling of Fig. 3.1 D. For reader's convenience, time cycle is segmented according to Fig. 3.2

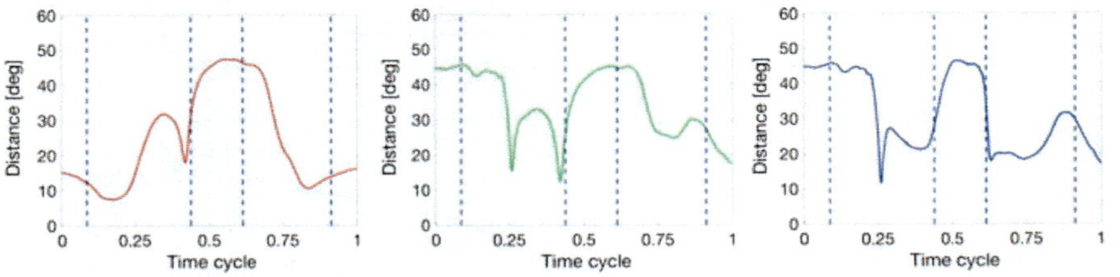

Fig. 3.4 Angular deviation between individual PCs computed through R-PCA and the corresponding global static PCs. From left to right, we report the values – in degrees – over the time for the first, the second and the third one, respectively. For reader's convenience, time cycle is segmented according to Fig. 3.2

tact/manipulation/fine gesture control phase. This aspect appears to be coherent with the observation we discussed earlier on the variance that components explain in time. In other terms, results show that, even though single PCs present a high variability in time, the subspace shaped by the first three PCs remains rather invariant during the pure motion phases. It is worth noticing that such effect is not present when we compare sub-manifolds with higher or lower dimensionality. We believe that, for two-dimensional manifolds (i.e. $S_1 + S_2$), this is due to the fact that two PCs are not sufficient to describe the dataset variability in a sufficiently accurate manner and, then, the third one is needed to effectively take into account upper limb variability and control. Regarding higher-dimensional manifolds (number of components between 4 and 6), this observation remains an open point. One possible explanation could be that, while the first three synergies are used to shape the gross movement, higher order synergies account for very reduced contribution mainly used to refine the motion and, hence, can show different pattern during the temporal evolution.

3.5.1 Robustness Across Subjects and Validation

To prove the validity of our results, we investigated their robustness across subjects by repeating the aforementioned analyses 10 times, while randomly selecting a subset of 20 subjects for further analyses. Results show that the norm of three-dimensional distance between subspaces presents a shape similar to the one reported in Fig. 3.5 (related plots are omitted here for the sake of space), and it is always below 25 deg during the motion phases.

Moreover, to verify that the time-varying angular deviation between individual PCs (see Fig. 3.4) is actually due to the non-stationarity of the upper-limb motor control, with no biases from the proposed R-PCA methodology, we performed the analyses on a synthetic dataset generated through an autoregressive (AR) model identified on real data. In particular, the following AR model:

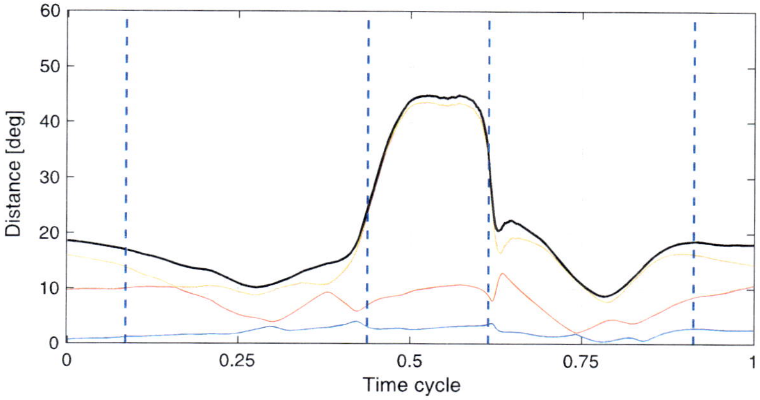

Fig. 3.5 Norm of 3D angular deviation between the subspace of the first three PCs computed through R-PCA and the subspace of the first three global static synergies. Yellow, red and blue lines show the three principal angles between the two subspaces as defined in [24], while in black we report their norm

$$y(t) = a_1 y(t-1) + \cdots + a_{n_a} y(t - n_a) + e(t), \tag{3.1}$$

with $e(t)$ i.i.d. Gaussian white noise with zero mean and standard deviation equal to 1, was fitted on real data (selecting one random trial from the whole dataset) at different model order. We selected $n_a = 4$ in accordance with the Akaike information criterion and the minimal root mean square error. Parameters a_i are then estimated through a least-square method. Then, we generated 1000 repetitions of a 7x500 realization of the AR process, and applied our R-PCA method. As the model is time-invariant, it ensures the stationarity of the output. As expected, we found negligible values of the norm of angular deviation between the first three PC-based subspace over time as compared to the global static one, with a resulting norm of the angular deviation as low as 1.3684 ± 0.4760 (Max $= 2.338$).

3.5.2 *Principal Component Description*

Results presented so far suggest that a static three-dimensional description of upper limb kinematics is sufficient to explain a high percentage of dataset variance during different motion phases. In addition, it is possible to affirm that such approximation can be considered effective during the free movements of the upper limb. Considering the movement phases in which the process can be intended as time-invariant, we can define a set of three vectors that are contained in the span of the first three global static synergies and use it to describe the upper limb movements through a time-invariant basis. In Fig. 3.6 we report the reconstruction of an upper limb kinematic model while moving around the mean pose along the first three global static synergies. Of note,

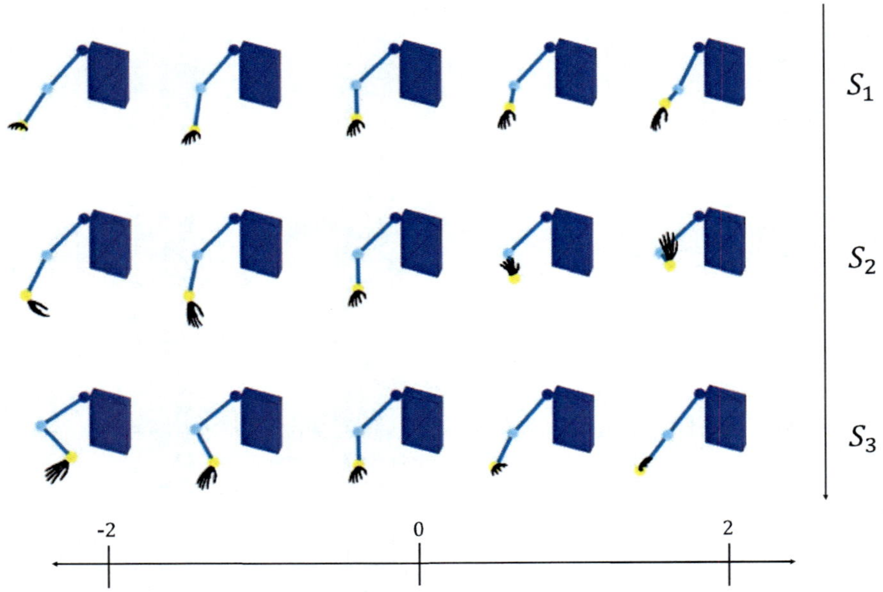

Fig. 3.6 Different upper limb shapes along the first three synergies. For each row, the central figure shows the mean pose, while the others report the effect of the corresponding synergy modulated by a coefficient σ varying between −2 and 2

Table 3.1 Coefficients of the first three global static synergies

	S_1	S_2	S_3
$\mathbf{q}_1$	−0.5280	0.4504	0.2619
$\mathbf{q}_2$	0.2898	−0.1376	0.0219
$\mathbf{q}_3$	−0.1011	0.1454	0.0705
$\mathbf{q}_4$	−0.2067	0.2296	−0.4727
$\mathbf{q}_5$	0.3999	−0.3577	−0.0880
$\mathbf{q}_6$	0.0967	−0.0520	0.8335
$\mathbf{q}_7$	0.6442	0.7574	−0.0063

the three main movements are not in general required to be physically meaningful. However, it is possible to observe that the first global static synergy appears to be an extension/flexion of the arm in the sagittal plane, while the second is more involved in modulating the level of the hand along the vertical axis. Numerical values of the first three global static synergies are listed in Table 3.1.

3.6 Discussions

The results presented in this study suggest that a synergistic coupling between upper limb joints is noticeable during daily living activities. These synergies define a three dimensional sub-manifold that appears to be stable during the motion phase. We do believe that these outcomes can shade light on the neural mechanisms underpinning synergistic motion generation over time and potentially impact the rehabilitation field too. Indeed, neurological disorders, e.g. stroke and focal dystonia, affect the ability to coordinate multi-joint movements [42]. Having a robust and simplified description of upper-limb kinematics in physiological conditions could serve as a baseline to evaluate eventual discrepancies induced by the pathology, which can affect the PC-based low-dimensional representation of motions (different number and/or type of PCs involved), its time-invariance, and the robustness across subjects. For the latter point, it is indeed known that there is a large phenomenological variety in subjects with motor disorders (related to different levels of impairment) [42], which can be compared with the physiological case to assess the effects of rehabilitative procedures.

However, the implications of this chapter are not limited to the neuroscientific field only, but rather we envision potential applications in robotics as well. The concept of synergies and its intrinsic connection with dimensionality reduction has found fertile application fields in engineering, with special focus on artificial hands and grasping, see the Introduction and e.g. [39] for a review on these topics. The time-invariant model described in this chapter can be used to couple (and under-actuate) the DoFs of an artificial upper limb, using mechanical implementation or though software control. This can enable the execution of a large variety of actions while maintaining the actuation complexity low. The same concept, translated in the mechanical domain, could be combined with soft robotics technologies, as it was done e.g. in [9], to instruct the design of simplified robotic manipulators, or prostheses, able to safely and effectively interact with the external environment. At the same time, the PC-based description can be used for simplifying the control of fully-actuated anthropomorphic robotic arms, e.g. for a safe human-robot communication and interaction [17].

3.7 Conclusions

In this chapter, we analyzed human upper limb movements during activities of daily living and described the underlying synergistic behavior. Time-varying Principal Components have been extracted using a frame-by-frame PCA approach. Results suggest that large part of upper limb kinematic trajectory variability can be approximated, with a good level of accuracy, using a time-invariant low-dimensional model identified by the first principal components of angular joints. Such description has been proven to be preserved for a large part of the motion execution. Variations of this time-unvarying synergistic description appear in specific portions of the task. We believe that this effect may be related to a fine—likely feedback—control of the limb in the proximity of goal reaching. Further works will follow two main directions: (i) include additional tasks or different task set to further strengthen the here reported outcomes, (ii) analyze the linkage between these results and the muscular counterpart. Indeed, the relation between kinematic and muscle synergies is still an open point in literature. Recently, some studies attempted to correlate the two domains. In particular, in [37] the authors argued that the relation between muscle and joint-torque patterns is not fixed in general because of the non-linear relation between the two domains, also considering the intrinsic redundancy of the muscular-skeletal system. Similar findings were discussed in [45] considering hand motion and EMG measurements during reach, grasp and pull tasks. The authors observed that muscle synergies are correlated although shifted in phase whit respect to the kinematic counterpart. The authors also hypothesized that kinematic synergies may have in part origin from a synergistic muscle activation. It is possible to speculate that the description essentially time-invariant of upper limb synergies we reported in this chapter may be related to principal muscle patterns, while changes observed during the interaction phase may be also introduced by the enrollment of different, time-dependent, muscle synergies in response to environment adaptation.

References

1. d Avella, A., Tresch, M.C.: Modularity in the motor system: decomposition of muscle patterns as combinations of time-varying synergies. Adv. Neural Inf. Process. Syst. **1**, 141–148 (2002)
2. Averta, G., Angelini, F., Bonilla, M., Bianchi, M., Bicchi, A.: Incrementality and hierarchies in the enrollment of multiple synergies for grasp planning. IEEE Robot. Autom. Lett. **3**(3), 2686–2693 (2018)
3. Averta, G., Della Santina, C., Battaglia, E., Felici, F., Bianchi, M., Bicchi, A.: Unvealing the principal modes of human upper limb movements through functional analysis. Front. Robot. AI **4**, 37 (2017)
4. Averta, G., Valenza, G., Catrambone, V., Barontini, F., Scilingo, E.P., Bicchi, A., Bianchi, M.: On the time-invariance properties of upper limb synergies. IEEE Trans. Neural Syst. Rehabil. Eng. **27**(7), 1397–1406 (2019)
5. Bernstein, N.A.: The control and regulation of movements (1967)
6. Butler, E.E., Ladd, A.L., Louie, S.A., LaMont, L.E., Wong, W., Rose, J.: Three-dimensional kinematics of the upper limb during a reach and grasp cycle for children. Gait Posture **32**(1), 72–77 (2010)
7. Casini, S., Tincani, V., Averta, G., Poggiani, M., Santina, C.D., Battaglia, E., Catalano, M.G., Bianchi, M., Grioli, G., Bicchi, A.: Design of an under-actuated wrist based on adaptive synergies (accepted). ICRA (2017)
8. Castellini, C., van der Smagt, P.: Evidence of muscle synergies during human grasping. Biol. Cybern. **107**(2), 233–245 (2013)
9. Catalano, M.G., Grioli, G., Farnioli, E., Serio, A., Piazza, C., Bicchi, A.: Adaptive synergies for the design and control of the Pisa/IIT softhand. Int. J. Robot. Res. **33**(5), 768–782 (2014)
10. Catrambone, V., Greco, A., Averta, G., Bianchi, M., Valenza, G., Scilingo, E.P.: Predicting object-mediated gestures from brain activity: An eeg study on gender differences. IEEE Trans. Neural Syst. Rehabil. Eng. **27**(3), 411–418 (2019). https://doi.org/10.1109/TNSRE.2019.2898469

11. Coradeschi, S., Ishiguro, H., Asada, M., Shapiro, S.C., Thielscher, M., Breazeal, C., Mataric, M.J., Ishida, H.: Human-inspired robots. IEEE Intell. Syst. **4**, 74–85 (2006)
12. d'Avella, A., Portone, A., Fernandez, L., Lacquaniti, F.: Control of fast-reaching movements by muscle synergy combinations. J. Neurosci. **26**(30), 7791–7810 (2006)
13. d'Avella, A., Portone, A., Fernandez, L., Lacquaniti, F.: Control of fast-reaching movements by muscle synergy combinations. J. Neurosci. **26**(30), 7791–7810 (2006)
14. d'Avella, A., Saltiel, P., Bizzi, E.: Combinations of muscle synergies in the construction of a natural motor behavior. Nat. Neurosci. **6**(3), 300 (2003)
15. Della Santina, C., Bianchi, M., Averta, G., Ciotti, S., Arapi, V., Fani, S., Battaglia, E., Catalano, M.G., Santello, M., Bicchi, A.: Postural hand synergies during environmental constraint exploitation. Front. Neurorobot. **11**, 41 (2017)
16. Dipietro, L., Krebs, H.I., Fasoli, S.E., Volpe, B.T., Stein, J., Bever, C.T., Hogan, N.: Changing motor synergies in chronic stroke. J. Neurophysiol. (2007)
17. Duffy, B.R.: Anthropomorphism and the social robot. Robot. Auton. Syst. **42**(3), 177–190 (2003)
18. Flash, T., Hogan, N.: The coordination of arm movements: an experimentally confirmed mathematical model. J. Neurosci. **5**(7), 1688–1703 (1985)
19. Gasparri, G.M., Manara, S., Caporale, D., Averta, G., Bonilla, M., Marino, H., Catalano, M., Grioli, G., Bianchi, M., Bicchi, A., et al.: Efficient walking gait generation via principal component representation of optimal trajectories: application to a planar biped robot with elastic joints. IEEE Robot. Autom. Lett. **3**(3), 2299–2306 (2018)
20. Heidari, O., Roylance, J.O., Perez-Gracia, A., Kendall, E.: Quantification of upper-body synergies: a case comparison for stroke and non-stroke victims. In: ASME 2016 International Design Engineering Technical Conferences and Computers and Information in Engineering Conference, pp. V05AT07A032–V05AT07A032. American Society of Mechanical Engineers (2016)
21. Ivanenko, Y.P., Poppele, R.E., Lacquaniti, F.: Five basic muscle activation patterns account for muscle activity during human locomotion. J. Physiol. **556**(1), 267–282 (2004)
22. Jolliffe, I.: Principal component analysis. Springer (2011)
23. Kim, S., Laschi, C., Trimmer, B.: Soft robotics: a bioinspired evolution in robotics. Trends Biotechnol. **31**(5), 287–294 (2013)
24. Knyazev, A.V., Argentati, M.E.: Principal angles between subspaces in an a-based scalar product: algorithms and perturbation estimates. SIAM J. Sci. Comput. **23**(6), 2008–2040 (2002)
25. Krishnamoorthy, V., Goodman, S., Zatsiorsky, V., Latash, M.L.: Muscle synergies during shifts of the center of pressure by standing persons: identification of muscle modes. Biol. Cybern. **89**(2), 152–161 (2003)
26. Latash, M.L., Scholz, J.P., Schöner, G.: Toward a new theory of motor synergies. Mot. Control **11**(3), 276–308 (2007). https://doi.org/10.1123/mcj.11.3.276
27. Loeb, G.E.: Optimal isn't good enough. Biol. Cybern. **106**(11–12), 757–765 (2012)
28. Maier, M., Hepp-Reymond, M.: Emg activation patterns during force production in precision grip. ii. muscular synergies in the spatial and temporal domain. Exp. Brain Res. **103**(1), 123–136 (1995)
29. Maselli, A., Dhawan, A., Russo, M., Cesqui, B., Lacquaniti, F., d'Avella, A.: A whole body characterization of individual strategies, gender differences, and common styles in overarm throwing. J. Neurophysiol. **122**(6), 2486–2503 (2019)
30. Mason, C.R., Gomez, J.E., Ebner, T.J.: Hand synergies during reach-to-grasp. J. Neurophysiol. **86**(6), 2896–2910 (2001)
31. McDaid, A., Kora, K., Xie, S., Lutz, J., Battley, M.: Human-inspired robotic exoskeleton (hurex) for lower limb rehabilitation. In: 2013 IEEE International Conference on Mechatronics and Automation (ICMA), pp. 19–24. IEEE (2013)
32. Newell, K.M., Vaillancourt, D.E.: Dimensional change in motor learning. Hum. Mov. Sci. **20**(4–5), 695–715 (2001)
33. Park, Y.L., Chen, B.r., Pérez-Arancibia, N.O., Young, D., Stirling, L., Wood, R.J., Goldfield, E.C., Nagpal, R.: Design and control of a bio-inspired soft wearable robotic device for ankle–foot rehabilitation. Bioinspiration Biomim. **9**(1), 016007 (2014)

34. Poston, B., Danna-Dos Santos, A., Jesunathadas, M., Hamm, T.M., Santello, M.: Force-independent distribution of correlated neural inputs to hand muscles during three-digit grasping. Am. J. Physiol.-Hear. Circ. Physiol. (2010)
35. Reisman, D.S., Scholz, J.P.: Aspects of joint coordination are preserved during pointing in persons with post-stroke hemiparesis. Brain **126**(11), 2510–2527 (2003)
36. Roh, J., Rymer, W.Z., Perreault, E.J., Yoo, S.B., Beer, R.F.: Alterations in upper limb muscle synergy structure in chronic stroke survivors. Am. J. Physiol.-Hear. Circ. Physiol. (2012)
37. Russo, M., D'Andola, M., Portone, A., Lacquaniti, F., d'Avella, A.: Dimensionality of joint torques and muscle patterns for reaching. Front. Comput. Neurosci. **8**, 24 (2014)
38. Saltiel, P., Wyler-Duda, K., D'Avella, A., Tresch, M.C., Bizzi, E.: Muscle synergies encoded within the spinal cord: evidence from focal intraspinal NMDA iontophoresis in the frog. J. Neurophysiol. **85**(2), 605–619 (2001)
39. Santello, M., Bianchi, M., Gabiccini, M., Ricciardi, E., Salvietti, G., Prattichizzo, D., Ernst, M., Moscatelli, A., Jörntell, H., Kappers, A.M., et al.: Hand synergies: integration of robotics and neuroscience for understanding the control of biological and artificial hands. Phys. Life Rev. (2016)
40. Santello, M., Flanders, M., Soechting, J.F.: Postural hand synergies for tool use. J. Neurosci. **18**(23), 10105–10115 (1998)
41. Santello, M., Flanders, M., Soechting, J.F.: Patterns of hand motion during grasping and the influence of sensory guidance. J. Neurosci. **22**(4), 1426–1435 (2002)
42. Santello, M., Lang, C.E.: Are movement disorders and sensorimotor injuries pathologic synergies? when normal multi-joint movement synergies become pathologic. Front. Hum. Neurosci. **8**, 1050 (2015)
43. Scholz, J.P., Schöner, G.: The uncontrolled manifold concept: identifying control variables for a functional task. Exp. Brain Res. **126**(3), 289–306 (1999). https://doi.org/10.1007/s002210050738
44. Schwarz, A., Averta, G., Veerbeek, J.M., Luft, A., Held, J.P., Valenza, G., Bicchi, A., Bianchi, M.: A functional analysis-based approach to quantify upper limb impairment level in chronic stroke patients: A pilot study. In: Engineering in Medicine and Biology Society, 2019. EMBC'19. 41th Annual International Conference of the IEEE. IEEE (2019)
45. Tagliabue, M., Ciancio, A.L., Brochier, T., Eskiizmirliler, S., Maier, M.A.: Differences between kinematic synergies and muscle synergies during two-digit grasping. Front. Hum. Neurosci. **9**, 165 (2015)
46. Thakur, P.H., Bastian, A.J., Hsiao, S.S.: Multidigit movement synergies of the human hand in an unconstrained haptic exploration task. J. Neurosci. **28**(6), 1271–1281 (2008)
47. Ting, L.H., Macpherson, J.M.: A limited set of muscle synergies for force control during a postural task. J. Neurophysiol. (2005)
48. Tresch, M.C., Cheung, V.C., d'Avella, A.: Matrix factorization algorithms for the identification of muscle synergies: evaluation on simulated and experimental data sets. J. Neurophysiol. (2006)
49. Tresch, M.C., Saltiel, P., Bizzi, E.: The construction of movement by the spinal cord. Nat. Neurosci. **2**(2), 162 (1999)
50. Tuller, B., Turvey, M.: The bernstein perspective: ll. the concept of muscle linkage or coordinative. Human Motor Behavior: An Introduction, p. 253 (1982)
51. Valero-Cuevas, F.J.: Predictive modulation of muscle coordination pattern magnitude scales fingertip force magnitude over the voluntary range. J. Neurophysiol. **83**(3), 1469–1479 (2000)
52. Vinjamuri, R., Sun, M., Chang, C.C., Lee, H.N., Sclabassi, R.J., Mao, Z.H.: Temporal postural synergies of the hand in rapid grasping tasks. IEEE Trans. Inf Technol. Biomed. **14**(4), 986–994 (2010)
53. Weiss, E.J., Flanders, M.: Muscular and postural synergies of the human hand. J. Neurophysiol. **92**(1), 523–535 (2004). https://doi.org/10.1152/jn.01265.2003

Chapter 4
Evidences on the Hierarchical Control of Human Hands

Humans are able to intuitively exploit the shape of an object and environmental constraints to achieve stable grasps and perform dexterous manipulations. In doing that, a vast range of kinematic strategies can be observed. However, in this work we formulate the hypothesis that such ability can be described in terms of a synergistic behavior in the generation of hand postures, i.e., using a reduced set of commonly used kinematic patterns. This is in analogy with previous studies showing the presence of such behavior in different tasks, such as grasping. We investigated this hypothesis in experiments performed by six subjects, who were asked to grasp objects from a flat surface. We quantitatively characterized hand posture behavior from a kinematic perspective, i.e., the hand joint angles, in both pre-shaping and during the interaction with the environment. To determine the role of tactile feedback, we repeated the same experiments but with subjects wearing a rigid shell on the fingertips to reduce cutaneous afferent inputs. Results show the persistence of at least two postural synergies in all the considered experimental conditions and phases. Tactile impairment does not alter significantly the first two synergies, and contact with the environment generates a change only for higher order Principal Components. A good match also arises between the first synergy found in our analysis and the first synergy of grasping as quantified by previous work. The present study is motivated by the interest of learning from the human example, extracting lessons that can be applied in robot design and control. Thus, we conclude with a discussion on implications for robotics of our findings.

G. Averta, *Human-Aware Robotics: Modeling Human Motor Skills for the Design, Planning and Control of a New Generation of Robotic Devices*, Springer Tracts in Advanced Robotics 145, https://doi.org/10.1007/978-3-030-92521-5_4

4.1 Kinematic Domain

4.1.1 Introduction

The human hand is a remarkably complex system, with many joints, ligaments, muscles, and sensory receptors contributing to its wide dexterity. The control of such an abundance, or redundancy, is classically referred to as Bernstein's problem [6]. Several neuroscientific findings [31, 38, 41, 44, 54] suggest that the human nervous system is able to cope with such complexity leveraging on a control space of reduced dimensionality. Under this regard, synergies were defined as principal patterns of actuation. Through them, the human Central Nervous System, also leveraging upon peripheral constraints, can generate movements by combining pre-organized patterns, i.e. by simultaneously activating different degrees of freedom, instead of acting separately on each joint or muscle. These patterns are implemented at various level of the motor control architecture, i.e. either they are represented by cortical mechanisms, hard-coded in low level neural circuits, or generated by the mechanical organization of human musculo-skeletal system [34, 45, 56]. The synergy concept has been investigated in different tasks. Examples regarding hand control are grasping imagined objects [47], reach-to-grasp [36], precision grip [25].

Another key aspect in the generation of meaningful motor actions was recognized in the interplay between synergistic components and the external environment [20]. Indeed, from a sensory point of view, it plays a crucial role to build up our knowledge of the world. Under this regard, a well-known characterization of hand movements during the exploration of the environment is represented by the Exploratory Procedures identified by Lederman and Klatzky [33]. Exploratory Procedures are stereotyped hand motions characteristic of human haptic exploration of the environment. By applying these procedures, subjects try to maximize the amount of haptic information acquired from external objects. Thakur et al. [55] analyzed Exploratory Procedures through Principal Component Analysis, identifying a reduced set of postural synergies generating the observed behavior.

Focusing instead on the use of constraints for effective motor task accomplishment, Environmental Constraint Exploitation (ECE) primitives are defined in [19]. We will refer in the following to ECE strategies as actions which aim to manipulate or grasp an object, and involve the use of a constraint as central element of the strategy. Figure 4.1 shows some examples reproducing experimentally observed behaviors, where a subject exploits a flat surface. In [43] a taxonomy of ECE primitives is proposed. In [19], authors also show that impairing the use of the environment during grasping and manipulation sensibly decreases subjects performance, suggesting the central role of these strategies in humans. This way of manipulating and grasping is very different from the classical strategies used in robotics, where the contact with the environment is avoided as much as possible [10]. Thus, it is authors' opinion that a deeper understanding of ECE can directly affect the performance of robotic manipulators.

Fig. 4.1 Photo sequences of some examples of experimental constraint exploitation (one for each row). In the first two rows, the subject uses the surface to guide the fingertips position and robustly grasp the object (pinch grasp in the first row, power grasp in the second). In the third row the subject uses the surface to re-orientate the object. In rows four and five, two thin objects are grasped by flipping them using the surface as pivot. In the last row, the subject shifts the object by sliding it through the surface to the edge of the table, where it can be efficiently grasped. From these few examples, we can observe a large variety of strategies and related hand postures

In this chapter we investigate the presence of a synergistic behavior underlying the generation of hand postures during ECE, execution and planning. We also investigate the effect that cutaneous impairment might have on such synergies. We were motivated in doing this by the many studies in literature highlighting the central role of tactile information in grasping [28, 40, 58].

We performed experiments with six participants, who were asked to grasp a set of objects from a flat surface. We selected this task since it represents a good trade-off between analytic complexity and richness of kinematic behavior induced by ECE, as accounted e.g. by Fig. 4.1 and in [19]. Two experimental conditions were considered: with and without cutaneous impairment. In the first case, participants were requested to wear rigid shells at their fingertips. Results were processed through Principal Component Analysis, as it is common in literature [45], in order to check the existence of a set of Principal Components describing the recorded data. We performed the analysis during the actual contact with the environment and on pre--shaping postures, i.e., during the approaching movement that precedes the hand-object-table interaction.

A marked synergistic behavior results in all the considered experiments. Both the interaction with the external environment and the cutaneous impairment appear to modify only the higher order Principal Components. Furthermore the first synergy exhibits a good resemblance with the one observed in grasping of imagined objects (reported in [47]).

In recent years, the idea of hand synergies (with special focus on grasping) has been successfully applied in robotics for a simplified yet effective design, sensing, and control of artificial systems. For a detailed review, we refer the interested reader to [1, 46]. In accordance with this previous successful experience of mutual inspiration between neuroscience and robotics, we believe that the here proposed results can inform the design and control of artificial hands able to take full advantage of the Exploitation of Environmental Constraints. The content of this chapter is adapted from [4, 16].

4.1.2 Materials and Methods

4.1.2.1 Participants

We tested six able-bodied volunteers (three females, three males; age range: 23–27 years, mean 25.17 years). All subjects were tested on their dominant hand (right hand; self-reported hand dominance). All participants were naive to the experimental purpose of the study and had no history of neuromuscular disorders. Before data collection, subjects signed an informed consent to participate in the experiment. The experimental protocols were approved by the Institutional Review Board of University of Pisa, in accordance to the declaration of Helsinki.

4.1.2.2 Apparatus

For the present investigation we employed a multimodal acquisition setup (see Fig. 4.2), composed of

- Phase Space Motion Capture System; we recorded kinematic data using a commercial system for 3D motion tracking with active LED markers, the Phase Space.[1] Ten stereo-cameras working 480 Hz tracked 3D positions of 24 markers fastened to hand and phalanges as shown in Fig. 4.2a. The LED frequency is in the visible red.
- A fingertip shell (Fig. 4.2b); we used the shell of ThimbleSense sensor [5]. Subjects wore Thimblesenses in all fingertips during the tactile impairment experiments, as in Fig. 4.3. Each shell is connected to the correspondent fingertip as in classical thimbles. Please refer to [5] for a more extensive description of the impairment

[1] http://www.phasespace.com/, website visited on April 2017.

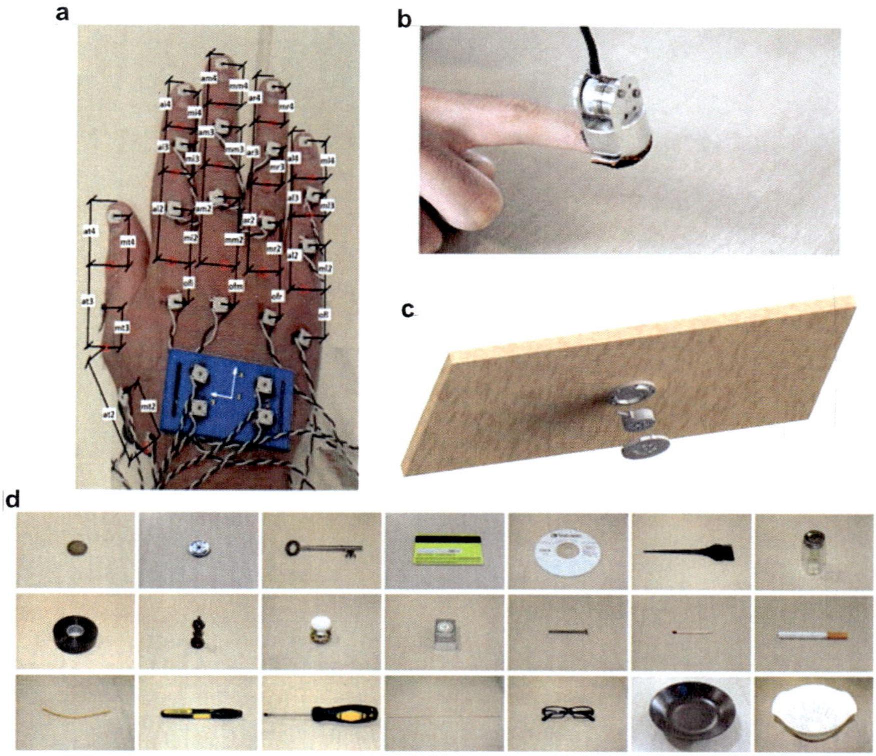

Fig. 4.2 Experimental apparatus. Kinematic measures are kept through an active marker motion capture system. Impaired condition is imposed through ThimbleSenses' rigid shells. Force and torque data were acquired from the sensorized table. **a** LED distribution, **b** ThimbleSense, **c** sensorized surface, and **d** set of objects used in the experiment

effects of the shells. Also force information is acquired, to be used for further analyses. Subjects wore Thimblesenses on all fingertips, during the tactile impairment experiments.

- A set of 21 objects (Fig. 4.2d); this is composed of 2 euro coin, button badge, key, credit card, CD, comb hair color, salt shaker, tape, chessman (queen), knob, matchbox, screw, match, cigarette, rubber band, marker, screw driver, shashlik, glasses, coffee mug, plat.
- Sensorized platform (600 $\times$ 400 mm), which includes the force-torque sensor ATI mini45E mounted as in Fig. 4.2c. It allows the sensing of interaction forces/torques with the table where objects were placed.
- Two cameras (Logitech hd 1080p) to record from two viewpoints the experiments, as in Fig. 4.4b. The recordings were used to visually check the effectiveness of the segmentation. We used blue filters to improve the quality of the recording, which

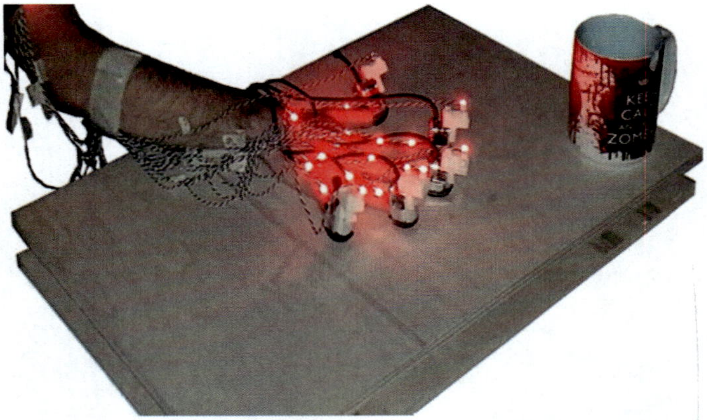

Fig. 4.3 Hand wearing ThimbleSense and optical markers supported by the sensorized table

was reduced by the poor light conditions and the red light emittance due to the motion capture system.

In contrast with previous work on grasping real objects (e.g. [36]), we chose objects that are difficult to grasp and therefore necessitate exploitation of the flat surface constraint. A similar set of objects was used in [19] in an analogous table-top scenario, demonstrating the effectiveness of eliciting a rich set of ECE strategies.

The acquisition is implemented through a custom application developed in C++ employing:

- Boost libraries [50] to perform the synchronization between Phase Space data and force/torque sensors;
- Phase Space OWL library to get the optical tracking system data
- A custom library providing an interface to acquire the force/torque sensor data [52].

The acquisition system allows to organize and synchronize data with respect to an absolute clock with period 0.025 s.

4.1.2.3 Experiments

The experiments were designed with the aim of identifying kineto-static primitives used by humans in grasping and manipulation tasks involving ECE. The subject was asked to comfortably seat in front of the table, as in Fig. 4.4a. The subject was instructed to pose her/his hand in the initial hand position location, positioned at the right side of the sensorized surface as in Fig. 4.4b. The distance between the subject's hand starting position and the site where the object was placed was 60cm. For each trial, subjects were asked to reach the object posed in the center of the sensorized surface. Once the hand had reached the object, subjects were asked to

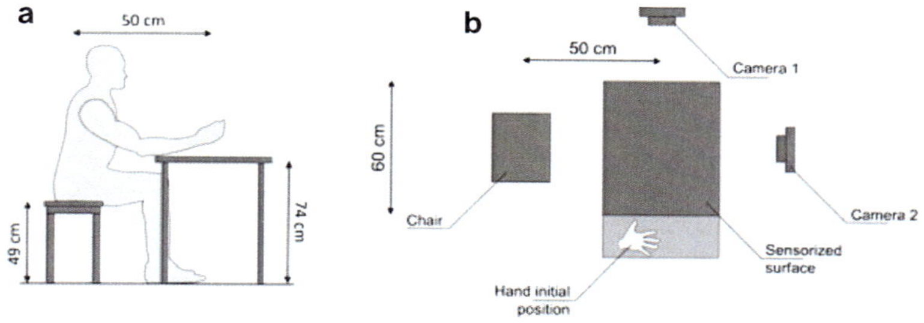

Fig. 4.4 Experimental setup. The subjects sat comfortably in front of the table. In the starting position, subject's hand was located over the drawing on right side of the sensorized surface. In addition to the ten cameras used for the motion capture, two cameras are included in the setup to record the scene. **a** Setup side view. **b** Setup from above

grasp, lift (∼21 cm height), hold (∼1 s), put it back on the table, and place the hand back to its starting position. The experimenter gave the starting signal to subjects. In the instructions, the experimenter emphasized that the whole movement should be performed in a natural fashion, i.e. the object should be grasped as if the subject was about to use it, in accordance with [47]. Two trials were performed for any of the 21 objects grasped. The object order was randomized for every subject. The sequence of trials was repeated two times, with and without tactile impairment, for a total of 84 trials per subject. Each subject performed the whole experiment in a single day. The experiments with tactile impairment were performed in the morning, the experiments without tactile impairment in the afternoon, in agreement with previous experiments [5]. Each subject performed the experiment in an independent day w.r.t the other ones.

4.1.2.4 Data Pre-Processing and Analysis of Joint Values

Data processing is organized in two phases. First, data is pre-processed to reduce the noise and to evaluate joint angles and contact points. Second, we perform statistical analyses to identify the presence of a synergistic behavior in ECE. We perform the analyses in pre-shaping, and during the whole contact with the environment. To quantify the role of tactile impairment, we consider separately the impaired and unimpaired cases. The analysis was performed considering the whole group of participants. In Sect. 4.1.3.4 we report the CI for the dot products between the synergies' vectors.

Data Processing
Raw data collected from the experimental setup were force and torque from the six axes F/T sensor ATImini45 and 24 LED positions from the motion capture system, both with and without tactile impairment.

a

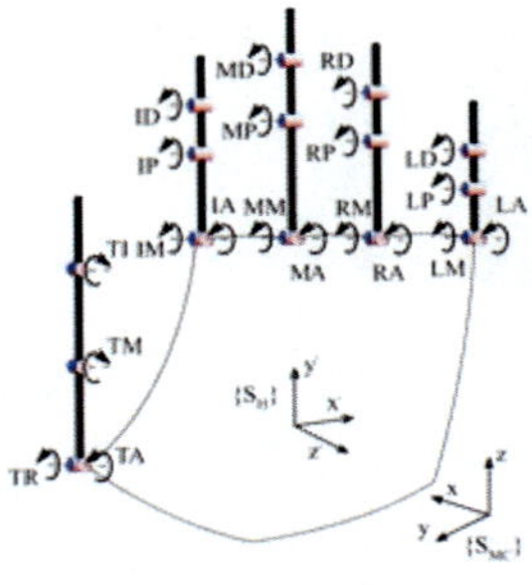

DoFs	Description
TA	Thumb Abduction
TR	Thumb Rotation
TM	Thumb Metacarpal
TI	Thumb Interphalangeal
IA	Index Abduction
IM	Index Metacarpal
IP	Index Proximal
ID	Index Distal
MM	Middle Metacarpal
MA	Middle Abduction
MP	Middle Proximal
MD	Middle Distal
RA	Ring Abduction
RM	Ring Metacarpal
RP	Ring Proximal
RD	Ring Distal
LA	Little Abduction
LM	Little Metacarpal
LP	Little Proximal
LD	Little Distal

b

Thumb (4 axis, 4R)

j	θ	d	a	α	offset
1	q_1	0	0	$-\frac{\pi}{2}$	0
2	q_2	0	a_2	$-\frac{\pi}{2}$	0
3	q_3	0	a_3	0	0
4	q_4	0	a_4	0	0

Long Finger (4 axis, 4R)

j	θ	d	a	α	offset
1	q_1	0	0	$-\frac{\pi}{2}$	0
2	q_2	0	a_2	0	0
3	q_3	0	a_3	0	0
4	q_4	0	a_4	0	0

Fig. 4.5 Kinematic model of the human hand considered in this chapter. The model has 20 DoFs, which include the 15 DoFs of the hand model used in [47]. The left panel graphically describes the kinematics, while the right one specifies the Denavit-Hartenberg parametrization of thumb and the long fingers. We denoted with j the joint index, starting from the proximal joint to the distal, while q_1, q_2, q_3, q_4 are the joint angles, and a_2, a_3, a_4 are the phalanx lengths. **a** Kinematic model. **b** DH parameterization

Kinematic Model of the Human Hand

An accurate description of the human hand is challenging due to the high number of bones and joints composing the human hand. As a trade-off between accuracy and complexity, in this chapter we considered a 20 DoF kinematic model of the human hand (Fig. 4.5a. Each long finger is described by a set of four angles: two DoFs for flexion-extension and abduction-adduction in metacarpo-phalangeal joints, one DoF for flexion-extension in proximal and distal intra-phalangeal joints. The thumb is described with four angles: two DoFs for the trapezio-metacarpal, one DoF for the metacarpo-phalangeal, and one DoF for the inter-phalangeal. For the sake of space, we do not report here the mathematical form of the kinematics,[2] which can be easily derived from the Denavit-Hartenberg parametrization in Fig. 4.5b (see e.g. [37]). In the Appendix, we report the direct kinematics derivation for the led positioned on a long finger nail. A key characteristic of the model is that is shares the 15 DoFs of the model used in [47], allowing an easy comparison with the classical postural synergies of grasp, as done e.g. in [23].

Model Calibration

To reconstruct realistic values of the joint angles from marker data, the kinematic model employed should reproduce as closely as possible the actual kinematics of

[2] Note that here we use the term kinematics as it is commonly done in robotics and biomechanics, to refer to the mapping of Lagrangian variables to a different point of the structure. In this context, the evolution in time is referred to as differential kinematics.

the subject being recorded. To achieve this goal, we implement an identification procedure that follows what was done in [23]. The dataset is acquired by asking the subject to perform the Kapandji test [30], i.e. touching the four long fingers with the tip of the thumb. The test is repeated two times.

The parameters to be identified are: $a_{\mathrm{B}} \in \mathbb{R}^{30}$ collecting the length of each phalanx and the three space positions of the abduction joints (TR/TA, IM/IA, MM/MA, RM/RA e LM/LA in Fig. 4.5a) w.r.t. the local frame, and $a_{\mathrm{G}} \in \mathbb{R}^{45}$ collecting the location of each marker placed on the phalanxes w.r.t. the joint to which the LED is connected.

Their values are evaluated as ones minimizing the Root Mean Square Error (RMSE) between measured marked positions y_k and the estimated ones from the hand kinematics $f(x_k; a_{\mathrm{G}}, a_{\mathrm{B}})$, in a set of reference time stamps $\mathbb{K}_{\mathrm{id}}$

$$\begin{aligned} \arg \min_{x,\, a_{\mathrm{G}},\, a_{\mathrm{B}}} & \sqrt{\sum_{k \in \mathbb{K}_{\mathrm{id}}} (y_k - f(x_k; a_{\mathrm{G}}, a_{\mathrm{B}}))^T (y_k - f(x_k; a_{\mathrm{G}}, a_{\mathrm{B}}))} \\ \text{s.t. } & x_k > 0 \ \forall k \,, \end{aligned} \tag{4.1}$$

where $x_k \in \mathbb{R}^{20}$ is the vector collecting the estimation of joint angles at the $k-$th time step, $x_k > 0$ is to be considered element-wise, x is the vector collecting all the joint estimations in the considered time steps. WE selected $\mathbb{K}_{\mathrm{id}}$ as 60 equidistant frames. The constraint $x_k > 0$ accounts for biomechanical joint limits, in order to achieve a more robust estimation.

The optimization problem is solved through the MatLab function *fmincon*. The initial guess for the parameters a_{G} and a_{B} are evaluated through direct measurements, done with a caliper on each subject's hand before the experiment, while the initial value of x is the null vector. Since we do not have direct access to the real joint angles, the quality of the calibration is evaluated as the RMSE between the measured led positions, and the ones resulting from the identification, as e.g. in [23]. The average error across trials and subjects is 2.2mm for the unimpaired case and 3.1mm for the impaired case.

Joint Angles Estimation

Articulated hand postures typically produce marker occlusion to cameras. We use here Piecewise Cubic Hermite Interpolating Polynomials [15] to interpolate missing values. To estimate the hand postures from marker position we propose a Constrained Extendend Kalman Filter [53]. Kalman filtering was already employed for a similar scope in [21]. We rely on the identified hand model, representing joint evolution as a random walk. In each time step the filter estimates the hand posture from the measure of the marker position in space and previous state estimation. Joint limits are considered as constraints. To increase the robustness against marker losses (we had an average of 2.47 marker loss per frame, and a maximum number of consecutive marker loss of $\sim$ 15), we multiply at each step the Kalmann Filter observation noise covariance matrix by the number of consecutive missing measures. The filter

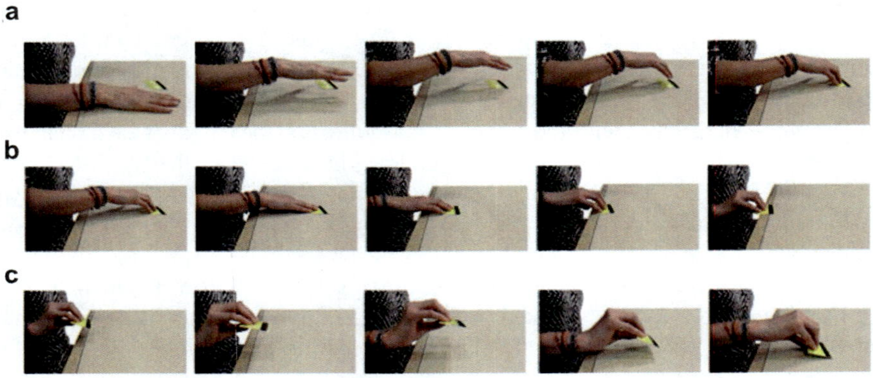

Fig. 4.6 Photo-sequences of experimental constraint exploitation. According to our contact based classification, **a** presents the *pre-shaping* phase, **b** shows the *during contact* phase, and **c** the *post-contact* phase

is initialized with the open hand posture, and a null state covariance matrix. The observation noise covariance matrix is 0.001 $I_{20\times20}$ and the state noise covariance matrix is 0.0005 $I_{24\times24}$. Both the matrices are heuristically tuned. The quality of the reconstruction is evaluated as the RMSE between measured and estimated LED positions. The mean values is 2.9mm for the unimpaired case and 3.2mm for the impaired case.

Force Data

The force/torque data from ATI Mini45 (i.e. sensorised surface) are filtered through a moving average filter based on Savitzky-Golay method [49]. The window width is heuristically tuned as the 1.5% of total data length. We then use the knowledge of surface form, to evaluate the centroid of contact of Force/Torque data.[3]

Analysis

We segment pre-processed data into three main phases, also described in Fig. 4.6: (i) pre-shaping, where the object is reached and the hand posture is shaped in order to purposefully interact with the environment, (ii) contact, in which the constraint is exploited in order to manipulate and grasp the object, and (iii) post-contact, when the object is grasped and lifted from the table.

By considering non-adhesive interactions with the environment, we can assume any change occurring to the force orthogonal to the surface as due to an interaction. We thus segmented the actions searching for a change in the corresponding force

[3] The contact points are computed through tactile toolbox, as in [52]. A portable version of this tool as well as raw data of our experiments will be released through handcorpus.

measured by the sensorized surface (see Sec. 4.1.2.2). The cut off from the first and the second phase is identified by the first contact with the table, when the force starts to increase. To accurately detect this point we consider both to the signal and to its derivative. The cut off identifying the end of the contact phase is taken as the first time in which the contact force returns to zero.

The aim of the analysis is to identify a subspace of reduced dimensionality embedding the hand postures, to test the hypothesis of presence of a synergistic behavior in Environmental Constraint Exploitation. Principal Component Analysis (PCA) is a valuable tool to achieve this goal [36, 45, 47, 55]. Given a set of data, described by a correlation matrix C and a mean m, PCA derives an orthonormal base of the data space, whose first element S_1 indicates the direction where data present the greatest variability. In turn, each successive component S_i has the highest variability under orthogonality constraint. The $i-$th element of the base is referred as Principal Component of the dataset. The normalized percentage of the data variability projected on each Principal Component is called explained variance of the component.

We evaluate PCA as the singular value decomposition of the data correlation matrix C, i.e. by finding an orthonormal matrix Σ, which brings C in Jordan form through the similitude transformation $\Sigma^T\, C\, \Sigma$. In that case the Principal Components are the column of the matrix $\Sigma = [S_1, \ldots, S_n]$, and the explained variances are the corresponding eigenvalues. For further details on PCA we refer the interested reader to [29].

If PCA is used to analyze hand postures in joint space, explained variances can be used to understand if the hand moves on a reduced set of the configurations by looking if there are few principal directions that explain the major part of the data. If this is the case we refer to such principal directions as synergies.

It is worth noticing that the use of the same calibrated kinematic model for every subject enables a coherent description of the hand configuration space as $\mathbb{R}^{20}$ for all the experimental conditions. We consider cosine of the angle between synergy's directions as the metric to compare results of the different analyses. We evaluate it as the absolute value of the normalized dot product between the synergy vectors.

We also compare synergies resulting from our analysis with those employed for the execution of grasping, as shown in [47]. The kinematic model in [47] takes into account a subset of the joints considered in this chapter (see Sect. 4.1.2.4). Thus the configuration space of grasping synergies is a sub-set of dimension 15 of the hand configuration space considered in this chapter. To compare vectors, we projected full hand configurations in the corresponding sub-space. This is equivalent to simply neglect the values corresponding to joints ID, MD, RD and LD in Fig. 4.5a.

Since our main focus is to extract lessons that can inform the robotic design (as discussed in Sec. 4.1.4.2), a complete study of inference is beyond the scope of the present chapter. However we considered t-student method to infer the general validity of some of our results. Please refer to [27] for an exhaustive introduction.

Pre-shaping Analysis

The pre-shaping analysis is done with the purpose of identifying kinematic regularities in the generation of hand postures for the ECE exploitation. In [47] PCA is performed for constant postures acquired in grasping. There authors took out effects due to interaction with the objects, by asking subjects to grasp imagined ones. In this analysis we aim at achieving the same goal by performing PCA on a dataset composed of the last poses before the contact with the environment in each trial, i.e. the last pose in pre-shaping phases when a purposeful interaction with the environment is planned. Unimpaired and impaired cases are separately analyzed. Both datasets are composed of 252 poses (six subjects, 21 objects, two trials)

Contact Analysis

In order to evaluate if there are some kinematic regularities in the strategies employed by the subjects during actual exploitation of the surface (see e.g. Fig. 4.1), we perform PCA on data collected during the contact phase. We perform the analysis separately in the impaired and unimpaired case. To characterize the effect of cutaneous impairment we also evaluated the mean amount of time in which subjects stay in contact with the table, mean time for task accomplishment, and the mean norm of interaction forces, by averaging the correspondent values for every subject and every object.

The datasets for both impaired and unimpaired conditions are composed of a variable number of poses depending on the strategy execution time. Each ECE generates an amount of postures equal to 40 times the execution time (see acquisition rate in Sect. 4.1.2.2). All 21 objects, two trials and six subjects were considered.

Differences Between Pre and During Contact

We investigate the persistence of the same basic ingredients of hand posture before and during the contact with the environment for both bare fingers and cutaneous impairment conditions. The dot product of synergies evaluated in previous sections was computed in order to quantify their similarities.

4.1.2.5 Unsupervised Clustering of Hand Configurations

To inspect human grasping strategies we also performed an unsupervised clustering on hand configurations. For each subject and for each task, clusters are identified on the first frame of contact. This frame was obtained looking at force data from the sensorized table, as the one for which the norm of force exceeded a threshold (0.1 N).

To increase clustering robustness and include also the configuration after contact, we selected also a second frame, one second after the first one. After this selection, we used k-means algorithm as unsupervised clustering with the goal to find principal hand configurations for the contact frame and the delayed frame.

4.1.3 *Results*

4.1.3.1 Pre-shaping Analysis

Figure 4.7 shows the explained variance associated to the PCs on the poses during the pre-shaping phase. For the unimpaired case, the first Principal Component explains about 54% of the variance, while the first three pre-shaping synergies explain more than 72%. Figure 4.8 shows the graphical representation of the first three resulting postural synergies. The same analysis in the impaired case shows that the first synergy explains about 42% of the variance, while three Principal Components explain

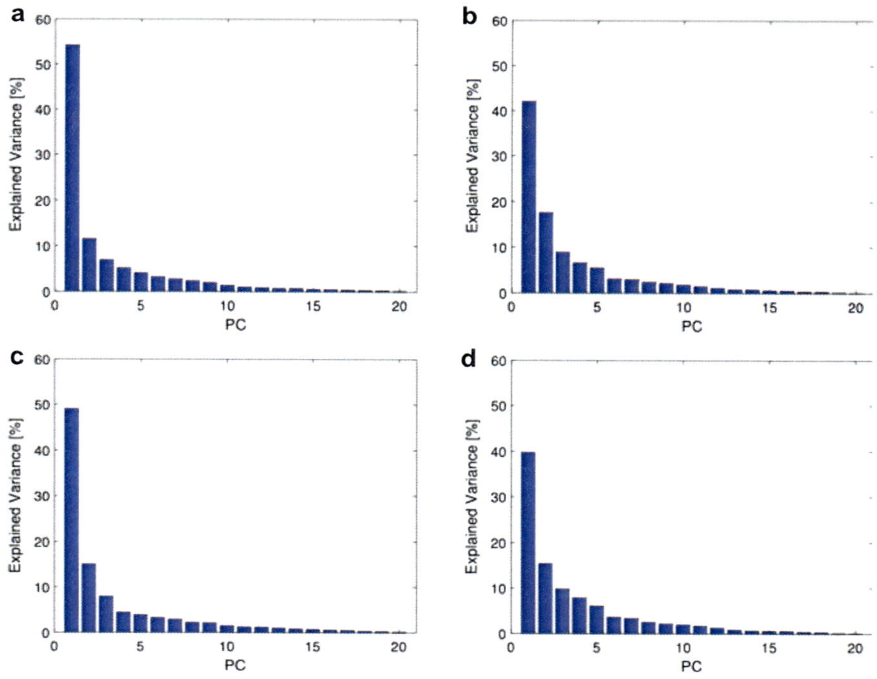

Fig. 4.7 Explained variance resulting from PCA analysis of postures in pre-shaping and during contact, in tactile impaired and unimpaired case. A marked predominance of the first Principal Component is present in all the cases, showing a synergistic behavior. **a** Unimpaired, pre-shape; **b** impaired, pre-shape; **c** unimpaired, contact; and **d** impaired, contact

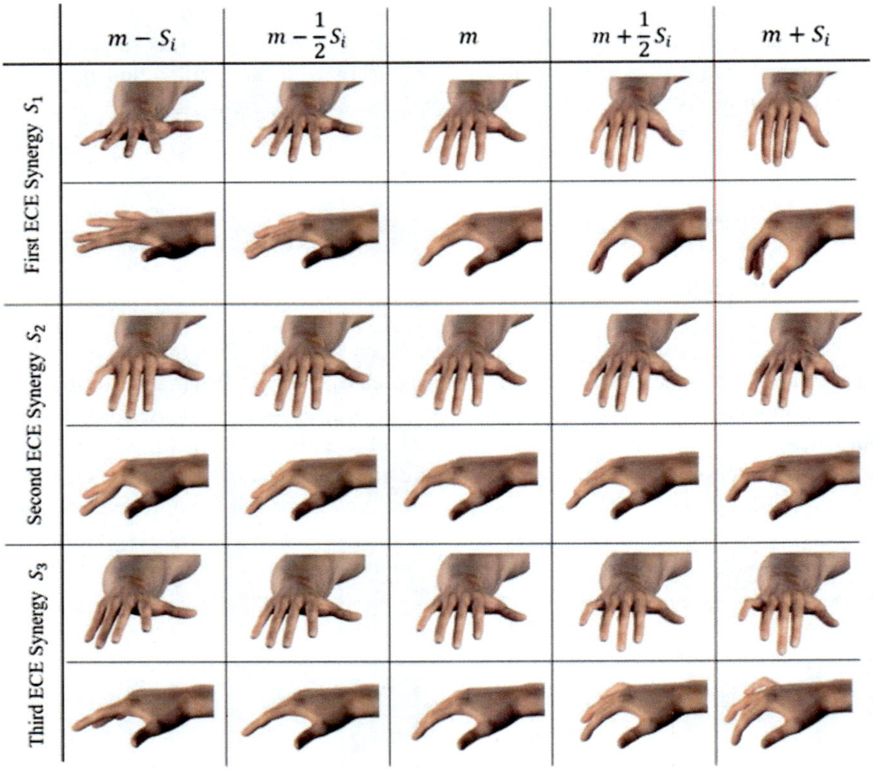

Fig. 4.8 Graphical representation of the movements w.r.t. the mean hand posture, associated to the first three synergies during pre-shaping in unimpaired condition. Each column presents a different stage of the synergistic movement, obtained by summing the hand mean configuration m, to the synergy vector S_i of the ith synergy

more than 68% of the variance. Thus, for our dataset, the variance explained by the first main synergies is lower in the impaired condition. In Fig. 4.8 we present the graphical representations of the movements corresponding to the first three synergies of pre-shaping without tactile impairment. Table 4.1 presents the numerical values of the first ECE synergy of pre-shaping with and without impairment, in comparison with the first synergy of grasp [47]. Figure 4.9 presents the movement corresponding to the second synergy of pre-shaping with tactile impairment.

In all the considered cases, the first synergy describes a opening-closing behavior of the whole hand, while the second synergy corresponds to a closure of the distal joints mostly of index and medial fingers, and a closing of the thumb. The third synergy is similar to the second, but concerning the little and ring fingers.

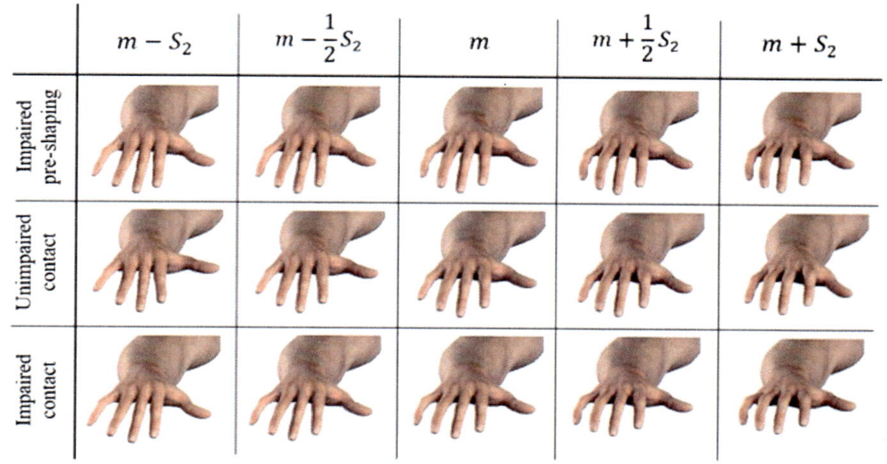

Fig. 4.9 Graphical representation of the movements associated to the second ECE synergy. Pre--shaping impaired, contact unimpaired and contact impaired conditions are considered. The mean posture is referred as m, the second synergy as S_2. We do not report here the first synergies, since they do not present visible discrepancies from each others. The present figure shows also a good coherence in the behavior described by the second synergy, among the considered conditions

Figure 4.10a shows the scalar product between the first grasp synergy in [47] and the ones found in this chapter for both sensory conditions. What is noticeable is that there is a high level of consistency between the main synergy of grasping and pre-shaping of human hand in impaired and unimpaired condition (≥ 0.9). The similarity is reduced for the second synergy, as shown in the same Fig. 4.10b, and so on for the other orders. Figure 4.10 reports also a high correlation between pre-shaping synergies with and without tactile impairment. The presence of impairment does not alter the first two synergies during the pre-shaping phase. However the similarity strongly drops when the synergy order increases further, reaching 0.36 for the third and 0.003 for the forth.

4.1.3.2 Contact Analysis

In the unimpaired condition, the first Principal Component explains about the 49% of the variance, and the first three synergies more than 73%. The same analysis in the impaired case returns a first synergy explaining about the 39% of the variance, and the three Principal Components explaining more than 65% of the variance. Also in this case, for our dataset the percentage of variance explained in the impaired condition is lower. Table 4.1 presents the numerical values of first ECE synergy during the contact with the environment, with and without impairment, in comparison with the first synergy of grasp [47]. Figure 4.9 presents the movement corresponding to the second synergy with and without tactile impairment.

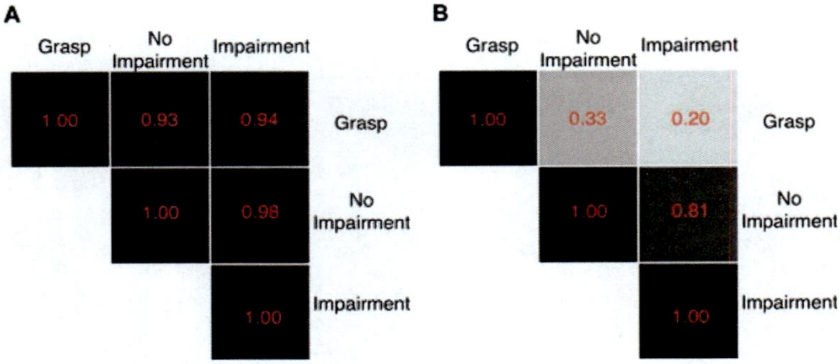

Fig. 4.10 Dot products between first and second grasp synergies and first and second ECE synergies evaluated during pre-shaping, with and without tactile impairment. The gray scale graphically codes the product value: black is 1, i.e. very similar, white is 0, i.e. very different. A high correlation between the first grasping and ECE synergies is evidenced. The ECE synergies with and without impairment present high similarity, which however drops for higher order synergies. **a** First synergies and **b** second synergies

The analysis also demonstrates that subjects with tactile impairment are in contact with the table for an average time of 4.2 ± 3.1s, while for the unimpaired case the average time is 2.4 ± 2.4s. The complete task is performed in 13.9 ± 2.7s for the tactile impairment case, while it is performed in 11.5 ± 2.0s in the other case. Finally, the contact force is different for the experiments considered. Mean value of norm of contact forces for the tactile impairment case is 23.2 ± 8.6 N, while mean value is 12.3 ± 5.7 N in the other case.

4.1.3.3 Differences Between Pre and During Contact

The result shows that the first Principal Components in unimpaired and impaired conditions are very similar w.r.t. the corresponding ones of the pre-shaping analysis. The similarity tends to decrease with the increase of synergy order (correlation ≥ 0.75 till 9th synergy). Dot products are graphically reported in Fig. 4.11 for the unimpaired case. Thus ECs induce only changes for the high order synergies, leaving unaltered the main ones, regardless of availability of tactile input. Note that the first synergy is still equivalent to the one in [47].

4.1.3.4 Inference and Statistical Relevance

In the previous sections we presented results from 6 subjects. Despite such a moderate number of participants, findings and analyses are in line with the existing literature in the field, see e.g. [36, 39, 47]. To generalize, we here report additional statistical analyses that provide t–Student based confidence intervals (CI) with 95% probability.

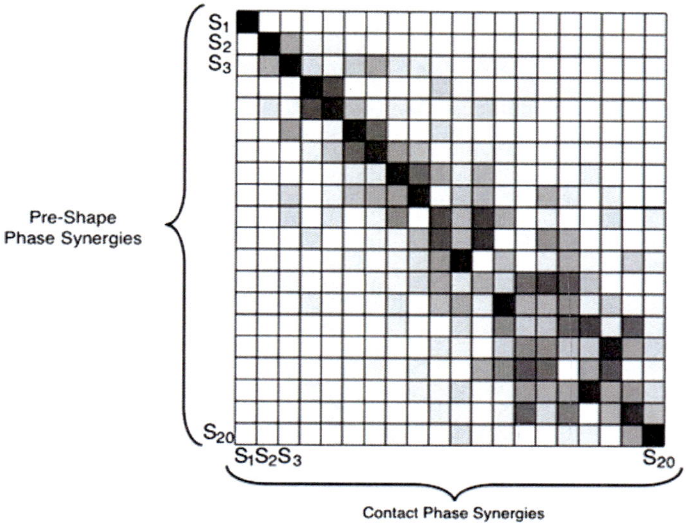

Fig. 4.11 Dot products between the ECE synergies in pre-shaping and in contact with the environment, in the bare finger case. The gray scale graphically codes the product value: black is 1, i.e. very similar, white is 0, i.e. very different. A tendency to maintain the first main components before and after the contact results clearly from this analysis. The results for the impaired case are analogous

CI refers to dot products performed on the PCs extracted for the different conditions, and the ones obtained in grasping.

CI for the dot product between the first synergy for impaired and unimpaired conditions is [0.81, 0.96]; CI for the dot product between the first grasping synergy and the unimpaired first ECE synergy is [0.88, 0.94]; CI for the dot product between the first grasping synergy and the impaired first ECE synergy is [0.79, 0.9]. Regarding the contact analysis of Sect. 4.1.3.3, the dot product between the first Principal Component before and during contact results in a CI of [0.97, 0.99] (unimpaired case).

4.1.3.5 Main Clusters of Hand Configurations

We found that for both the impaired and unimpaired case three clusters allow to increase the compactness of clusters by more than 80% (see Fig. 4.12). Both for the unimpaired and impaired case, we identified three main hand configurations:

- Power grasp (Cluster #1), characterized by higher closure of the long fingers;
- Sliding configuration (Cluster #2), in which fingers are spread out along the table;
- Precision grasp (Cluster #3), with long fingers slightly extended.

The centroids of the three clusters are plotted in Fig. 4.13. We found that for the unimpaired case the cardinality of the first cluster is 68, the cardinality of the second

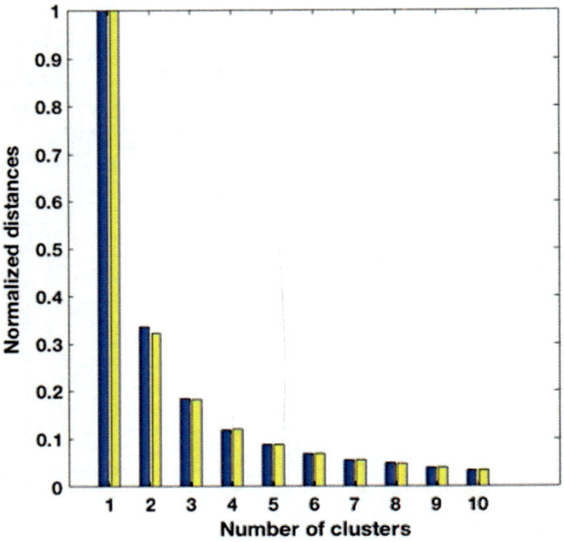

Fig. 4.12 Mean values of the within-cluster sums of point-to-centroid distances w.r.t. the number of clusters used. On blue the unimpaired case, on yellow the impaired case

cluster is 92 and the cardinality of the third cluster is 82. For the impaired case, instead, the cardinality of the first cluster is 80, the cardinality of the second cluster is 117 and the cardinality of the third cluster is 43. What is noticeable is that for the latter case we found an increase of sliding and power grasp approaches. This suggests that precision grasp is penalized by the absence of tactile feedback. Under a robotic point of view, this result can suggest main planning strategies for ECE with soft robotic hands but also can inform the design of robotic hands sensorization. For example, IMUs could be used, as a first approximation of haptic information, to detect contacts with the environment and to be employed in refining and completing the ECE strategies reported in the previous section, e.g. as a combination of feedback and feedforward actions.

4.1.3.6 Analysis of Contact Forces and Times of Task Execution

The ATI sensor placed under the table allowed to measure the force/torque applied by the subject while interacting with the environment. We found that subjects with tactile impairment required more time to complete the task w.r.t. the unimpaired case. In particular, the complete task is performed in $13.9 \pm 2.7s$ for the tactile impairment case, while it is performed in $11.5 \pm 2.0s$ in the other case. Also the amount of time required for the interaction with the table increased for the tactile impairment case. In fact, subjects with impairment was in contact with the table for an average time of $4.2 \pm 3.1s$, while for the unimpaired case the average time is $2.4 \pm 2.4s$. Finally, the contact force was different for the experiments considered. In fact, mean value

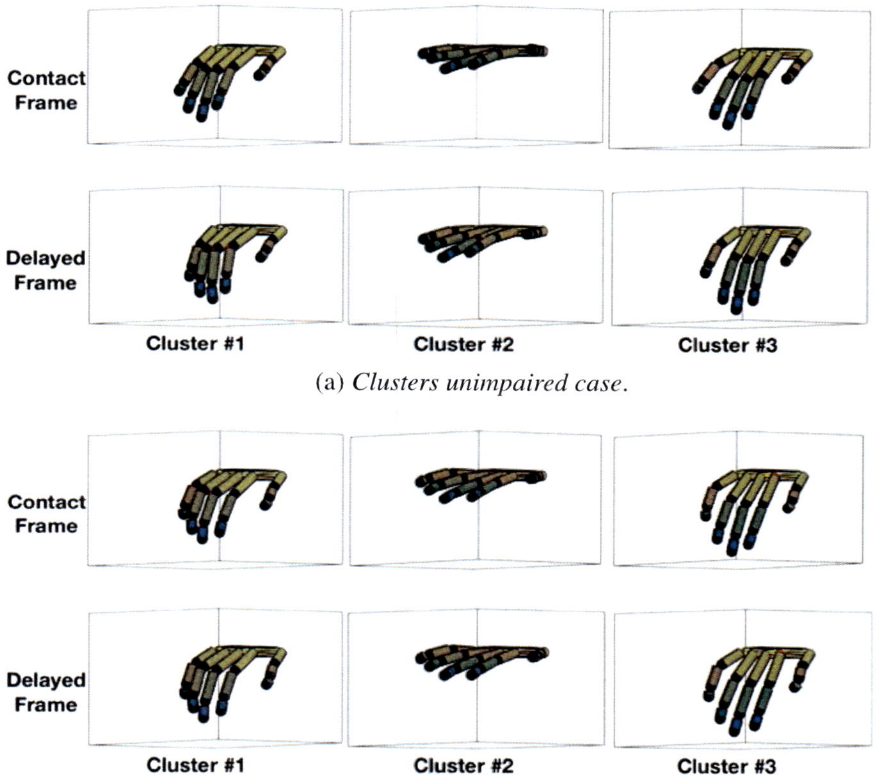

(a) *Clusters unimpaired case.*

(b) *Clusters impaired case.*

Fig. 4.13 Centroids of the three principal clusters of hand configurations in impaired and unimpaired cases. For each one we report the configuration in the contact time frame and the configuration in a time frame delayed by 1 sec. In both the conditions we can identify one cluster for power grasp (Cluster #1), one cluster for sliding configuration (Cluster #2) and one cluster for precision grasp (Cluster #3)

of norm of contact forces for the tactile impairment case is $23.2 \pm 8.6N$, while is $12.3 \pm 5.7N$ in the other case.

4.1.4 *Discussion and Conclusions*

4.1.4.1 Implications for Motor Control

The explained variance reported in Fig. 4.7 suggests the presence of an underlying synergistic behavior in the purposeful exploitation of environmental constraints.

Indeed, the dimensionality of the space required to approximate hand posture is considerably smaller than the number of degrees of freedom. In particular, the first Principal Component shows a marked predominance, with a maximum total variance explained of 54% in the unimpaired pre-shaping case. The three first synergies explain more than 65% of the total variance in all the considered conditions. In the considered experiments, during contact with the environment, the amount of variability accounted by higher order synergies increases. This could be explained by observing that the interaction can shape the subject hand beyond its nominal kinematics. This behavior is also in agreement with the findings in [48], where the synergistic analysis of [47] was performed on grasped real objects instead of imagined ones.

Despite our inference analysis (Sect. 4.1.3.4) is limited to the first synergy, a series of characteristics of our dataset can be pointed out, which are in accordance with existing neuroscientific findings. Future works will focus on different experimental procedures and tasks to further investigations. In our dataset, the first two synergies in the impaired and unimpaired conditions are very similar, as shown in Figs. 4.9 and 4.10. This suggests that the presence of tactile impairment, while modifying the strategies themselves, does not substantially modify the most basic kinematic ingredients commonly used to generate hand postures. Subjects are aware of the presence of the tactile impairment, so it is reasonable to expect that they might have changed their planning in accordance to that. Indeed, the drop of such similarity after the third synergy suggests that cutaneous impairment affects posture refinement, which can be likely ascribed to higher order synergies (as described in [47]). However the amount of data collected does not allow a sound statistical characterization of this behavior. To assess whether higher order synergies are primarily noise or they actually contribute to hand postures, we will resort to the usage of discriminant analysis and information theory, as in [47], as future works.

Analogously, Fig. 4.11 shows a high similarity in the first Principal Components during contact and pre-shaping, while differences can be observed for higher order synergies. Uncontrolled Manifold theory [32, 51] suggests that the central nervous system selects in the space of joint angles a subset of variables of interest, which are regulated, purposefully leaving free the remaining variables. The persistence of main postural synergies of pre-shaping during the contact with the environment can be interpreted in the light of this theory by considering the first set of ECE synergies as the variables of interest for the considered task, which remain constant when an external disturbance occurs. The sub-space individuated by the higher order synergies is instead left free to adapt to the external environment. Moreover such behaviour could also be due to peripheral constraints embedded in the musculoskeletal system, as discussed e.g. in [45].

Data shown in Fig. 4.10 and Table 4.1 demonstrate that a strong resemblance exists between the first synergy, resulting from the analysis of ECE strategies, and the first synergy of grasp as found in [47]. This could suggest the presence of underlying synergies, which are integrated with task specific ones. This was proposed e.g. in [24], and it is in agreement with experimental results presented in [55], where task independent synergies are estimated by a set of unconstrained tasks (see CI estimates in Sect. 4.1.3.4).

4.1.4.2 Implications for Robotics

As it can be widely observed in literature, neuroscientific results of synergistic behavior of human hands have been successfully translated and applied to robotics to inform the design, control and sensing of artificial systems, with special focus on grasping, see [7, 9]. One of the first notable application of synergies to robotics was in [11], where authors propose to use grasp synergies to derive actuation patterns for an underactuated robotic hand. In [22], the use of hand synergies for the choice of grasping forces is discussed. In [14] and later in [2, 35, 57], a synergy based low-dimensional synergistic space is considered to obtain effective pre-grasp shapes for fully actuated robotic hands.

Recently, synergy-inspired actuation has been combined with the introduction of compliance in the structure [12, 13, 17, 59] (according to the soft synergy framework [9]). The availability of robotic hands embedding elasticity in their mechanics has also led to a shift in their control philosophy, accounted e.g. in [10, 18]. In the classical planning, suitable points are selected on the object to be grasped or manipulated, generating a nominal grasp of good quality. Trajectories are then executed, to correctly position the fingertips while avoiding contacts with the environment. On the contrary, soft manipulation has changed this scheme. The hand-environment contacts are no more avoided but exploited to successfully shape the hand around the object. Under this regard, the study of environmental exploitation in humans could inform the design, planning, and control of robotic hands to take full advantage from the external environment.

The most direct implication of the presented results could leverage upon the observation that the first synergy of grasping is very similar to the first synergy for Environmental Constraint Exploitation (ECE). Thus the implementation of the first synergy of grasp as degree of actuation can target the twofold goal of realizing under-actuated robotic hands that can effectively grasp objects and, at the same time, are able to exploit Environmental Constraints. To increase hand functionalities beyond the first degree of actuation, we could implement additional ECE synergies, possibly in combination with the grasp synergies. For some examples of the implementation of synergies for the design of under-actuated robotic hands, we refer the interested reader to [17, 26, 42].

Looking at the differences between the impaired and unimpaired conditions, the key kinematic ingredients seem to remain unaltered at least for the gross movements. However, the time for task accomplishment and the force exchanged with the environment is higher for the impaired case. This result could indicate possible sensing strategies for soft robotic hands, i.e. to detect contact with the environment e.g. through IMU sensors, which can lead to the development of planning and control laws aiming at minimizing force execution on external objects.

Appendix A: Numerical Values of ECE Postural Synergies

Table 4.1 Numerical values of the first synergy of Grasp [47] and of Enviromental Constraint Exploitation, with and without impairment, before and after contact. We indicate with 'x' the DoFs that were not considered in [47]

DoFs	Grasp	Unimpaired pre-shape	Impaired pre-shape	Unimpaired contact	Impaired contact
TA	–0.43	–0.14	–0.15	–0.12	–0.15
TR	0.29	0.31	0.35	0.30	0.34
TM	0.14	0.14	0.17	0.17	0.16
TI	0.03	0.04	0.05	0.05	0.09
IA	–0.13	–0.08	–0.12	–0.08	–0.11
IM	0.33	0.39	0.35	0.40	0.34
IP	0.15	0.16	0.16	0.17	0.20
ID	x	0.01	0.03	0.03	0.05
MA	x	–0.03	–0.08	–0.02	*–0.06
MM	0.33	0.38	0.33	0.38	0.32
MP	0.16	0.27	0.27	0.27	0.30
MD	x	0.04	0.06	0.05	0.08
RA	0.06	0.00	–0.02	0.02	–0.02
RM	0.40	0.44	0.37	0.43	0.32
RP	0.20	0.22	0.27	0.22	0.35
RD	x	0.04	0.06	0.05	0.11
LA	0.14	0.05	0.1	0.08	0.09
LM	0.37	0.43	0.42	0.41	0.37
LP	0.27	0.12	0.21	0.17	0.24
LD	x	0.02	0.04	0.02	0.08

Appendix B: Finger Kinematics

The forward kinematics of a serial chain with n degrees of freedom is described by post-multiplicating different rotation-translation matrices with the following form

$$T_{i-1,i} = \begin{pmatrix} R_{i-1,i} & h_{i-1,i} \\ 0_{1\times 3} & 1 \end{pmatrix},$$

where $R_{i-1,i}$ is the rotation matrix between frame $i-1$ and frame i and $h_{i-1,i}$ is the translation vector between frame $i-1$ and frame i. Then the whole forward kinematics is calculated as

$$T_{0,n} = T_{0,1} T_{1,2} \dots T_{n-1,n}.$$

For the long finger (its kinematics is introduced in Sect. 4.1.2.4 and parametrized with Denavit-Hartenberg in Fig. 4.5b) we have

$$T_{01} = \begin{pmatrix} C_1 & 0 & -S_1 & 0 \\ S_1 & 0 & C_1 & 0 \\ 0 & -1 & 0 & 0 \\ 0 & 0 & 0 & 1 \end{pmatrix}, \qquad T_{12} = \begin{pmatrix} C_2 & -S_2 & 0 & a_2 C_2 \\ S_2 & C_2 & 0 & a_2 S_2 \\ 0 & 0 & 1 & 0 \\ 0 & 0 & 0 & 1 \end{pmatrix},$$

$$T_{23} = \begin{pmatrix} C_3 & -S_3 & 0 & a_3 C_3 \\ S_3 & C_3 & 0 & a_3 S_3 \\ 0 & 0 & 1 & 0 \\ 0 & 0 & 0 & 1 \end{pmatrix}, \qquad T_{34} = \begin{pmatrix} C_4 & -S_4 & 0 & a_4 C_4 \\ S_4 & C_4 & 0 & a_4 S_4 \\ 0 & 0 & 1 & 0 \\ 0 & 0 & 0 & 1 \end{pmatrix}.$$

where C_i stands for $\cos(q_i)$, S_i stands for $\sin(q_i)$, a_i is the length of the phalanx $i-1$ and q_i is the ith joint rotation. Then, the forward kinematics is obtained as

$$T_{04} = T_{01} T_{12} T_{23} T_{34} = \begin{pmatrix} R_{0,4} & h_{0,4} \\ 0_{1\times 3} & 1 \end{pmatrix}.$$

The extended expression of T_{04} is neglected for simplicity and can be easily calculated by the reader. The final position $\left(x_{LF}\ y_{LF}\ z_{LF}\right)^T$ of the external phalanges can be extracted from T_{04} by selecting $h_{0,4}$:

$$\begin{pmatrix} x_{LF} \\ y_{LF} \\ z_{LF} \end{pmatrix} = \begin{pmatrix} C_1(a_2 C_2 + a_3 C_{23} + a_4 C_{234}) \\ S_1(a_2 C_2 + a_3 C_{23} + a_4 C_{234}) \\ -a_2 S_2 - a_3 S_{23} - a_4 S_{234} \end{pmatrix},$$

where C_{23} stands for $\cos(q_2 + q_3)$, C_{234} stands for $\cos(q_2 + q_3 + q_4)$, S_{23} stands for $\sin(q_2 + q_3)$, S_{234} stands for $\sin(q_2 + q_3 + q_4)$. The same procedure can lead to the description of the thumb forward kinematics. We have

$$T_{01} = \begin{pmatrix} C_1 & 0 & -S_1 & 0 \\ S_1 & 0 & C_1 & 0 \\ 0 & -1 & 0 & 0 \\ 0 & 0 & 0 & 1 \end{pmatrix}, \qquad T_{12} = \begin{pmatrix} C_2 & 0 & -S_2 & a_2C_2 \\ S_2 & 0 & C_2 & a_2S_2 \\ 0 & -1 & 0 & 0 \\ 0 & 0 & 0 & 1 \end{pmatrix},$$

$$T_{23} = \begin{pmatrix} C_3 & -S_3 & 0 & a_3C_3 \\ S_3 & C_3 & 0 & a_3S_3 \\ 0 & 0 & 1 & 0 \\ 0 & 0 & 0 & 1 \end{pmatrix}, \qquad T_{34} = \begin{pmatrix} C_4 & -S_4 & 0 & a_4C_4 \\ S_4 & C_4 & 0 & a_4S_4 \\ 0 & 0 & 1 & 0 \\ 0 & 0 & 0 & 1 \end{pmatrix},$$

and then

$$\begin{pmatrix} x_T \\ y_T \\ z_T \end{pmatrix} = \begin{pmatrix} a_4S_4(C_3S_1 - C_1C_2S_3) + a_2C_1C_2 + a_3S_1S_3 + a_4C_4(S_1S_3 + C_1C_2C_3) + a_3C_1C_2C_3 \\ a_2C_2S_1 - a_4S_4(C_1C_3 + C_2S_1S_3) - a_3C_1S_3 - a_4C_4(C_1S_3 - C_2C_3S_1) + a_3C_2C_3S_1 \\ a_4S_2S_3S_4 - a_3C_3S_2 - a_4C_3C_4S_2 - a_2S_2 \end{pmatrix}. \tag{4.2}$$

It is worth to be noticed that, while linear w.r.t. the parameters a_2, a_3, a_4, Eq. 4.2 defines a strongly nonlinear relationship between joint angles q_1, q_2, q_3, q_4 and LED position.

4.2 Force Synergies in Environmental Constraint Exploitation

While a lot of effort has been devoted to describing the synergistic control that modulates hand movements during grasp [48], the relationship between postural hand synergies and contact force distribution and how the interaction with the external environment shapes synergistic control is still debated [3, 16, 45]. The previous section investigates how kinematic synergistic control is modified when the hand interacts with the environment to grasp an object [16], and proposes a consequent unsupervised clustering of principal hand approaches [4]. In this section, we extend the analysis presented in previous ones to the force domain.

Our analysis is based on force and torque (F/T) measurements at fingertip level for thumb, index and middle fingers [5] during grasping of a set of objects from a flat surface, and uses Principal Component Analysis (PCA) to identify the principal force directions or synergies. This investigation is motivated by the need of developing new tools for the evaluation of physiological and pathological behavior in object grasping and manipulation. More specifically, some pathologies such as stroke can alter subject grasping capabilities in everyday-life, eventually resulting in different synergistic behavior in force distribution. At the same time, the study reported in this work could inspire simplified sensing guidelines for robotic hands and prostheses. The latter point is particularly important when considering soft robotic devices, which deform with the environment as humans actually do.

4.2.1 *Materials and Methods*

4.2.1.1 Pre-Processing

—We extrapolated two ranges of interest from the whole task execution: (a) in which the hand is in contact with the table and (b) in which the hand has lifted the object (see Fig. 4.15). Regarding phase (a), during which the interaction with the object has the goal of obtaining a stable grasp, I analyze the last five frames, which contain information on the force exchange immediately before lift. These F/T vectors are then averaged to get a single vector for each acquisition. Regarding phase (b), we extract a single F/T vector as mean between all the frames. With this procedure we extrapolate, for each task, for each subject and for each experimental phase (a, b), one representative F/T vector $\in \mathbf{R}^m$, containing 3D forces at fingertip level. Note that, although we recorded forces and torques during the experiments, for the analysis we used only the force components (i.e. $m = 9$), since the torques appear to be negligible w.r.t. the forces.

4.2.1.2 Data Analysis

—For each experimental phase (a, b), we collected the F/T vectors in a matrix $X_{A,B} \in \mathbf{R}^{n,m}$, where $n = 252$ is the number of trials. Principal Components (PCs) are calculated as eigenvectors of the covariance matrix $S = X_{A,B}X^T_{A,B} - \mu\mu^T$, where $\mu = \text{mean}(X_{A,B})$. The variance explained by each PC is calculated as the normalized corresponding eigenvalue.

4.2.2 *Results*

Principal Component Analysis (PCA) on the two experimental phases shows that one predominant Principal Component exists and explains over the 60% of the total dataset variability (see Fig. 4.16). In Fig. 4.14 we plot the first PC as direction of forces insisting on thumb, index and middle fingers (in blue the contact phase (a), in red the flight phase (b)). As expected the first PC appears with a strong dominance of direction along the axis normal to the fingertip surface. What is noticeable is that the coordination between fingers appears maintained in both the experimental phases, with slight variations mainly on index and middle finger probably due to the effect of the object weight.

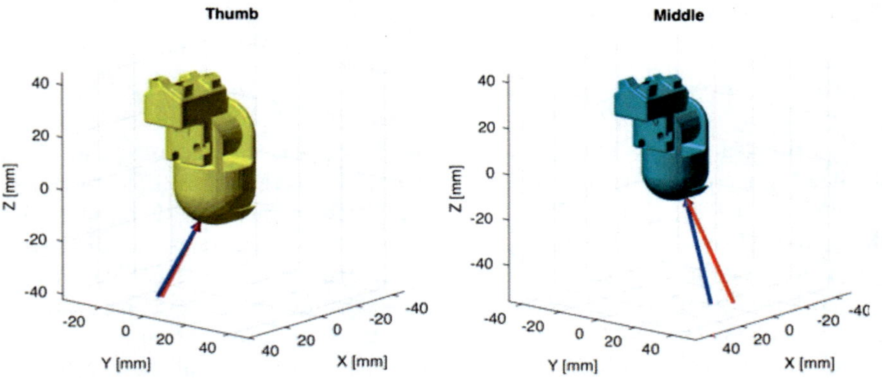

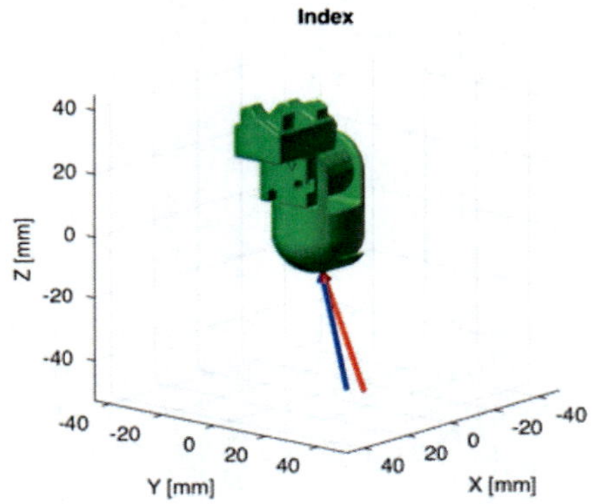

	F_x	F_y	F_z
T	-0.32	0.22	0.75
I	0.11	-0.11	0.49
M	0.01	-0.06	0.14

	F_x	F_y	F_z
T	-0.34	0.27	0.77
I	0.07	-0.03	0.39
M	-0.01	-0.08	0.24

Fig. 4.14 First force synergy for the three fingers. In blue the contact phase, in red the flight phase. On the right-bottom, in top table we report the numerical values of the first synergy in the contact phase while in the bottom table we report the values for the flight phase. Each row refers to one finger, each column refers to the direction. Please note that all coefficients are dimensionless

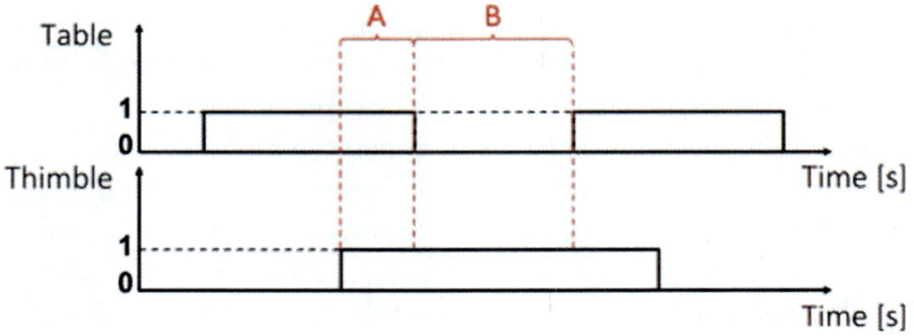

Fig. 4.15 In the top graph we show an ideal plot of signal presence on table sensor (1 if over a treshold, 0 otherwise), while in the bottom graph we report the analogous for ThimbleSense recordings. We identify two main regions of interest for our analysis: **a** in which the hand is in contact with the table; **b** in which the hand has lifted the object

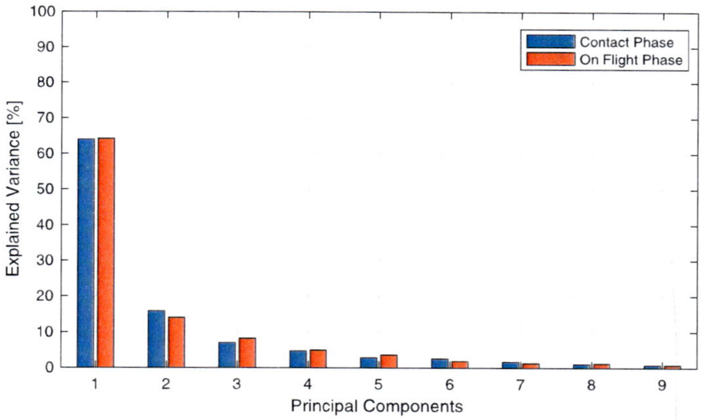

Fig. 4.16 Variance explained by each PC. In blue the contact phase, in red the flight phase. A predominant first synergy explaining more than the 60% of the data can be observed in both phases

4.2.3 Discussions and Conclusions

These results suggest that a coordination between finger forces exists in grasping tasks, and that this behavior is preserved also in case of interaction with the environment. This hypothesis is further confirmed by the fact that the first PC plays a crucial role in explaining the force distribution, with an explained variance higher than 60% in both the experimental cases. We believe that these finding could be used to develop new assessment procedures, e.g. in post-stroke patients, where the pathology could alter the synergistic behavior thus resulting in different principal components. At the same time, the co-variation in force distribution could be employed to suggest optimal sensing strategies for soft robotic hands, that deform as human hands actually do , when constraints on the number and quality of sensors are present, similarly to what was done in [8] for kinematic under-sensing techniques of hand poses relying on postural synergies.

References

1. Alessandro, C., Delis, I., Nori, F., Panzeri, S., Berret, B.: Muscle synergies in neuroscience and robotics: from input-space to task-space perspectives. Front. Comput. Neurosci. **7**, 43 (2013). https://doi.org/10.3389/fncom.2013.00043
2. Amor, H.B., Kroemer, O., Hillenbrand, U., Neumann, G., Peters, J.: Generalization of human grasping for multi-fingered robot hands. In: 2012 IEEE/RSJ International Conference on Intelligent Robots and Systems (IROS), pp. 2043–2050. IEEE (2012). https://doi.org/10.1109/iros.2012.6386072
3. Averta, G., Angelini, F., Bonilla, M., Bianchi, M., Bicchi, A.: Incrementality and hierarchies in the enrollment of multiple synergies for grasp planning. IEEE Robot. Autom. Lett. **3**(3), 2686–2693 (2018)

4. Averta, G., Della Santina, C., Battaglia, E., Ciotti, S., Arapi, V., Fani, S., Bianchi, M.: From humans to robots: The role of cutaneous impairment in human environmental constraint exploitation to inform the design of robotic hands. In: 2017 9th International Congress on Ultra Modern Telecommunications and Control Systems and Workshops (ICUMT), pp. 179–184. IEEE (2017)
5. Battaglia, E., Bianchi, M., Altobelli, A., Grioli, G., Catalano, M.G., Serio, A., Santello, M., Bicchi, A.: Thimblesense: a fingertip-wearable tactile sensor for grasp analysis. IEEE Trans. Haptics **9**(1), 121–133 (2016). https://doi.org/10.1109/toh.2015.2482478
6. Bernstein, N.A.: The control and regulation of movements (1967)
7. Bianchi, M., Moscatelli, A.: Human and Robot Hands: Sensorimotor Synergies to Bridge the Gap Between Neuroscience and Robotics. Springer (2016)
8. Bianchi, M., Salaris, P., Bicchi, A.: Synergy-based hand pose sensing: optimal glove design. Int. J. Robot. Res. **32**(4), 407–424 (2013)
9. Bicchi, A., Gabiccini, M., Santello, M.: Modelling natural and artificial hands with synergies. Philos. Trans. R. Soc. B: Biol. Sci. **366**(1581), 3153–3161 (2011)
10. Bonilla, M., Farnioli, E., Piazza, C., Catalano, M., Grioli, G., Garabini, M., Gabiccini, M., Bicchi, A.: Grasping with soft hands. In: 2014 14th IEEE-RAS International Conference on Humanoid Robots (Humanoids), pp. 581–587. IEEE (2014). https://doi.org/10.1109/HUMANOIDS.2014.7041421
11. Brown, C.Y., Asada, H.H.: Inter-finger coordination and postural synergies in robot hands via mechanical implementation of principal components analysis. In: IEEE/RSJ International Conference on Intelligent Robots and Systems, 2007. IROS 2007, pp. 2877–2882. IEEE (2007). https://doi.org/10.1109/iros.2007.4399547
12. Catalano, M.G., Grioli, G., Farnioli, E., Serio, A., Piazza, C., Bicchi, A.: Adaptive synergies for the design and control of the Pisa/IIT softhand. Int. J. Robot. Res. **33**(5), 768–782 (2014)
13. Chen, W., Xiong, C.: On adaptive grasp with underactuated anthropomorphic hands. J. Bionic Eng. **13**(1), 59–72 (2016). https://doi.org/10.1016/s1672-6529(14)60160-8
14. Ciocarlie, M., Goldfeder, C., Allen, P.: Dexterous grasping via eigengrasps: A low-dimensional approach to a high-complexity problem. In: Robotics: Science and Systems Manipulation Workshop-Sensing and Adapting to the Real World (2007)
15. De Boor, C., De Boor, C., Mathématicien, E.U., De Boor, C., De Boor, C.: A practical guide to splines, vol. 27. Springer, New York (1978). https://doi.org/10.1007/978-1-4612-6333-3
16. Della Santina, C., Bianchi, M., Averta, G., Ciotti, S., Arapi, V., Fani, S., Battaglia, E., Catalano, M.G., Santello, M., Bicchi, A.: Postural hand synergies during environmental constraint exploitation. Front. Neurorobot. **11**, 41 (2017)
17. Della Santina, C., Grioli, G., Catalano, M., Brando, A., Bicchi, A.: Dexterity augmentation on a synergistic hand: the pisa/iit softhand+. In: 2015 IEEE-RAS 15th International Conference on Humanoid Robots (Humanoids), pp. 497–503. IEEE (2015). https://doi.org/10.1109/humanoids.2015.7363595
18. Eppner, C., Brock, O.: Planning grasp strategies that exploit environmental constraints. In: 2015 IEEE International Conference on Robotics and Automation (ICRA), pp. 4947–4952. IEEE (2015). https://doi.org/10.1109/icra.2015.7139886
19. Eppner, C., Deimel, R., Alvarez-Ruiz, J., Maertens, M., Brock, O.: Exploitation of environmental constraints in human and robotic grasping. Int. J. Robot. Res. **34**(7), 1021–1038 (2015)
20. Feldman, A.G., et al.: Referent control of action and perception. Challenging Conventional Theories in Behavioral Neuroscience. Springer, New York (2015). https://doi.org/10.1007/978-1-4939-2736-4
21. Fu, Q., Santello, M.: Tracking whole hand kinematics using extended kalman filter. In: 2010 Annual International Conference of the IEEE, Engineering in Medicine and Biology Society (EMBC), pp. 4606–4609. IEEE (2010). https://doi.org/10.1109/iembs.2010.5626513
22. Gabiccini, M., Bicchi, A., Prattichizzo, D., Malvezzi, M.: On the role of hand synergies in the optimal choice of grasping forces. Auton. Robot. **31**(2–3), 235–252 (2011)
23. Gabiccini, M., Stillfried, G., Marino, H., Bianchi, M.: A data-driven kinematic model of the human hand with soft-tissue artifact compensation mechanism for grasp synergy analysis. In:

2013 IEEE/RSJ International Conference on Intelligent Robots and Systems, pp. 3738–3745. IEEE (2013)

24. Gorniak, S.L., Zatsiorsky, V.M., Latash, M.L.: Hierarchies of synergies: an example of two-hand, multi-finger tasks. Exp. Brain Res. **179**(2), 167–180 (2007). https://doi.org/10.1007/s00221-006-0777-z
25. Grinyagin, I.V., Biryukova, E.V., Maier, M.A.: Kinematic and dynamic synergies of human precision-grip movements. J. Neurophysiol. **94**(4), 2284–2294 (2005)
26. Grioli, G., Catalano, M., Silvestro, E., Tono, S., Bicchi, A.: Adaptive synergies: an approach to the design of under-actuated robotic hands. In: 2012 IEEE/RSJ International Conference on Intelligent Robots and Systems, pp. 1251–1256. IEEE (2012)
27. Hogg, R.V., Craig, A.T.: Introduction to mathematical statistics. (5th edn.) Upper Saddle River, New Jersey, Prentice Hall (1995)
28. Johansson, R., Westling, G.: Roles of glabrous skin receptors and sensorimotor memory in automatic control of precision grip when lifting rougher or more slippery objects. Exp. Brain Res. **56**(3), 550–564 (1984). https://doi.org/10.1007/bf00237997
29. Jolliffe, I.: Principal component analysis. Springer (2011)
30. Kapandji, A.: Clinical test of apposition and counter-apposition of the thumb. Annales de chirurgie de la main: organe officiel des societes de chirurgie de la main **5**(1), 67–73 (1985). https://doi.org/10.1016/S0753-9053(86)80053-9
31. Latash, M.L.: Synergy. Oxford University Press (2008). https://doi.org/10.1093/acprof:oso/9780195333169.001.0001
32. Latash, M.L., Scholz, J.P., Schöner, G.: Toward a new theory of motor synergies. Mot. Control **11**(3), 276–308 (2007). https://doi.org/10.1123/mcj.11.3.276
33. Lederman, S.J., Klatzky, R.L.: Haptic classification of common objects: Knowledge-driven exploration. Cogn. Psychol. **22**(4), 421–459 (1990). https://doi.org/10.1016/0010-0285(90)90009-s
34. Leo, A., Handjaras, G., Bianchi, M., Marino, H., Gabiccini, M., Guidi, A., Scilingo, E.P., Pietrini, P., Bicchi, A., Santello, M., et al.: A synergy-based hand control is encoded in human motor cortical areas. Elife **5** (2016)
35. Malhotra, M., Rombokas, E., Theodorou, E., Todorov, E., Matsuoka, Y.: Reduced dimensionality control for the act hand. In: 2012 IEEE International Conference on Robotics and Automation (ICRA), pp. 5117–5122. IEEE (2012). https://doi.org/10.1109/icra.2012.6224651
36. Mason, C.R., Gomez, J.E., Ebner, T.J.: Hand synergies during reach-to-grasp. J. Neurophysiol. **86**(6), 2896–2910 (2001)
37. Murray, R.M., Li, Z., Sastry, S.S., Sastry, S.S.: A mathematical introduction to robotic manipulation. CRC press (1994)
38. Mussa-Ivaldi, F.A.: Modular features of motor control and learning. Curr. Opin. Neurobiol. **9**(6), 713–717 (1999). https://doi.org/10.1016/s0959-4388(99)00029-x
39. Naceri, A., Santello, M., Moscatelli, A., Ernst, M.O.: Digit position and force synergies during unconstrained grasping. In: Human and Robot Hands, pp. 29–40. Springer (2016). https://doi.org/10.1007/978-3-319-26706-7_3
40. Nowak, D.A., Glasauer, S., Hermsdörfer, J.: How predictive is grip force control in the complete absence of somatosensory feedback? Brain **127**(1), 182–192 (2004). https://doi.org/10.1093/brain/awh016
41. Overduin, S.A., d'Avella, A., Carmena, J.M., Bizzi, E.: Muscle synergies evoked by microstimulation are preferentially encoded during behavior. Front. Comput. Neurosci. **8**, 20 (2014). https://doi.org/10.3389/fncom.2014.00020
42. Piazza, C., Della Santina, C., Catalano, M., Grioli, G., Garabini, M., Bicchi, A.: Softhand pro-d: Matching dynamic content of natural user commands with hand embodiment for enhanced prosthesis control. In: 2016 IEEE International Conference on Robotics and Automation (ICRA), pp. 3516–3523. IEEE (2016)
43. Puhlmann, S., Heinemann, F., Brock, O., Maertens, M.: A compact representation of human single-object grasping. In: 2016 IEEE/RSJ International Conference on Intelligent Robots and Systems (IROS), pp. 1954–1959. IEEE (2016). https://doi.org/10.1109/iros.2016.7759308

44. Saltiel, P., Wyler-Duda, K., D'Avella, A., Tresch, M.C., Bizzi, E.: Muscle synergies encoded within the spinal cord: evidence from focal intraspinal nmda iontophoresis in the frog. J. Neurophysiol. **85**(2), 605–619 (2001)
45. Santello, M., Baud-Bovy, G., Jörntell, H.: Neural bases of hand synergies. Front. Comput. Neurosci. **7**, 23 (2013)
46. Santello, M., Bianchi, M., Gabiccini, M., Ricciardi, E., Salvietti, G., Prattichizzo, D., Ernst, M., Moscatelli, A., Jörntell, H., Kappers, A.M., et al.: Hand synergies: integration of robotics and neuroscience for understanding the control of biological and artificial hands. Phys. Life Rev. (2016)
47. Santello, M., Flanders, M., Soechting, J.F.: Postural hand synergies for tool use. J. Neurosci. **18**(23), 10105–10115 (1998)
48. Santello, M., Flanders, M., Soechting, J.F.: Patterns of hand motion during grasping and the influence of sensory guidance. J. Neurosci. **22**(4), 1426–1435 (2002)
49. Schafer, R.W.: What is a savitzky-golay filter? [lecture notes]. IEEE Signal Process. Mag. **28**(4), 111–117 (2011). https://doi.org/10.1109/msp.2011.941097
50. Schäling, B.: The boost C++ libraries. Boris Schäling (2011)
51. Scholz, J.P., Schöner, G.: The uncontrolled manifold concept: identifying control variables for a functional task. Exp. Brain Res. **126**(3), 289–306 (1999). https://doi.org/10.1007/s002210050738
52. Serio, A., Bianchi, M., Gabiccini, M., Bicchi, A.: [d94] the tactile toolbox. In: Haptics Symposium (HAPTICS), 2014 IEEE, pp. 1–1. IEEE (2014). https://doi.org/10.1109/haptics.2014.6775573
53. Sircoulomb, V., Hoblos, G., Chafouk, H., Ragot, J.: State estimation under nonlinear state inequality constraints. a tracking application. In: 2008 16th Mediterranean Conference on Control and Automation, pp. 1669–1674. IEEE (2008). https://doi.org/10.1109/med.2008.4602024
54. Stratmann, P., Lakatos, D., Albu-Schäffer, A.: Neuromodulation and synaptic plasticity for the control of fast periodic movement: energy efficiency in coupled compliant joints via PCA. Front. Neurorobotics **10** (2016). https://doi.org/10.3389/fnbot.2016.00002
55. Thakur, P.H., Bastian, A.J., Hsiao, S.S.: Multidigit movement synergies of the human hand in an unconstrained haptic exploration task. J. Neurosci. **28**(6), 1271–1281 (2008)
56. Tresch, M.C., Saltiel, P., Bizzi, E.: The construction of movement by the spinal cord. Nat. Neurosci. **2**(2), 162 (1999)
57. Villani, L., Ficuciello, F., Lippiello, V., Palli, G., Ruggiero, F., Siciliano, B.: Grasping and control of multi-fingered hands. In: Advanced Bimanual Manipulation, pp. 219–266. Springer (2012). https://doi.org/10.1007/978-3-642-29041-1_5
58. Westling, G., Johansson, R.S.: Responses in glabrous skin mechanoreceptors during precision grip in humans. Exp. Brain Res. **66**(1), 128–140 (1987). https://doi.org/10.1007/bf00236209
59. Xu, K., Liu, H., Du, Y., Zhu, X.: Design of an underactuated anthropomorphic hand with mechanically implemented postural synergies. Adv. Robot. **28**(21), 1459–1474 (2014)

Part II
On the Design of Nature-Inspired Prostheses and the Assessment of Motion Impairment

Chapter 5
Using Nature-Inspired Principles to Design of Robotic Limbs: The Soft Wrist

Although hands represent the primary tool for humans and humanoid robots to interact with the surrounding environment, their capability are strongly limited in the absence of a companion wrist, responsible for a correct orientation of the palm. This aspect is much more evident in robotic prostheses. In this chapter, we propose an under-actuated wrist system, compliant to gently adapt to the environment, and capable to implement different under-actuation schemes. Our approach leverages on the concept of soft synergies—in particular the design method of adaptive synergies—as it derives from the field of robot hand design. First, we introduce the design principle and its implementation in a configurable test bench prototype, which can be used to demonstrate the feasibility of our idea. Then, we report on results from preliminary experiments with humans, targeting the identification of the most probable wrist pose during the pre-grasp phase in activities of daily living. Based on these outcomes, we calibrate our wrist prototype accordingly and demonstrate its effectiveness to accomplish grasping and manipulation tasks.

5.1 Introduction

In humans, the wrist is responsible for fine adjustments of hand pose [2, 21, 36] and plays a fundamental role in hand mobility and manipulation in everyday actions [3]. The development of effective robotic wrists is also very important in robotic manipulation, since the wrist enables the control of the end-effector pose. This is crucial in many applications, ranging from assistive robotics and prosthetics, human-robot interaction, micro-manipulation for medical procedures, orienting machine tools, see e.g. [3, 17, 19]. Furthermore, within the novel paradigm of soft manipulation, i.e. robot embodied ability to comply and adapt to features of the environment, an effective design of robotic wrists is an essential component to simplify the grasp planning phase. Indeed, leveraging upon the intrinsic capabilities of soft end effectors to

G. Averta, *Human-Aware Robotics: Modeling Human Motor Skills for the Design, Planning and Control of a New Generation of Robotic Devices*, Springer Tracts in Advanced Robotics 145, https://doi.org/10.1007/978-3-030-92521-5_5

continuously deform in different shapes, an approximation of object properties and hand pose definition, which is ultimately determined by the wrist, can be sufficient to generate successful candidate poses to grasp an external objects [9]. Among the applications mentioned, prosthetics probably represents one of the most challenging, due to the significant contribution that a robotic wrist can provide in terms of users quality of life and prosthesis embodiment [3], and to the additional constraints that must be taken into account during the design phase. Interestingly, while so far the effort of the community provided notable results for the development of dexterous hands for prosthetic use, the same is not mirrored for the development of robotic wrists. Indeed, only few examples of commercially available prosthetic wrists can be found, including both motorized solutions for hand closure and pronation/supination (P-S), flexion/extension (F-E) and finger closing, as well as passive solutions (see [24] for a review on these topics). Considering also research developments, we can find several architectures, including passive wrists (see e.g. [23]), 2 DOFs fully-motorized wrists (see e.g. [20, 35]), 3 DOFs fully-motorized wrist (see e.g. [31]), rotators [1], body-powered wrists [28], among the others (refer to [3] for a complete review).

In this chapter, we present the mechanical design of a soft and adaptable under-actuated wrist prototype with 2 DOFs (see Fig. 5.1). This system allows to implement different under-actuation patterns and represents a useful test-bed to shape the development of future developments of our implementations with reduced encumbrance. Since the absence of P-S is very penalizing to be compensated for (because it requests

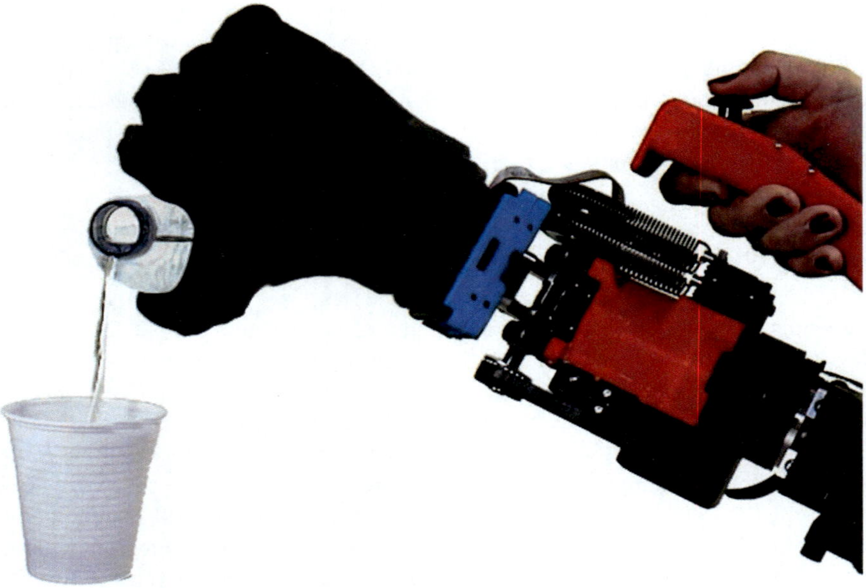

Fig. 5.1 The two DOFs adaptive under-actuated wrist pouring water

a rotation of the shoulder around the axis of the arm), we decided to implement this DOF, while our architecture allows to choose between F-E and add-abduction (A-A), or a combination of the two, as the second DOF. At the same time, we also developed a kinematic model of human upper limb and performed preliminary experiments with human participants to identify upper limb joint angles and extract the most probable wrist pose during the pre-grasp phase. This pattern was then implemented in our robotic wrist. We conducted experiments in which the wrist was used in conjunction with the Pisa/IIT SoftHand [13] to grasp different items to demonstrate the effectiveness of our solution in terms of grasp success, opening interesting perspectives for robotic, prosthetic and rehabilitative applications [32]. The content of this chapter is adapted from [12].

A companion video is available at the following link: https://www.dropbox.com/s/mxnjkxh0g7xw7r5/Underactuated_wrist.mp4?dl=0.

5.2 A Wrist with Synergies

The theoretical basis for the proposed modular under-actuation moves from the concept of adaptive synergies [13], which combines synergy-based actuation and mechanical adaptability for the development of soft robotic hands. This approach can be applied also to the simplification of the design of modular wrists. In addition, the proposed modular approach enables the implementation and testing of different under-actuation strategies, which in turn can take inspiration from human example.

In neuroscience the concept of synergies, i.e. multiple elements working together towards a common goal, has been extensively studied and developed as a principled simplification approach to the well-known Bernstein's problem of motor redundancy [5]. Although synergies can be defined at different levels, such as kinematic, muscular, neural and also sensory (see e.g. [33, 34]), the main idea shared across these definitions is the *geometric* concept of dimensionality reduction, i.e. multiple degrees of freedom that can be controlled in a coordinated manner, resulting in a dimensionality of the actuation domain lower than the number of available degrees of freedom. The formalism of motor synergies has been applied to study how the central nervous system can control the large number of sensory-motor elements of the hand, such as ligaments, muscles, receptors, joints and bones. At the same time, the existence of a co-variation (i.e. a synergy) for the joint angles of the whole upper limb has been also experimentally proven in literature. For example, previous works demonstrated that movements of the arm and shaping of the hand during reach-to-grasp are highly coordinated [10], and hence that a global control strategy for the coordination of the fingers, wrist and arm can be observed [22, 25]. Focusing on grasping and manipulation, the control of hand position and orientation w.r.t external objects represents a key factor for enabling a correct task accomplishment and can be regarded as the result of coordination and contribution of different upper-limb parts, such as the shoulder, the elbow, and finally, the wrist.

At the same time, robotics research has deeply leveraged upon these neuroscientific insights for the design of artificial hands and sensing systems with a reduced number of actuators, control inputs, and sensors (see e.g. [6, 11, 13, 34]).

More specifically, soft synergies [7, 15] were proposed as a tool for modeling human motor behavior and driving the design of artificial hands. According to this framework, a reference configuration can be generated using a reduced sub-set of all the possible degrees of freedom (called postural synergies [34]), while the real limb is compliantly attracted to it and, at the same time, repelled by the forces of interaction with the external world. When an equilibrium is reached, we have

$$q = S\sigma - CJ^T f_{\text{ext}}, \tag{5.1}$$

where $\sigma \in \mathbb{R}^s$ are the synergy variables, $q \in \mathbb{R}^n$ are the joint variables. $S \in \mathbb{R}^{n\times s}$ represents the synergistic *basis* of the joints configuration space, mapping σ into q. The matrix $C \in \mathbb{R}^{n\times n}$ is the compliance and $J^T f_{ext} \in \mathbb{R}^n$ collects the effect of all the external forces acting on the system.

Starting from these considerations, adaptive synergies [13, 18] were proposed as a design paradigm to implement soft synergies leveraging upon the adaptability of differential mechanism. In this case, let be $R \in \mathbb{R}^{s\times n}$ the transmission distribution matrix and $E \in \mathbb{R}^{n\times n}$ the joint space stiffness matrix. The equilibrium of the joint torques, which includes a linear elastic force in joint space $-Eq \in \mathbb{R}^n$ and the contribution of external forces $J^T f_{\text{ext}} \in \mathbb{R}^n$, is

$$\begin{cases} J^T f_{\text{ext}} = R^T f - Eq \\ Rq = \sigma. \end{cases} \tag{5.2}$$

where $f \in \mathbb{R}^s$ is the generalized force produced by the motor. Solving Eq. (5.2) leads to Eq. (5.1) with the following choice of matrices (see [18] for more details)

$$\begin{aligned} S &= E^{-1}R^T(RE^{-1}R^T)^{-1} \\ C &= E^{-1} - E^{-1}R^T(RE^{-1}R^T)^{-1}RE^{-1}. \end{aligned} \tag{5.3}$$

Through this relation the matrices S and C are related to the mechanical elements represented by E and R. Thus both un-perturbed movements and elastic behavior of the robot can be independently designed by proper choice of elastic constant of the springs and the radii of the pulleys. This framework was successfully applied in the design of robotic hands [13, 14, 27]. Naturally conforming to the external environments in a human-like manner, such hands are able to perform complex grasps and basic in-hand manipulations with a small set of control inputs and an intuitive interface.

As described above, such characteristics are highly attractive for a robotic wrist. In the authors' opinion, passive adaptability of the wrist should help the user providing an intuitive way to take advantages from environment, in order to achieve more stable hand pre-orientations, and to perform manipulation tasks. For these reasons, in this chapter we propose an adaptive synergy-based robotic wrist, which allows us to explore these ideas.

Furthermore, the modularity of our prototype enables the implementation and testing of different under-actuation patterns.

In our first prototype, we considered only the two DOFs q_1, q_2 rotating around axes J_1, J_2, instead of the three DOFs which are usually considered to model the human wrist. The two DOFs endowed in the wrist are P-S and F-E in the configuration shown in Fig. 5.2. Main motivation for this relies on results from preliminary human studies, described more in detail in Sect. 5.4, which revealed that a suitable combination of these two DOFs is enough to explain around 95% of wrist pose variability during the pre-grasp phase in daily life activities. However, it is worth noticing that the orientation of the hand w.r.t. the wrist can be changed, which causes J_2 to switch between F-E and A-A, as better explained in Sect. 5.3. Figure 5.2 shows the main components of the proposed system.

The elastic field $E\,q$ is here implemented through the springs k_1, k_2, while the transmission ratio R is implemented by the pulleys D_{p2}, D_{pm}. The next section pro-

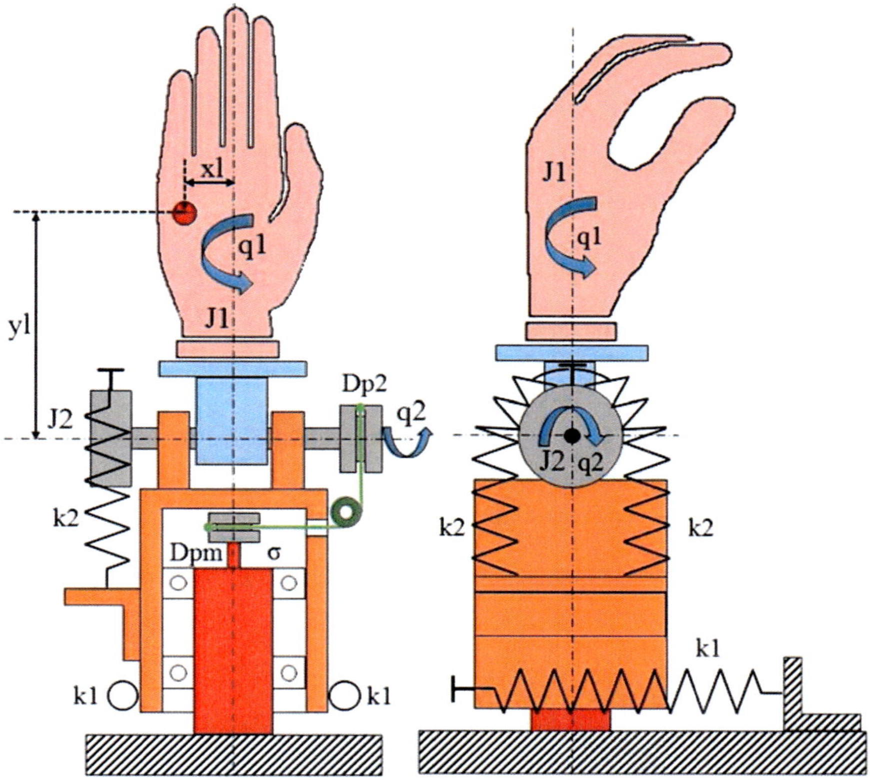

Fig. 5.2 Wrist model used for the mechanical design

vides a more accurate description of the actual mechanics implementing the described components.

We call x_G and x_L respectively the distance between $J1$ and the center of mass G of the hand and an applied load L normal to the palm of the hand and similarly y_G and y_L the distance between $J2$ and G and f_L. Then Eq. (5.2) for the proposed wrist is

$$\begin{bmatrix} x_G & x_L \\ y_G & y_L \end{bmatrix} \begin{bmatrix} M_h g \\ f_L \end{bmatrix} = \begin{bmatrix} 1 \\ \frac{D_{p2}}{D_{pm}} \end{bmatrix} f - \begin{bmatrix} k_1 & 0 \\ 0 & k_2 \end{bmatrix} \begin{bmatrix} q_1 \\ q_2 \end{bmatrix}, \tag{5.4}$$

where M_h is the mass of the hand and k_1 and k_2 the elastic constants of the springs of joint $J1$ and $J2$ respectively. Thus (5.3) leads to the following description of wrist behavior

$$\begin{aligned} q = & \begin{bmatrix} D_{pm}\, k_2 \\ D_{p2}\, k_1 \end{bmatrix} \frac{\sigma\, D_{pm}}{k_1\, D_{p2}^2 + k_2\, D_{pm}^2} + \\ & \begin{bmatrix} \frac{D_{p2}^2}{k_1\, D_{p2}^2 + k_2\, D_{pm}^2} & \frac{-D_{p2}\, D_{pm}}{k_1\, D_{p2}^2 + k_2\, D_{pm}^2} \\ \frac{-D_{p2}\, D_{pm}}{k_1\, D_{p2}^2 + k_2\, D_{pm}^2} & \frac{D_{pm}^2}{k_1\, D_{p2}^2 + k_2\, D_{pm}^2} \end{bmatrix} \begin{bmatrix} x_G & x_L \\ y_G & y_L \end{bmatrix} \begin{bmatrix} M_h g \\ f_L \end{bmatrix}. \end{aligned} \tag{5.5}$$

Among the many ways to link the DOFs of the wrist, the proposed device reproduces wrist synergies with a direct proportional relationship between q_1 and q_2. From Eq. (5.4), it can be noted that, for null external loads, the ratio of joint variables is given by the equation $\frac{q_1}{q_2} = \frac{k_2}{k_1}\frac{D_{pm}}{D_{p2}}$. Acting on the ratio of the elastic constants of the springs and on the diameters of the driver and driven pulleys, we can tune a coordinated motion of the DOFs of the wrist in absence of external loads (as during the grasp acquisition stage). In the following section, we discuss the mechanical implementation and thoroughly analyze the possible choices for spring constants.

5.3 Mechanical Design

Figure 5.3 shows an overall 3D render of the wrist. The majority of the parts are made in ABS plastic with a FDM machine, whereas the most stressed parts are in Aluminum Alloy or Stainless Steel. The device comprises three main groups:

Frame

It is the yellow part located at the bottom of the device. It constitutes the fixed frame on which a joystick can be mounted to handle the wrist (see the pictures of Sect. 5.5) and contains the electronic components to drive the motors of the hand and the wrist.

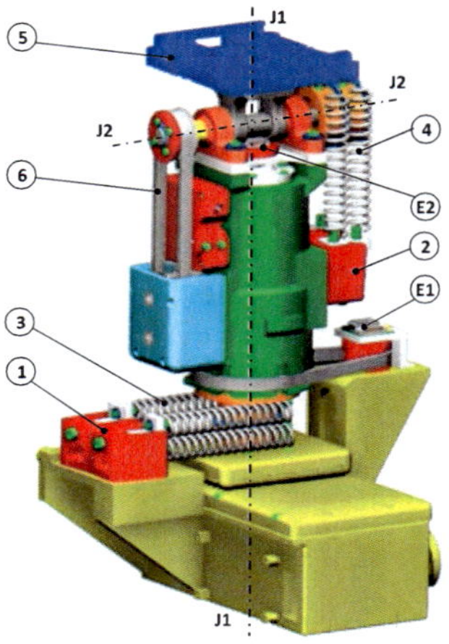

Fig. 5.3 3D render of the wrist with the main components of the prototype

Group (1), fixed to the frame, enables the loading of springs (3) at axis J_1. Encoder (E1) directly measures the P-S angle q_1 with a transmission ratio 2.5 : 1.

Pronation-Supination Group

It is the main cylindrical body which confers the P-S movement to the wrist (axis J_1), with the motor embedded. Group (2) enables the loading of springs (4) wrapping around axis J_2 while toothed belt (6) transmits the power from the motor to the J_2. On the top of the cylindrical body, encoder (E2) measures the relative angle between the motor shaft and the cylindrical body $\sigma - q_1$. As shown in Sect. 5.2, this reading is equal to $\frac{q_2 D_{p2}}{D_{pm}}$, thanks to which we can get a direct reading of A-A or F-E. A third redundant encoder (E3) housed in the cylindrical body, has the purpose of allowing a direct measurement of the rotation of the motor without being dependent from other encoder readings.

Flexion-Extension Group (or Abduction-Adduction Group)

It is located at the top of the device and confers a rotation around axis J_2. At the top of the wrist, a flange (5) allows to mount the hand with different orientations at 90°. If the hand is mounted as in Fig. 5.2 the rotation around axis J_2 is F-E. Conversely, if the hand is rotated 90° on the flange with respect to Fig. 5.2, the rotation is A-A.

5.4 Wrist Synergies in Humans

One of the motivations of having a modular adaptive wrist is to enable a suitable under-actuation strategy which takes inspiration from the human example. For this reason, we performed a set of preliminary experiments aiming at understanding which are the most probable poses of the human wrist during activities of daily living. The experimental setup is described in detail in Chaps. 2 and 3 of this thesis. In the following, a brief recap of the major technical details are reported to improve the clarity of presentation of this Chapter.

To perform the experimental campaign, we defined a kinematic model for the whole upper limb, and an estimation strategy to obtain joint angles from a marker-based optical motion capture system (PhaseSpace, working frequency 480 Hz). While the wrist represents the main interest of this chapter, tracking is extended to the whole upper limb in order to improve performances in optical motion tracking. Inspired by [4, 8, 26, 29], we modeled the upper limb using three rigid segments and seven DOFs: three DOFs for the shoulder (A-A, F-E and external-internal rotation of the humerus w.r.t the scapula), two for the elbow (F-E of the ulna-radius w.r.t the humerus and P-S due to the relative movement of radius and ulna), and two for the wrist (F-E and A-A of hand w.r.t the forearm). The axes for the shoulder, elbow and wrist joints are assumed to be orthogonal and intersecting.

We used rigid supports with attached markers to define a reference frame on the chest, and local frames in the middle of forearm, arm and hand dorsum as in Fig. 5.4a. The segments model upper arm, forearm and hand, respectively, and are connected through revolute joints. We chose to use the *Product of Exponentials* (POE) parametrization to model kinematics [16]. Importantly, in this model the elbow prono-supination degree of freedom, that ultimately determines the pose of the hand, is attributed to the elbow, as was done in [4]. It is worth noticing that in our robotic prototype this degree of freedom is considered as part of the wrist. Under a kinematic point of view these two approaches are equivalent in terms of hand pose control. However, our mechanical solution is preferable looking at prosthetic applications, e.g. for trans radial amputees, where hand P-S is controlled at wrist level. In Fig. 5.4b, we show twist axes for each degree of freedom and other descriptive parameters of the model used during the calibration phase.

Referring to [16], we implemented a calibration procedure to estimate model parameters based on a constrained least-squares minimization between marker positions and model-related marker positions. Joint angle were retrieved implementing a recursive estimation procedure based on an iterative Extended Kalman Filter.

5.4.1 Experimental Procedure

10 right-handed subjects (6 M, 4F, age ranging from 20 to 30 y.o.) gave their informed consent to take part to the experiment. No one had any physical limitation that could

Table 5.1 Daily living activities

1	Brush/comb hair
2	Stir in a bowl
3	Cut meat with knife and fork
4	Pour from a bottle/can
5	Write the word "LETTER" legibly
6	Use a pair of scissors
7	Use a key in lock
8	Dial a touch tone phone
9	Use a hammer and nail

have affected the experimental outcomes. During the experiment, participants wore the previously described marker supports, and were comfortably seated. They were asked to grasp to use a set of objects as in Table 5.1 and to perform the related activities of daily living tasks extracted from [30].

Each task was repeated three times with an order of actions randomized. In particular, for each action, the object was placed on a table ($\approx$ 80 cm height) in front of the participant, at a distance of $\approx$ 50 cm, while the starting hand position was palm-down on the table. This list of actions was performed as trade-off between experimental time and human workspace description, with no claim of exhaustiveness. Study of a more complete list of human actions will be carried out in future works. Note that, although we are interested in the pose of the wrist in the pre-grasp phase, we asked participants to complete all the action phases to have more natural data for our analysis.

5.4.2 Data Analysis

Once joint angles were estimated, we focused on the degrees of freedom corresponding to the ones we can implement in our robotic wrist, i.e. q_5, q_6 and q_7 as in Fig. 5.4. To identify the most common and probable pre-grasp wrist pose, we performed Principal Component Analysis on the data. Wrist orientations were collected in a matrix $A \in \mathbb{R}^{90\times 3}$, where the number of rows comes from the 10 participants and 9 actions, with the three repetitions being averaged, and the number of columns corresponds to the three degrees of freedom considered. The principal components (PCs) were then computed from the eigenvalues and eigenvectors of the matrix of the covariance coefficients between each of the 90 samples. Results of the PC analysis are reported in Table 5.2.

Although preliminary, these results provided us useful indications on how to control the wrist, i.e. to constrain the degrees of freedom, according to the first common actuation pattern PC1 of the wrist poses in human pre-grasping, which takes

Table 5.2 Wrist synergies extracted from the experiments

	P-S	A-A	F-E	Variance (%)
PC1	0.36	$4 \cdot 10^{-3}$	0.93	74
PC2	0.93	0.013	–0.36	21
PC3	–0.013	~1	$9.3 \cdot 10^{-4}$	5

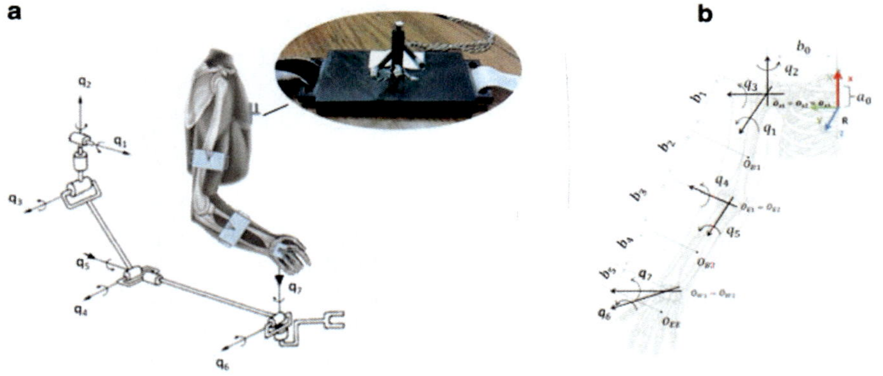

Fig. 5.4 Upper limb model used in the identification. Joint variables model shoulder abduction-adduction (q_1), shoulder external-internal rotation (q_2), shoulder flexion-extension (q_3), elbow flexion-extension (q_4), elbow prono-supination (q_5), wrist abduction-adduction (q_6) and wrist flexion-extension (q_7), respectively. **a** Kinematic Model (adapted from [4]) and marker supports (in light gray and in the zoomed view) used for the experiments. **b** Geometric Parameters

into account a large part of pose variability. What is noticeable is that according to PC1 F-E plays the most important role: this is not surprising given the set of actions we considered. In the future these results will be further investigated, varying both the location of the object in front of the participant and the typologies of the items used for the experiments.

5.5 Experiments

In this Section, we describe the hardware and software setup used to prove the device effectiveness in P-S and F-E or A-A movements, while grasping different objects with the Pisa/IIT SoftHand. The 2-DOF under-actuated wrist has been provided with a 2-axes joystick hardware interface to handle both the hand and the wrist motor. The thumb-joystick controls the device through a velocity-based control loop, i.e. the farther is the joystick movement from the rest position the faster is the motor rotation. The joystick is at rest when the thumb is centred on both axes. From this position the SoftHand closes when the joystick is moved upward, while the wrist rotates if it is moved laterally.

Two different sets of experiments have been carried out with this setup: the first targeting a qualitative demonstration of the wrist capabilities, and the second with the purpose of showing the implementation of the first principal component from the previous section.

5.5.1 System Demonstration

Figures 5.5 and 5.6 show the wrist free movement in two different configurations. In Fig. 5.5 from the neutral position (center), it is possible to have either pronation and extension (left) or supination and flexion (right). On the contrary, Fig. 5.6 refers to the case in which the hand is mounted on the wrist toothed flange to have pronation and adduction (left) or supination and abduction (right). It is of relevance also to have a look on the movement of the device while a DOF is blocked to notice how a pure P-S or a F-E / A-A movement can be obtained (see Figs. 5.7 and 5.8). Figure 5.9 shows the under-actuated wrist device while it is used with the thumb-joystick interface for a typical daily living activity. The here proposed device is also capable of exploiting the environment and of passively adapting to improve objects approach and grasping (Fig. 5.10). Figure 5.11 shows an example of this kind of adaptability, while the user presses the under-actuated wrist on a table to block F-E and to convey only P-S movement.

In the following figures it is possible to see the wrist in conjunction with the Pisa/IIT SoftHand [13]. What is noticeable is that the integrated system enables advanced manipulation capabilities, see e.g. Figure 5.9. Furthermore, it is worth to observe that the orientation of the arm of the user maneuvering our device is minimized during task execution. This is extremely important in view of future prosthetic implementations, to reduce the need of compensatory movements in amputees during the execution of activities of daily living. Finally, in Fig. 5.11 the adaptability of the wrist-hand system is used to exploit environmental constraints: this aspect is

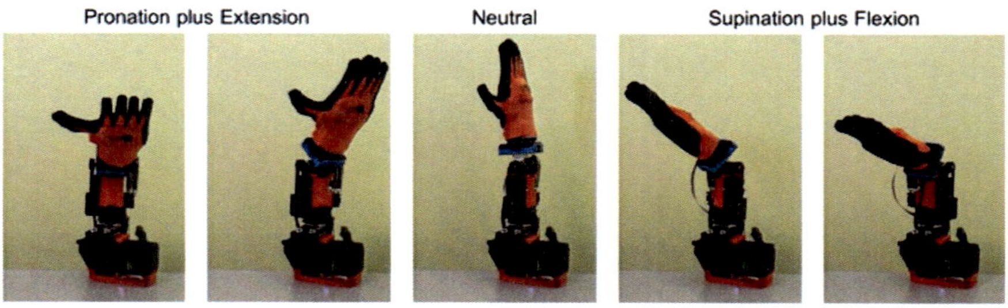

Fig. 5.5 Free movement of the under-actuated system wrist-SoftHand ($\pm 90^\circ$ excursion for both P-S and F-E)

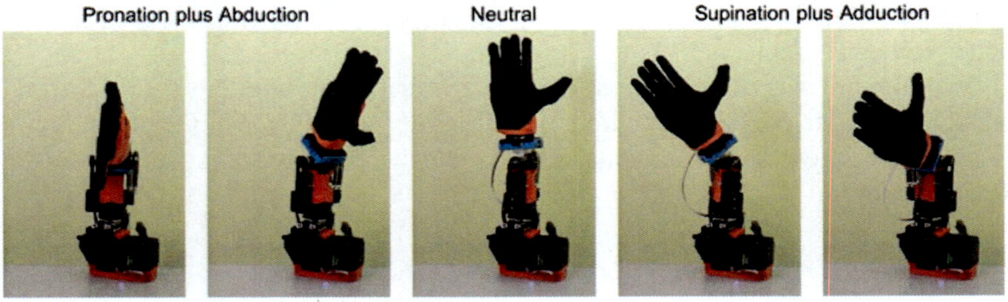

Fig. 5.6 Free movement of the under-actuated system wrist-SoftHand (±90° excursion for both P-S and A-A)

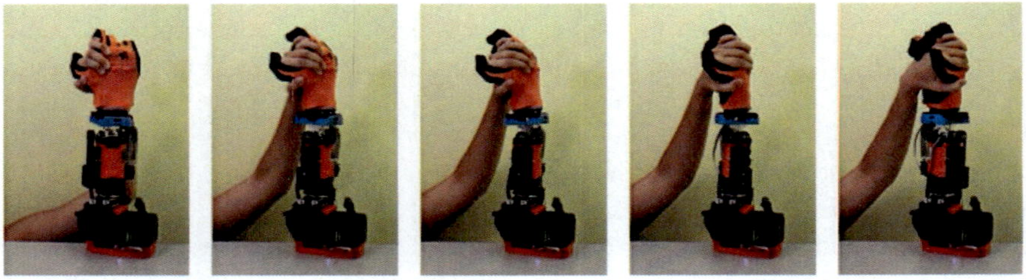

Fig. 5.7 P-S full range movement while F-E is blocked by an external constraint

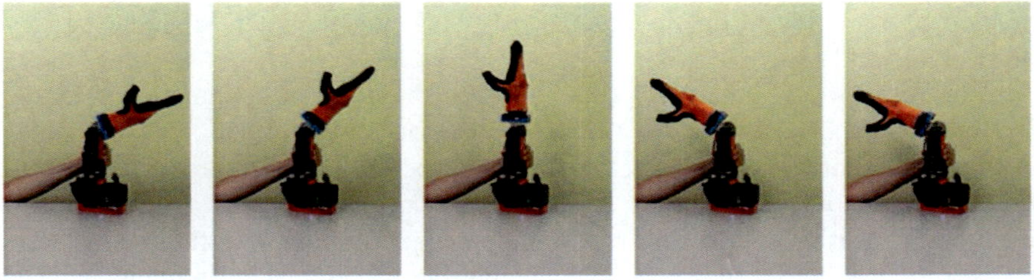

Fig. 5.8 F-E full range movement while P-S is blocked by an external constraint

Fig. 5.9 The coordinate motions of the wrist are used to approach a bottle and pour water into a glass

Fig. 5.10 Wrist motions are exploited to grasp a glass and rotate it upside down

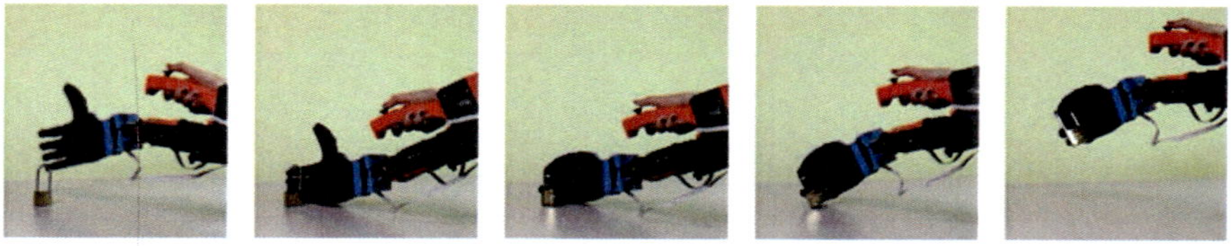

Fig. 5.11 A typical usage of the here realized device where passive adaptability to environment is effective during task accomplishment

fundamental not only for a natural usage in assistive robotic applications, but also for soft robotic manipulation, in general.

5.5.2 *Implementation of PCs*

In Sect. 5.4, we reported on a set of preliminary experiments aiming at understanding which are the most probable poses of the human wrist during activities of daily living, which led to the individuation of a set of principal components such that the first accounts for more than 74% of the total dataset variability. In this section, we will show an implementation of such PC1 in our device. Referring to Eq. (5.4), the problem of implementing PC1 through hardware is reduced to finding k_1 and k_2 such that the ratio between q_1 and q_2 is coherent with the ratio assigned by the first principal component. We recall that this PC has a dominance of the flex-extension joint, with P-S coming second and add-abduction being set to an almost null value. This makes it possible to implement PC1 by assembling the hand in the P-S/F-E configuration (Fig. 5.5) and choosing the elastic constants appropriately. In other words, such implementation results in requiring a target ratio $\frac{q_1}{q_2} = \frac{0.36157}{0.93234} \approx 0.39$. With our available springs at the moment of the experiments, we implemented a ratio $\frac{q_1}{q_2} = 0.42$ as the closest approximation ($k_1 = 1.8Nmm$ and $k_2 = 0.55Nmm$). In Fig. 5.12 we show a plot of the $\frac{q_1}{q_2}$ ratio from encoders measurements after 10 cycles of P-S and F-E full range movement, compared with the nominal ratio that would be required by the PC1 implementation. Two considerations can be done: (i) the hysteresis comes from clearances and static friction in the physical implementation of the wrist (ii) the slope of the experimental curves is close to the ideal one. Figure 5.13 shows a grasp performed with this PC1 implemented, with all the previ-

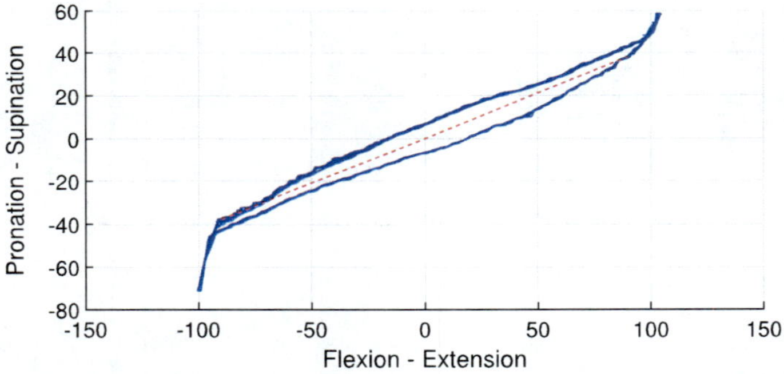

Fig. 5.12 Joint angles behaviour while PC1 is implemented in the artificial wrist: ideal ratio between F-E and P-S movements (red line) versus real ratio deduced by 10 cycles experiments (blue lines)

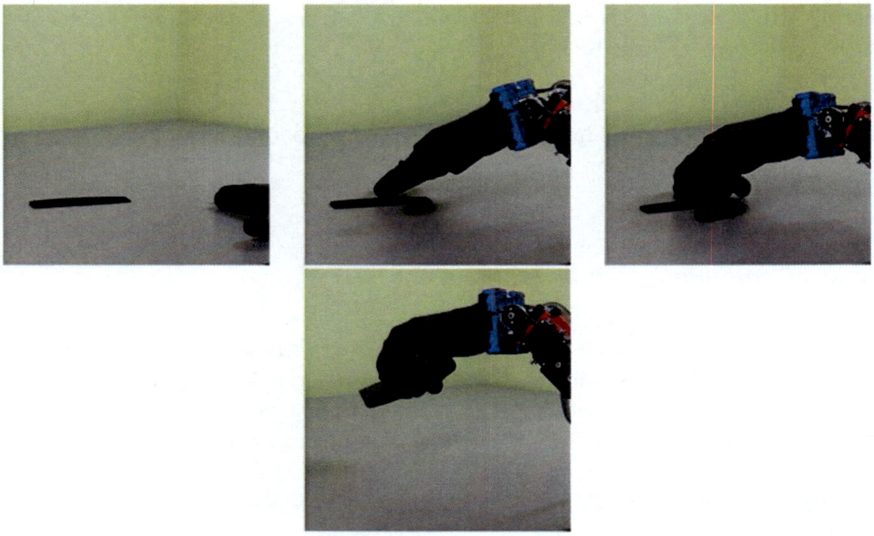

Fig. 5.13 Example of object grasping with the under-actuated wrist implementing PC1

ously discussed characteristics (i.e. adaptability, enhanced grasping capabilities and absence of compensatory movements) well depicted.

5.6 Conclusions

In this chapter we have presented a modular, soft, under-actuated wrist with two DOFs and only one motor. The mechanical design, whose theoretical foundations rely on the framework of soft and adaptive synergies, is modular in order to allow the implementation of different under-actuation patterns. In parallel, we also performed experiments with human participants to obtain a preliminary estimation of the most probable pose of the human wrist during the pre-grasp-phase in activities of daily living, i.e. a first estimation of wrist synergies. Finally, we have experimentally demonstrated the capabilities of our prototype and discussed a possible implementation of a preliminarly estimated first human wrist synergy, which was obtained thanks to a suitable choice of the elastic elements. Experiments where our device was used in conjunction with the Pisa/IIT SoftHand to grasp different items show the effectiveness of our solution as a multiplier of manipulation capabilities in humans, also thanks to the *softness* of the system.

The outcomes of this chapter opens promising scenarios for prosthetic applications, with the possibility to enhance grasping and manipulation performance in amputees, while keeping the complexity of control as simple as possible and still guaranteeing robustness and movement naturalness—the latter representing important novel contributions of soft robotic technologies such as the Pisa/IIT SoftHand and its prosthetic evolution.

Furthermore, these results can also be used within the soft manipulation paradigm in industrial applications and human-robot-interaction, allowing to further multiply the capabilities of soft manipulators.

Future works will address the more in depth investigation of wrist synergies in humans and, consequently, to reduce the dimensions of the wrist prototype. Of note, this aspect is especially important in trans-radial amputations for a successful integration of the wrist with the socket and the hand, which could be designed to implement a specific under-actuation pattern devised from human example. Additional experiments will be also performed to further evaluate the effectiveness of the wrist in grasping and manipulation tasks.

References

1. Hormer dorrance friction wrist. http://hosmer.com/
2. van Andel, C.J., Wolterbeek, N., Doorenbosch, C.A., Veeger, D.H., Harlaar, J.: Complete 3D kinematics of upper extremity functional tasks. Gait Posture **27**(1), 120–127 (2008)
3. Bajaj, N.M., Spiers, A.J., Dollar, A.M.: State of the art in prosthetic wrists: commercial and research devices. In: 2015 IEEE International Conference on Rehabilitation Robotics (ICORR), pp. 331–338. IEEE (2015)
4. Benati, M., Gaglio, S., Morasso, P., Tagliasco, V., Zaccaria, R.: Anthropomorphic robotics. Biol. Cybern. **38**(3), 125–140 (1980)
5. Bernstein, N.A.: The co-ordination and regulation of movements (1967). https://doi.org/10.1037/007805
6. Bianchi, M., Salaris, P., Bicchi, A.: Synergy-based hand pose sensing: optimal glove design. Int. J. Robot. Res. **32**(4), 407–424 (2013)
7. Bicchi, A., Gabiccini, M., Santello, M.: Modelling natural and artificial hands with synergies. Philos. Trans. R. Soc. B: Biol. Sci. **366**(1581), 3153–3161 (2011)
8. Biryukova, E., Roby-Brami, A., Frolov, A., Mokhtari, M.: Kinematics of human arm reconstructed from spatial tracking system recordings. J. Biomech. **33**(8), 985–995 (2000)

9. Bonilla, M., Farnioli, E., Piazza, C., Catalano, M., Grioli, G., Garabini, M., Gabiccini, M., Bicchi, A.: Grasping with soft hands. In: 2014 14th IEEE-RAS International Conference on Humanoid Robots (Humanoids), pp. 581–587. IEEE (2014). https://doi.org/10.1109/HUMANOIDS.2014.7041421
10. Bootsma, R.J., Marteniuk, R.G., MacKenzie, C.L., Zaal, F.T.: The speed-accuracy trade-off in manual prehension: effects of movement amplitude, object size and object width on kinematic characteristics. Exp. Brain Res. **98**(3), 535–541 (1994)
11. Brown, C.Y., Asada, H.H.: Inter-finger coordination and postural synergies in robot hands via mechanical implementation of principal components analysis. In: IEEE/RSJ International Conference on Intelligent Robots and Systems, 2007. IROS 2007, pp. 2877–2882. IEEE (2007). https://doi.org/10.1109/iros.2007.4399547
12. Casini, S., Tincani, V., Averta, G., Poggiani, M., Della Santina, C., Battaglia, E., Catalano, M.G., Bianchi, M., Grioli, G., Bicchi, A.: Design of an under-actuated wrist based on adaptive synergies. In: 2017 IEEE International Conference on Robotics and Automation (ICRA), pp. 6679–6686. IEEE (2017)
13. Catalano, M.G., Grioli, G., Farnioli, E., Serio, A., Piazza, C., Bicchi, A.: Adaptive synergies for the design and control of the Pisa/IIT softhand. Int. J. Robot. Res. **33**(5), 768–782 (2014)
14. Della Santina, C., Grioli, G., Catalano, M., Brando, A., Bicchi, A.: Dexterity augmentation on a synergistic hand: the Pisa/IIT softhand+. In: 2015 IEEE-RAS 15th International Conference on Humanoid Robots (Humanoids), pp. 497–503. IEEE (2015). https://doi.org/10.1109/humanoids.2015.7363595
15. Feldman, A.G.: Once more on the equilibrium-point hypothesis (λ model) for motor control. J. Mot. Behav. **18**(1), 17–54 (1986)
16. Gabiccini, M., Stillfried, G., Marino, H., Bianchi, M.: A data-driven kinematic model of the human hand with soft-tissue artifact compensation mechanism for grasp synergy analysis. In: 2013 IEEE/RSJ International Conference on Intelligent Robots and Systems, pp. 3738–3745. IEEE (2013)
17. Gosselin, C., Angeles, J.: The optimum kinematic design of a spherical three-degree-of-freedom parallel manipulator. J. Mech., Transm., Autom. Des. **111**(2), 202–207 (1989). https://doi.org/10.1115/1.3258984
18. Grioli, G., Catalano, M.G., Silvestro, E., Tono, S., Bicchi, A.: Adaptive synergies: an approach to the design of under-actuated robotic hands. In: International Conference of Intelligent Robots and Systems—IROS 2012, pp. 1251 – 1256. Vilamoura, Algarve, Portugal (2012). https://doi.org/10.1109/IROS.2012.6385881
19. Hammond, F.L., Howe, R.D., Wood, R.J.: Dexterous high-precision robotic wrist for micromanipulation. In: 2013 16th International Conference on Advanced Robotics (ICAR), pp. 1–8. IEEE (2013)
20. Goudreau, Louis, BASc, P., Greg Bush, B., et al.: Two-degree-of-freedom powered prosthetic wrist. J. Rehabil. Res. Dev. **48**(6), 609 (2011)
21. Ma, S., Feldman, A.: Two functionally different synergies during arm reaching movements involving the trunk. J. Neurophysiol. **73**(5), 2120–2122 (1995)
22. Mason, C.R., Gomez, J.E., Ebner, T.J.: Hand synergies during reach-to-grasp. J. Neurophysiol. **86**(6), 2896–2910 (2001)
23. Montagnani, F., Controzzi, M., Cipriani, C.: Preliminary design and development of a two degrees of freedom passive compliant prosthetic wrist with switchable stiffness. In: 2013 IEEE International Conference on Robotics and Biomimetics (ROBIO), pp. 310–315. IEEE (2013)
24. Montagnani, F., Controzzi, M., Cipriani, C.: Is it finger or wrist dexterity that is missing in current hand prostheses? IEEE Trans. Neural Syst. Rehabil. Eng. **23**(4), 600–609 (2015)
25. Paulignan, Y., MacKenzie, C., Marteniuk, R., Jeannerod, M.: The coupling of arm and finger movements during prehension. Exp. Brain Res. **79**(2), 431–435 (1990)
26. Peppoloni, L., Filippeschi, A., Ruffaldi, E., Avizzano, C.A.: A novel 7° of freedom model for upper limb kinematic reconstruction based on wearable sensors. In: 2013 IEEE 11th International Symposium on Intelligent Systems and Informatics (SISY), pp. 105–110. IEEE (2013)

27. Piazza, C., Santina, C.D., Catalano, M.G., Grioli, G., Garabini, M., Bicchi, A.: Softhand Pro-D: matching dynamic content of natural user commands with hand embodiment for enhanced prosthesis control. In: IEEE International Conference of Robotics and Automation (ICRA2016), pp. 3516 – 3523. IEEE, IEEE, Stockholm, Sweden, May 16–21 (2016)
28. Prout, W.: Universal wrist system (1974). US Patent 3,798,680
29. Rab, G., Petuskey, K., Bagley, A.: A method for determination of upper extremity kinematics. Gait Posture **15**(2), 113–119 (2002)
30. Resnik, L., Adams, L., Borgia, M., Delikat, J., Disla, R., Ebner, C., Walters, L.S.: Development and evaluation of the activities measure for upper limb amputees. Arch. Phys. Med. Rehabil. **94**(3), 488–494 (2013)
31. Resnik, L., Klinger, S.L., Etter, K.: The deka arm: Its features, functionality, and evolution during the veterans affairs study to optimize the deka arm. Prosthet. Orthot. Int. **38**(6), 492–504 (2014)
32. Rosen, J., Perry, J.C., Manning, N., Burns, S., Hannaford, B.: The human arm kinematics and dynamics during daily activities-toward a 7 DOF upper limb powered exoskeleton. In: ICAR'05. Proceedings., 12th International Conference on Advanced Robotics, 2005, pp. 532–539. IEEE (2005)
33. Santello, M., Baud-Bovy, G., Jörntell, H.: Neural bases of hand synergies. Front. Comput. Neurosci. **7**, 23 (2013)
34. Santello, M., Bianchi, M., Gabiccini, M., Ricciardi, E., Salvietti, G., Prattichizzo, D., Ernst, M., Moscatelli, A., Jörntell, H., Kappers, A.M., et al.: Hand synergies: integration of robotics and neuroscience for understanding the control of biological and artificial hands. Phys. Life Rev. (2016)
35. Takeda, H., Tsujiuchi, N., Koizumi, T., Kan, H., Hirano, M., Nakamura, Y.: Development of prosthetic arm with pneumatic prosthetic hand and tendon-driven wrist. In: 2009 Annual International Conference of the IEEE Engineering in Medicine and Biology Society, pp. 5048–5051. IEEE (2009)
36. Yang, N., Zhang, M., Huang, C., Jin, D.: Motion quality evaluation of upper limb target-reaching movements. Med. Eng. Phys. **24**(2), 115–120 (2002)

Chapter 6
A Novel Approach to Quantify Motion Impairment

The accurate assessment of upper limb motion impairment induced by stroke—which represents one of the primary causes of disability world-wide—is the first step to successfully monitor and guide patients' recovery. As of today, the majority of the procedures relies on clinical scales, which are mostly based on ordinal scaling, operator-dependent, and subject to floor and ceiling effects. In this work, we intend to overcome these limitations by proposing a novel approach to analytically evaluate the level of pathological movement coupling, based on the quantification of movement complexity. To this goal, we consider the variations of functional Principal Components applied to the reconstruction of joint angle trajectories of the upper limb during daily living task execution, and compared these variations between two conditions, i.e. the affected and non-affected arm. A Dissimilarity Index, which codifies the severity of the upper limb motor impairment with respect to the movement complexity of the non-affected arm, is then proposed. This methodology was validated as a proof of concept upon a set of four chronic stroke subjects with mild to moderate arm and hand impairments. As a first step, we evaluated whether the derived outcomes differentiate between the two conditions upon the whole data-set. Secondly, we exploited this concept to discern between different subjects and impairment levels. Results show that: (i) differences in terms of movement variability between the affected and non-affected upper limb are detectable and (ii) different impairment profiles can be characterized for single subjects using the proposed approach. Although provisional, these results are very promising and suggest this approach as a basis ingredient for the definition of a novel, operator-independent, sensitive, intuitive and widely applicable scale for the evaluation of upper limb motion impairment.

G. Averta, *Human-Aware Robotics: Modeling Human Motor Skills for the Design, Planning and Control of a New Generation of Robotic Devices*, Springer Tracts in Advanced Robotics 145, https://doi.org/10.1007/978-3-030-92521-5_6

6.1 Introduction

Human upper limb movements require an extraordinary coordination of numerous degrees of freedom (DoFs). Based on neuroscientific literature, this coordination is supposed to be organized according to covariation patterns, the so called *synergistic control*, which allow a successful interaction with the environment [10, 21, 29, 36]. After stroke, these upper limb movements can be affected by different types of sensorimotor impairments, such as weakness or loss of inter-joint coordination resulting in deficits ranging from paralysis, abnormal movement patterns or pathological synergies (e.g. coupling of shoulder abduction and elbow flexion) to a certain degree of inter-joint coordination [20, 30]. In consequence, stroke subjects with arm impairments can show a reduced adaptability to task demands [30], inefficient movement trajectories [24], higher energy and force-consumption [34] (when for example trying to perform goal-directed reaching movements), and the increased risk of frustration in case of unsuccessful movement attempts [26]. Hence, these impairments lead to long-term disabilities contributing in making stroke one of the main causes of disability world-wide, with a tremendous socio-economic impact [11]. Being able to evaluate stroke-specific upper limb movement patterns with sufficient detail is critical to properly monitor upper limb impairments after stroke, the recovery thereof and determining the effectiveness of different treatment approaches [18]. There exist multiple different standard clinical upper limb assessments in stroke research, among which the Fugl-Meyer Assessment of the Upper Extremity (FMA-UE) [13] is widely used to describe voluntary movement control [31]. This is assessed in a hierarchical structure; from within to out-of synergistic movements and from proximal to distal upper extremity segments. Despite the satisfactory measurement properties [14, 33], most of these clinical assessments exhibit floor and ceiling effects [35] and heavily rely on subjective observer scoring on broad ordinal scales (such as: 0 = not, 1 = partially, 2 = fully possible).

Novel technologies for kinematic signal detection and processing provide the opportunity for an objective and accurate motion analysis, which allows to overcome the limitations of standard clinical assessments in stroke research and rehabilitation [18]. Among the different approaches, it is worth mentioning (i) device-based assessments [6, 19] and (ii) less interfering wearable systems for bio-signal and motion capture processing [7, 23]. However, although promising, the state-of-the-art solutions need to face with important issues, such as operator and patient safety and costs of the procedure (i). Bio-signal measurements, especially kinematic recordings (ii), can offer a reliable way for motion evaluation during a set of natural movements. In particular, the wearable sensor-based approach has shown to be applicable to a variety of movement tasks and expendable to daily-living tasks [16], which increases the relevance of the assessment compared to the more abstract and stringent nature of movements performed during standard clinical assessments. Though some characteristics such as longer movement times and higher trunk displacement were described in stroke subjects [2], an exhaustive mathematical characterization on the level of loss of inter-joint coordination or pathological synergies after stroke still lacks in literature [30]. The most common way to quantify synergies is through Principal Component Analysis (PCA), a statistical method that allows the identification of dominant variation patterns in the data [3, 15]. Using this approach in a single-case study, fewer combinations of joint motions were identified in one stroke subject [28] as well as stronger synergistic coupling between shoulder, elbow and wrist motions in contrast to a healthy subject [15]. In [17], the authors applied a PCA on a larger dataset of stroke subjects (a grand total of 46 participants) to investigate the components of linear relations between the upper limb joints and the trunk at the end of reaching movements.

To extend this analysis to the temporal domain, in [5] the authors proposed functional Principal Component Analysis (fPCA) as a technique to investigate the dominant modes of time-dependent variation upper limb movements on a comprehensive set of upper limb daily-living activities in healthy subjects. The main advantage of this analysis is that, while classical PCA-based analysis consider single kinematic postures, with fPCA all the temporal evolution of the movement is considered (thus intrinsically including the dynamic aspects). Results showed that a reduced number of functional Principal Components (fPCs) can be used to describe and accurately reconstruct the complexity of upper limb activities in healthy subjects, at joint level. The authors also pointed to the possibilities of an automatic recognition of physiological and pathological movements in stroke research and rehabilitation by analyzing fPCs variations between the affected and non affected upper limb [5].

In the present study we investigate whether the outcomes discussed in [5] can be effectively exploited to characterize upper limb motor impairment in chronic stroke subjects. For that purpose, four chronic stroke subjects with moderate arm and hand impairments performed the same 30 tasks of daily living activities [5] with both arms (affected and non-affected) (Fig. 6.1) and a functional PCA analysis

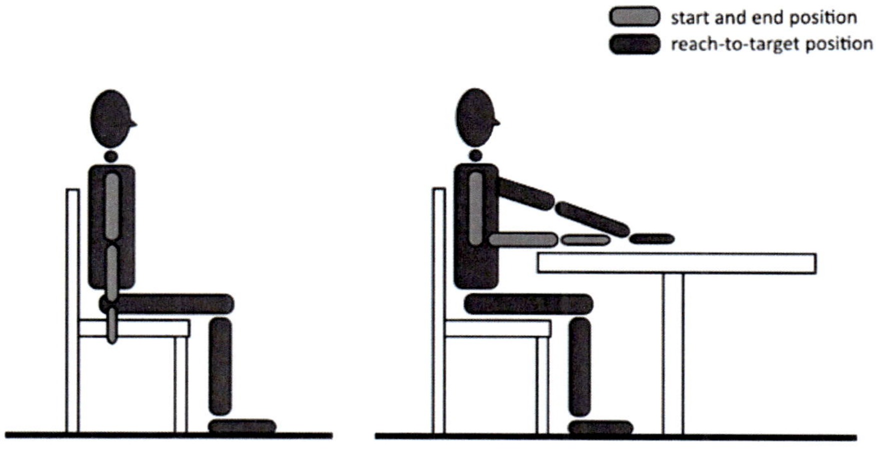

Fig. 6.1 Overview of the experimental setup for the intransitive task (left) and transitive and tool-mediated task (right)

was applied. Based on this data set, we evaluated whether the set of fPCs and the associated approximation error in reconstructing joint trajectories are appropriate outcome measures to differentiate (i) between the affected and the non-affected arm across stroke subjects and (ii) between different levels of upper limb impairments of the single subjects. The content of this chapter is adapted from [32].

6.2 Experimental Protocol and Setup

6.2.1 Set of Daily Living Tasks

In [5], the authors discussed a large set of tasks (i.e., 30 different actions), which were selected to excite the whole upper limb work-space [1, 22, 25] and to span all the major hand configurations, e.g. referring to the most common hand grasping taxonomies [9, 12]. Leveraging on this, we employed here the same protocol, with the twofold goal to: (i) providing a comparison between different groups of populations, including chronic stroke subjects in this case, and (ii) contributing to building up a large experimental data set of upper limb motion (acquired using different acquisition modalities), to enable comparison and benchmarking in future works. Under this regard, data are made freely available in [4]. The task-set is divided in three sub-groups of ten actions each: intransitive, transitive and tool-mediated tasks. Actions included in the first group are gestures with no contact with the external environment (i.e., requiring movement of the proximal part of the upper limb), while the second group contains actions that involve interaction with an object. The third group implies tasks in which one external object is used to act on the environment. A detailed description of the task-set can be found in [5, 8]. In contrast to standard clinical scales that mostly consist of abstract movement executions and postures, the execution of a set of daily living tasks increases the meaningfulness of arm and hand movements, and thereby the ecological validity of measurements.

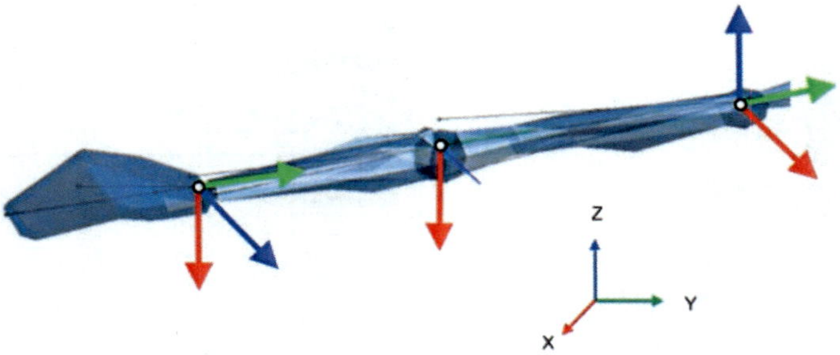

Fig. 6.2 Kinematic model used in this chapter. Three rigid links are connected by seven joints. Picture adapted from [37]

6.2.2 Experimental Setup for Data Acquisition

The data were recorded with a full-body worn IMU-based sensor suit (Xsens technologies B.V., Enschede, The Netherlands). The system consists of 17 inertial measurement units (IMUs) placed symmetrically on predefined body positions and fixed with Velcro straps and a size-fitting T-Shirt. The IMUs provide 3D angular velocity using rate gyroscopes, 3D acceleration using accelerometers, 3D earth magnetic field using magnetometers, as well as atmospheric pressure using the barometer in an operating frequency 2405-1475 MHz [37].

A calibration procedure was required to evaluate sensors drifts and internal parameters. This was achieved using information related to subjects' body dimensions and through data fusion. The calibrated model was then used to reconstruct the whole body kinematic description. In particular, 23 links (or segments) connected through 22 spherical joints are used to model the human body. Several relevant motion-related quantities are then calculated and provided as output with a recording frequency of 60 Hz, such as segments and joint position, velocities, and accelerations (see Fig. 6.2). More details can be found in [37]. Once the system is calibrated, the experimental procedure established the repetition of each task of the protocol three times. All measurements were performed in upright sitting position on a chair. Subjects were instructed to perform each movement task at comfortable speed, first using the non-affected and then the affected upper limb. Task actions including grasping and manipulating objects (e.g. transitive and tool-mediated task group, see [5, 8] for more details) were performed in front of an height-adjustable table (at forearm height in 90° elbow flexion and neutral shoulder position) with the targets placed at about 90% of the arm length (shoulder joint-axis until line of proximal interphalangeal joints).

Participant characteristics

Subject ID	Age (years)	Stroke localization	Affected arm	Dominant hand	Months since stroke	FMA-UE					
						Total (Max. 66)		Arm* (Max. 36)	Wrist* (Max. 10)	Hand* (Max. 14)	Coordination* (Max. 6)
						non-affected	affected	affected	affected	affected	affected
S1	73	Left, corona radiata, putamen	Right	Right	18	62	40	23	7	7	3
S2	61	Left, corona radiata	Right	Right	108	66	41	24	5	8	4
S3	59	Right, inferior frontal gyrus	Left	Right	20	63	46	25	6	12	3
S4	42	Right, total MCA territory	Left	Right	12	65	40	26	6	4	4
Median (IQR)	60.00 (54.75, 64.00)				20.00 (16.00; 64.00)	64.00 (62.75; 65.25)	40.50 (40.00; 42.30)	24.50 (23.80; 25.30)	6.00 (5.75; 6.25)	7.50 (6.25; 9.00)	4.00 (3.50; 4.00)

Fig. 6.3 Legend: FMA-UE, Fugl-Meyer assessment of the upper extremity; IQR, Interquartile range; Max., Maximal; MCA, Middle cerebral artery; *, Subscale of the FMA-UE

6.2.3 *Study Information*

In this chapter, we use data recorded from a subset of four chronic stroke subjects, which are part of the observational study *"Assessing pathological synergies of upper limb function and the relationship to visuospatial function after stroke"*. All subjects gave written consent in accordance with the current version of the Declaration of Helsinki and the Swiss regulatory authority requirements. The protocol was approved by the Cantonal Ethics Committee Northwest and Central Switzerland (BASEC-ID: 2016-02075) and registered on ClinicalTrials.gov (Identifier: NCT03135093). Subject-specific characteristics are reported in Fig. 6.3.

6.3 Data Analysis

6.3.1 *Modeling and Pre-processing*

We decided to use the XZY Euler parametrization to represent the data. In this way, we obtained nine angles in total to describe upper limb kinematics, three for the shoulder, three for the elbow and three for the wrist. Additional details on angles identification are provided in [37]. Without any loss of generality and to allow future comparisons, we chose to represent data with a 7 DoF model coherent with [5]. For these reasons, we considered only one DoF out of the three provided for the elbow (i.e. flexion-extension), three DoFs for the shoulder, and three DoFs for the wrist (of which one is related to forearm pronation-supination) (See Fig. 6.2). Recorded data were manually segmented, meaning that for every data stream, we selected initial and final frames of each task repetition. Then, we linearly warped in time all the segments through a re-sampling procedure with respect to a fixed number of time frames ($T = 300$). This was made to enable time-comparison of different movements and allow a proper implementation of functional analysis (see [5]).

6.3.2 *Evaluation-Index of Motion Complexity*

After the pre-processing phase, for each task, we got the temporal evolution of each joint normalized in time. To quantify the complexity of these movements under a functional point of view, a possible strategy is to evaluate how many basis functions are required to reconstruct the specific joint trajectories. Functional Principal Component Analysis (fPCA) represents a classic approach to identify the main modulating functions of one data set, and to order these in a descending order related to the explained variance. fPCA is a functional extension of Principal Component Analysis (PCA), typically used in several research fields to analyze multi-dimensional

Algorithm 1 General procedure to calculate Reconstruction Error Plots and Dissimilarity Index (I_D)

```
procedure FUNCTIONALPCA
    Resample Signals to T time frames;
    Calculate fPCs (ref. [5]);
end
procedure GETRECONSTRUCTIONERROR
    Load Dataset;
    N ← NumElements(Dataset);
    fPCs ← FunctionalPCA(Dataset);
    M ← MaxNumfPCs(fPCs);
    k ← 0;
    while k ≤ M do                                          ▷ For each fPC
        while i ≤ N do                          ▷ For each element in Dataset
            Load Dataset(i);
            Approx(i) ← Approximate Dataset(i) using the first k fPCs¹;
            error(k, i) ← rms(Dataset(i) − Approx(i));
            i = i + 1;
        end
        k = k + 1;
    end
    GlobalError ← rms(error)                                ▷ rms per columns
end
procedure CALCULATE DISSIMILARITY INDEX (I_D)
    Load Dataset1;
    Load Dataset2;
    MeanError1 ← GetReconstructionError(Dataset1);
    MeanError2 ← GetReconstructionError(Dataset2);
    I_D ← norm(MeanError1 − MeanError2);
end
```

time series [27]. More recently, this technique has been profitably applied in [5] to identify the main functions that modulate human movements.

In the following, we will briefly describe the main idea behind fPCA and how it has been applied for motion description. Let us assume, without any loss of generality, a 7 DoF kinematic model to represent upper limb joint trajectories $q(t) : \mathbb{R} \rightarrow \mathbb{R}^7$ where $t \in [0, 1]$ is the normalized time. In these terms, generic upper limb motion $q(t)$ can be decomposed in terms of the weighted sum of base elements $S_i(t)$, or functional Principal Components:

$$q(t) \simeq \bar{q}(t) + S_0(t) + \sum_{i=1}^{s_{\max}} \alpha_i \circ S_i(t), \tag{6.1}$$

where $\alpha_i \in \mathbb{R}^n$ is a vector of weights, $S_i(t) \in \mathbb{R}^n$—in our case n equals to 7—is the ith basis element or synergy and $s_{\max}$ is the number of basis elements. The operator $\circ$ is the element-wise product (Hadamard product). and $\bar{q} \in \mathbb{R}^7$ is the average posture of q

$$\bar{q} = \int_0^1 q(\tau) \mathrm{d}\tau, \tag{6.2}$$

while $S_0 : \mathbb{R} \rightarrow \mathbb{R}^7$ is the average trajectory, also called *zero-order* synergy. The output of fPCA, which is calculated independently for each joint, is a basis of functions $\{S_1, \ldots, S_{s_{\max}}\}$ that maximizes the explained variance of the movements in the collected dataset. Given a dataset with N elements collecting the trajectories recorded in a given joint j, the first fPC $S_{j,1}(t)$ is the function that solves the following problem

$$\begin{aligned} &\max_{S_{j,1}} \sum_{j=1}^{N} \left(\int S_{j,1}(t) q_j(t) dt \right)^2 \\ &\text{subject to} \quad ||S_{j,1}(t)||_2^2 = \int_0^1 S_{j,1}^2(t) dt = 1. \end{aligned} \tag{6.3}$$

Subsequent fPCs $S_{j,i}(t)$ are the functions that solve the following:

$$\begin{aligned} &\max_{S_{j,i}} \sum_{j=1}^{N} \left(\int S_{j,i}(t) q_j(t) dt \right)^2 \\ &\text{subject to} \quad ||S_{j,i}(t)||_2^2 = 1 \\ &\int_0^1 S_{j,i}(t) S_p(t) dt = 0, \ \forall p \in \{1, \ldots, i-1\}. \end{aligned} \tag{6.4}$$

A detailed implementation of this method—which bypasses the solution of the minimization problem—is detailed in Chap. 2. Here we used $s_{\max} = 15$.

It has been proved that the higher is the number of fPCs used to reconstruct a signal, the lower is the error obtained for reconstructing the real data. This observation implies that also the complexity of a sample could be quantified in terms of the number of functional components needed to provide reliable reconstruction. Indeed, given an arbitrary reconstruction error threshold, the higher is the variability of a time series, the higher is the number of fPCs required to approximate the signal with an error lower than the threshold.

In this chapter, we propose to exploit these characteristics to quantify the differences, in terms of functional complexity, between two different physical conditions, i.e. non-affected versus affected upper limb movements. The main hypothesis is that an affected motion, due to the loss of inter-joint coordination or—in other words—increased joint coupling, leads to more simple and less variable joint contributions than the normal condition [30], when analyzed on the same task-set of upper limb activities. In our analysis, this should be reflected in the fact that, given a specific number of fPCs used to reconstruct a signal, the reconstruction error will be lower in case of pathological movements. Given d_i as the ith element of the dataset, and given the first k synergies, the approximation error is evaluated as

$$e_k(i) = \min_{\alpha_{j,i}} \ [\text{rms}(q_i(t) - \hat{q}_i^k(t, \alpha_{j,i}))], \tag{6.5}$$

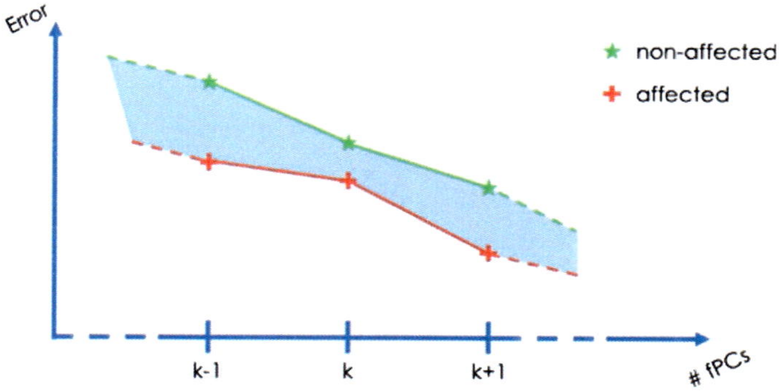

Fig. 6.4 Typical profiles of reconstruction error w.r.t. the number of fPCs enrolled. Red and green lines are the expected shapes in case of affected and healthy motion, respectively. The area between the two curves can be intended an index of dissimilarity between the two conditions

where $\hat{q}_i^k(t, \alpha_{j,i})$ is the approximation of $q_i(t)$ using the first k synergies, calculated as

$$\hat{q}_i^k(t) = \bar{q}(t) + S_0(t) + \sum_{j=1}^{k} \alpha_{j,i} \circ S_j(t), \tag{6.6}$$

and $\alpha_{j,i}$ are the optimal weights associated to the element d_i. Representative error values for each k are then calculated as the average $\bar{e}_k = mean([e_k(1), \ldots, e_k(N)])$ and collected in a vector $E = [\bar{e}_1, \ldots, \bar{e}_{s_{\max}}]$.

A schematics of this idea is depicted in Fig. 6.4, where the red plot refers to affected motion, while the green plot stands to the non-affected motion. The blue area within the two profiles can be regarded as an index of dissimilarity between the two conditions. The dissimilarity index (I_D) can be easily calculated as:

$$I_D = \sum_{i=1}^{s_{\max}} E_H(i) - \sum_{j=1}^{s_{\max}} E_S(i) = \sum_{i=1}^{s_{\max}} (E_H(i) - E_S(i)), \tag{6.7}$$

where E_H is the vector of reconstruction error for the healthy case, E_S is the vector of reconstruction error for the stroke case.

The procedure to obtain the plots theorized in Fig. 6.4 and to calculate I_D is reported in Algorithm 1. Please note that under a theoretical point of view, this methodology could be generalized to the analysis of different types of motion datasets with multiple subjects, and applied to intra-subject analysis.

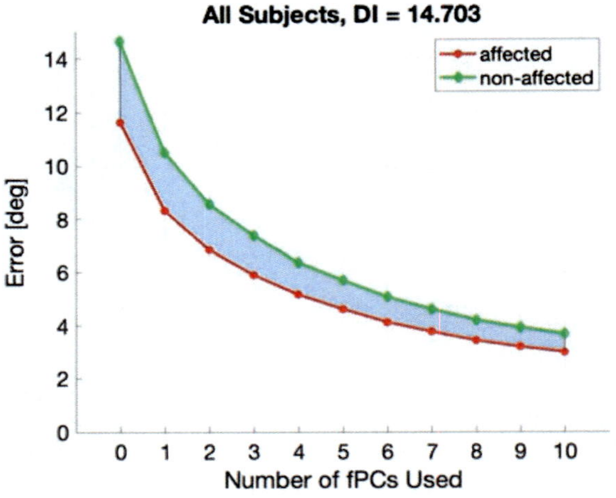

Fig. 6.5 Root mean square of reconstruction error versus number of fPCs used. All subject are considered for this analysis. $I_D = 14.703$

6.4 Results and Discussions

To verify whether our main hypothesis holds true, we calculated the dissimilarity index (I_D) defined in the previous section between the whole data-set of non-affected arm motions and the whole data-set of affected arm motions.

Results are depicted in Fig. 6.5 for all the stroke subjects. It is possible to observe that the reconstruction error is lower for the affected upper limb, as per research hypothesis discussed in the previous section. The blue area can be used as a gross difference index between the affected and non-affected conditions for all stroke subjects in the whole data sample ($I_D = 14.703$). This result supports our assumption that analysis of movement variability can be used to effectively distinguish between non-affected and affected conditions due to stroke.

To further characterize the proposed I_D as index of the impairment level, we performed the same analyses on a single-subject-level. In order to evaluate whether our approach is able to differentiate between subjects and their individual impairment severity, we repeated our procedure to each single-subject's data-set. Results, reported in Fig. 6.6, illustrate differences in terms of movement variability between the affected and non-affected side of each subject. While our main hypothesis of lower reconstruction errors in the affected arm holds true for S1, S2 and S3 (I_D ranging from 12.12 to 27.31), higher reconstruction errors are visible in the affected arm of S4 when compared to the non-affected side leading to a negative I_D equal to –6.91. We believe that, this higher variability in the affected arm is related to extra-movements, which S4 performed to compensate for the diminished grasp function (FMA-UE hand score: 4 out of 14) when performing the transitive and tool-mediated task actions. To verify this hypothesis, we performed the same approach for the sub-set of intransitive task actions, where finger and hand functions are not crucial for the task accomplishment, for all the considered subjects. Almost equal variability

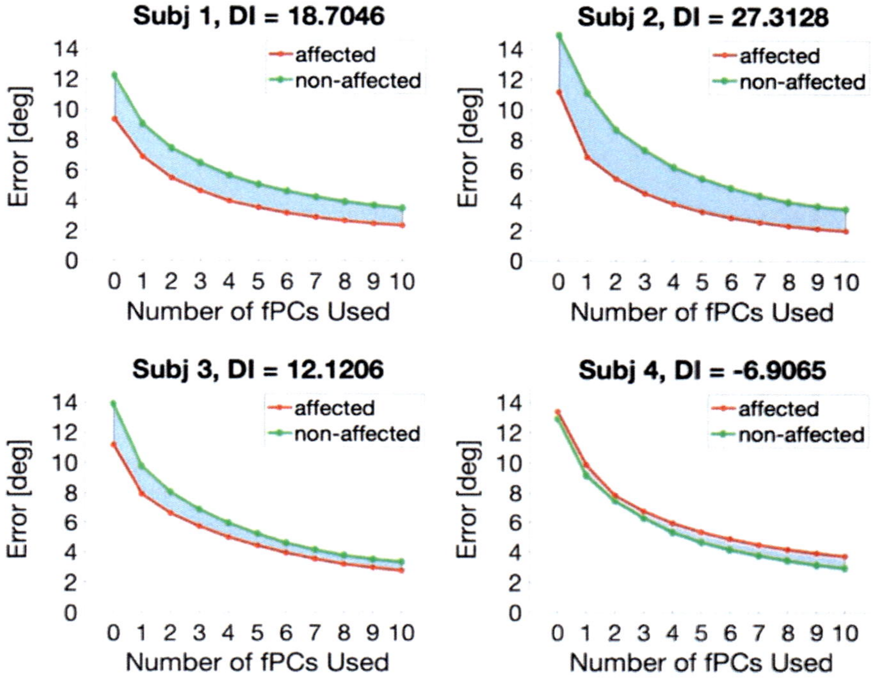

Fig. 6.6 Root mean square of reconstruction error versus number of fPCs used. Each sub-plot reports the analysis subject-specific. From top-left to bottom-right results for subjects from 1 to 4 are considered. Correspondent I_D is reported in figure headings

profiles between the affected and non-affected arm and a I_D of 1.27 were found in S4, supporting the idea of approximately normal movement behavior as long as no hand or finger function is needed, whereas the movement variability seems to be comparably diminished in S1 (I_D of 11.65), S2 (I_D of 24.33) and S3 (I_D of 12.60). Illustrations of these analyses are omitted for sake of space.

6.5 Implications and Conclusions

In this chapter, we demonstrated how stroke subjects' level of impairment may be assessed through the quantification of upper limb movement variability, in terms of the resulting root mean square of reconstruction error using a fPCs based description. The I_D calculated upon the whole dataset revealed differences in terms of motion variability between the two upper limb conditions. In this manner, novel measurement outcomes are provided to describe stroke-related upper limb movement impairments, such as the loss of inter-joint coordination, commonly associated with reduced variability of motion elements, when performing a comprehensive set of daily living activities. This opens fascinating perspectives toward the usage of this methodology as a tool for assessing motor impairment after a stroke, herein defined as reduced movement variability, in a quantitative, sensitive and operator-independent fashion. Although only preliminary, results seems confirming our hypothesis of generally lower profiles of reconstruction errors in the affected compared to the non-affected upper limb (as shown in Fig. 6.5 by comparing both conditions for the whole set of four subjects as shown in Fig. 6.6). The single-subject analysis revealed a I_D ranging from 27.31 in S2 to -6.91 in S4, that can partially be explained by clinical assessment outcomes as shown in Fig. 6.3. Studies on a larger data set are needed to confirm these preliminary findings, also including other motor impairments, such as ataxia, which would likely influence the results and interpretation of the presented outcomes. These findings suggest the usefulness of fPCA to study the motion variability and provide implications for sensitive outcomes of post-stroke upper limb impairment, which may be effectively used as suitable biomarkers to discriminate between pathological and physiological movement behavior in stroke research and rehabilitation [18]. In [17] the authors similarly found correct predictions of the presence or absence of basic limb synergies (defined by FMA-UE $\geq$34 and FMA-UE $<$34) for 38 of 46 patients (82.6%) using PCA on endpoint reaching kinematics. Here, we extend the PCA to the whole time set of motion data using shape analysis described in functions, which enables the investigation of dynamic aspects of movement behavior, including the spatio-temporal evolution of joint trajectories to precisely describe pathological joint coupling or pathological synergies due to stroke. However, we acknowledge that, in contrast to the classical PCA performed at one point in time, direct coupling between joints are not explicitly considered with our approach. As another limitation, we acknowledge that limiting the application of the approach to the upper limb except for the hand and fingers could lead to erroneous assessment and interpretation of the results. For this reason, model extensions to hand and finger motions will be considered to additionally include relevant kinetic information for grasping activities. Finally, our future works will also take into account possible deficits in the non-affected arm. Considering previous works on physiological movement data-sets [5], here, we could broaden the analysis to pathological movement data from mildly to moderately affected stroke subjects, when performing a set of daily living tasks. In a next step, differences between specific single task-items will be investigated using the same methodology. Finally, this methodology could be further exploited for an automatic recognition of physiological and pathological movements through machine learning and in terms of online evaluations of improvements related to rehabilitation procedures.

References

1. Abdel-Malek, K., Yang, J., Brand, R., Tanbour, E.: Towards understanding the workspace of human limbs. Ergonomics **47**(13), 1386–1405 (2004)
2. Alt Murphy, M., Häger, C.K.: Kinematic analysis of the upper extremity after Stroke-how far have we reached and what have we grasped? Phys. Ther. Rev. **20**(3), 137–155 (2015)

3. Averta, G., Angelini, F., Bonilla, M., Bianchi, M., Bicchi, A.: Incrementality and hierarchies in the enrollment of multiple synergies for grasp planning. IEEE Robot. Autom. Lett. **3**(3), 2686–2693 (2018)
4. Averta, G., Barontini, F., Catrambone, V., Haddadin, S., Handjaras, G., Held, J.P., Hu, T., Jakubowitz, E., Kanzler, C.M., Kühn, J., et al.: U-limb: a multi-modal, multi-center database on arm motion control in healthy and post-stroke conditions. GigaScience **10**(6), giab043 (2021)
5. Averta, G., Della Santina, C., Battaglia, E., Felici, F., Bianchi, M., Bicchi, A.: Unvealing the principal modes of human upper limb movements through functional analysis. Front. Robot. AI **4**, 37 (2017)
6. Basteris, A., Contu, S., Plunkett, T.K., Kuah, C.W., Konczak, I.J., Chua, K.S., Masia, L.: Robot-aided bimanual assessment of wrist proprioception in people with acute stroke. In: 2018 7th IEEE International Conference on Biomedical Robotics and Biomechatronics (Biorob), pp. 473–478. IEEE (2018)
7. Camardella, C., Murciego, L.P., Tang, S., Bertolucci, F., Chisari, C., Barsotti, M., Frisoli, A.: Simple tool for functional and physiological stroke patients assessment. In: International Conference on NeuroRehabilitation, pp. 779–782. Springer (2018)
8. Catrambone, V., Greco, A., Averta, G., Bianchi, M., Valenza, G., Scilingo, E.P.: Predicting object-mediated gestures from brain activity: an EEG study on gender differences. IEEE Trans. Neural Syst. Rehabil. Eng. **27**(3), 411–418 (2019). https://doi.org/10.1109/TNSRE.2019.2898469
9. Cutkosky, M.R.: On grasp choice, grasp models, and the design of hands for manufacturing tasks. IEEE Trans. Robot. Autom. **5**(3), 269–279 (1989)
10. Della Santina, C., Bianchi, M., Averta, G., Ciotti, S., Arapi, V., Fani, S., Battaglia, E., Catalano, M.G., Santello, M., Bicchi, A.: Postural hand synergies during environmental constraint exploitation. Front. Neurorobot. **11**, 41 (2017)
11. Feigin, V.L., Forouzanfar, M.H., Krishnamurthi, R., Mensah, G.A., Connor, M., Bennett, D.A., Moran, A.E., Sacco, R.L., Anderson, L., Truelsen, T., et al.: Global and regional burden of stroke during 1990–2010: findings from the global burden of disease study 2010. Lancet **383**(9913), 245–255 (2014)
12. Feix, T., Romero, J., Schmiedmayer, H.B., Dollar, A.M., Kragic, D.: The grasp taxonomy of human grasp types. IEEE Trans. Hum.-Mach. Syst. **46**(1), 66–77 (2016)
13. Fugl-Meyer, A.R., Jääskö, L., Leyman, I., Olsson, S., Steglind, S.: The post-stroke hemiplegic patient. 1. a method for evaluation of physical performance. Scand. J. Rehabil. Med. **7**(1), 13–31 (1975)
14. Gladstone, D.J., Danells, C.J., Black, S.E.: The Fugl-Meyer assessment of motor recovery after stroke: a critical review of its measurement properties. Neurorehabil. Neural Repair **16**(3), 232–240 (2002)
15. Heidari, O., Roylance, J.O., Perez-Gracia, A., Kendall, E.: Quantification of upper-body synergies: a case comparison for stroke and non-stroke victims. In: ASME 2016 International Design Engineering Technical Conferences and Computers and Information in Engineering Conference, pp. V05AT07A032–V05AT07A032. American Society of Mechanical Engineers (2016)
16. Held, J.P., Klaassen, B., Eenhoorn, A., Beijnum, B.J.F.V., Buurke, J.H., Veltink, P.H., Luft, A.R.: Inertial sensor measurements of upper-limb kinematics in stroke patients in clinic and home environment. Front. Bioeng. Biotechnol. **6**, 27 (2018)
17. van Kordelaar, J., van Wegen, E.E., Kwakkel, G.: Unraveling the interaction between pathological upper limb synergies and compensatory trunk movements during reach-to-grasp after stroke: a cross-sectional study. Exp. Brain Res. **221**(3), 251–262 (2012)
18. Kwakkel, G., Lannin, N.A., Borschmann, K., English, C., Ali, M., Churilov, L., Saposnik, G., Winstein, C., Van Wegen, E.E., Wolf, S.L., et al.: Standardized measurement of sensorimotor recovery in stroke trials: consensus-based core recommendations from the stroke recovery and rehabilitation roundtable. Neurorehabil. Neural Repair **31**(9), 784–792 (2017)

19. Lambercy, O., Fluet, M.C., Lamers, I., Kerkhofs, L., Feys, P., Gassert, R.: Assessment of upper limb motor function in patients with multiple sclerosis using the virtual peg insertion test: a pilot study. In: 2013 IEEE International Conference on Rehabilitation Robotics (ICORR), pp. 1–6. IEEE (2013)
20. Langhorne, P., Bernhardt, J., Kwakkel, G.: Stroke rehabilitation. Lancet **377**(9778), 1693–1702 (2011)
21. Latash, M.L.: Synergy. Oxford University Press (2008). https://doi.org/10.1093/acprof:oso/9780195333169.001.0001
22. Lenarcic, J., Umek, A.: Simple model of human arm reachable workspace. IEEE Trans. Syst. Man Cybern. **24**(8), 1239–1246 (1994)
23. Lorussi, F., Carbonaro, N., De Rossi, D., Paradiso, R., Veltink, P., Tognetti, A.: Wearable textile platform for assessing stroke patient treatment in daily life conditions. Front. Bioeng. Biotechnol. **4**, 28 (2016)
24. Osu, R., Ota, K., Fujiwara, T., Otaka, Y., Kawato, M., Liu, M.: Quantifying the quality of hand movement in stroke patients through three-dimensional curvature. J. Neuroeng. Rehabil. **8**(1), 62 (2011)
25. Perry, J.C., Rosen, J., Burns, S.: Upper-limb powered exoskeleton design. IEEE/ASME Trans. Mechatron. **12**(4), 408 (2007)
26. Raghavan, P.: Upper limb motor impairment after stroke. Phys. Med. Rehabil. Clin. **26**(4), 599–610 (2015)
27. Ramsay, J.O.: Functional data analysis. Wiley Online Library (2006)
28. Reisman, D.S., Scholz, J.P.: Aspects of joint coordination are preserved during pointing in persons with post-stroke hemiparesis. Brain **126**(11), 2510–2527 (2003)
29. Santello, M.: Synergistic control of hand muscles through common neural input. In: The Human Hand as an Inspiration for Robot Hand Development, pp. 23–48. Springer (2014)
30. Santello, M., Lang, C.E.: Are movement disorders and sensorimotor injuries pathologic synergies? when normal multi-joint movement synergies become pathologic. Front. Hum. Neurosci. **8**, 1050 (2015)
31. Santisteban, L., Térémetz, M., Bleton, J.P., Baron, J.C., Maier, M.A., Lindberg, P.G.: Upper limb outcome measures used in stroke rehabilitation studies: a systematic literature review. PLoS One **11**(5), e0154792 (2016)
32. Schwarz, A., Averta, G., Veerbeek, J.M., Luft, A., Held, J.P., Valenza, G., Bicchi, A., Bianchi, M.: A functional analysis-based approach to quantify upper limb impairment level in chronic stroke patients: a pilot study. In: Engineering in Medicine and Biology Society, 2019. EMBC'19. 41th Annual International Conference of the IEEE. IEEE (2019)
33. See, J., Dodakian, L., Chou, C., Chan, V., McKenzie, A., Reinkensmeyer, D.J., Cramer, S.C.: A standardized approach to the Fugl-Meyer assessment and its implications for clinical trials. Neurorehabil. Neural Repair **27**(8), 732–741 (2013)
34. Suresh, N.L., Zhou, P., Rymer, W.Z.: Abnormal emg-force slope estimates in the first dorsal interosseous of hemiparetic stroke survivors. In: Engineering in Medicine and Biology Society, 2008. EMBS 2008. 30th Annual International Conference of the IEEE, pp. 3562–3565. IEEE (2008)
35. Thrane, G., Sunnerhagen, K.S., Persson, H.C., Opheim, A., Alt Murphy, M.: Kinematic upper extremity performance in people with near or fully recovered sensorimotor function after stroke. Physiother. Theory Pract. 1–11 (2018)
36. Tresch, M.C., Jarc, A.: The case for and against muscle synergies. Curr. Opin. Neurobiol. **19**(6), 601–607 (2009)
37. Xsens Technologies, B.V.: MVN User Manual. MVN Awinda. Rev. March, User Guide MVN, MVN BIOMECH MVN Link (2017)

Chapter 7
A Novel Mechatronic System for Evaluating Elbow Muscular Spasticity Relying on Tonic Stretch Reflex Threshold Estimation

Muscular spasticity represents one of the most common motor disorder associated to lesions of the Central Nervous System, such as Stroke, and affects joint mobility up to the complete prevention of skeletal muscle voluntary control. Its clinical evaluation is hence of fundamental relevance for an effective rehabilitation of the affected subjects. Standard assessment protocols are usually manually performed by humans, and hence their reliability strongly depends on the capabilities of the clinical operator performing the procedures. To overcome this limitation, one solution is the usage of mechatronic devices based on the estimation of the Tonic Stretch Reflex Threshold, which allows for a quite reliable and operator-independent evaluation. In this work, we present the design and characterization of a novel mechatronic device that targets the estimation of the Tonic Stretch Reflex Threshold at the elbow level, and, at the same time, it can potentially act as a rehabilitative system. Our device can deliver controllable torque/velocity stimulation and record functional parameters of the musculo-skeletal system (joint position, torque, and multi-channel ElectroMyoGraphyc patterns), with the ultimate goals of: (i) providing significant information for the diagnosis and the classification of muscular spasticity, (ii) enhancing the recovery evaluation of patients undergoing through therapeutic rehabilitation procedures and (iii) enabling a future active usage of this device also as therapeutic tool. Clinical relevance- The contribution presented in this work proposes a technological advancement for a device-based evaluation of motion impairment related to spasticity, with a major potential impact on both the clinical appraisal and the rehabilitation procedures.

G. Averta, *Human-Aware Robotics: Modeling Human Motor Skills for the Design, Planning and Control of a New Generation of Robotic Devices*, Springer Tracts in Advanced Robotics 145, https://doi.org/10.1007/978-3-030-92521-5_7

7.1 Introduction and Motivation

The connective-musculo-skeletal system is a marvelous and extremely complex natural system, organized in rigid segments (bones) that are moved by a redundant number of muscles through specific connective structures (tendons, aponeuroses, etc.). In physiological conditions, this system is responsible for the extraordinary adaptability in motion execution. When a lesion affects the Central Nervous System (CNS), e.g. stroke, stereotyped movement patterns arise, which are the consequence of the pathological coupling between different Degrees of Freedom (DoFs) and of the loss of individual control over single joints [21]. This behavior can be observed both at the level of the upper limb, during reaching and task-oriented movements, and at the level of the lower limb, during gait.

Among the different neurological diseases, stroke has a considerable societal impact and represents one of the main causes of death and disability worldwide. For example, looking at the United States only, an American dies of stroke every 4 min (Centers for Disease Control and Prevention, 2015). In stroke patients, it is common to observe an impairment of the sensory-motor capabilities, which is typically associated to the reduction of the active range of motion, the loss of meaningful inter-joint coordination, muscle weakness and spasticity [3, 17, 22].

Spasticity is characterized by an involuntary velocity-dependent stretch reflex that causes inappropriate activation of the stretched muscle during passive and active movements. This leads to the abnormal increase of the resistance to the motion itself, as underlined in the classical definition of Lance [13]. It is hence clear that the clinical appraisal of muscle spasticity is highly relevant not only for the assessment of the level of impairment but also to evaluate the effects of the rehabilitation strategies in use.

To this aim, standard clinical approaches often rely on the application of a modified Ashworth scale (MAS) [6], which quantifies the resistance exhibited by a muscle that undergoes through passive movements imposed by the examiner. However, since muscle stretch resistance is a combination of multiple causes, which act both at the central neural level and locally (due to the modification of the viscoelastic properties of muscles, tendons and ligaments), MAS cannot distinguish between the neural and non-neural components. Furthermore, the MAS scale is not able to evaluate the velocity-dependent resistance to passive stretching, which is one of the core characteristics of spasticity. Last but not least, MAS evaluation is highly operator-dependent, since it is usually manually performed.

To overcome these limitations, alternative approaches based on the equilibrium point hypothesis (EPH) have been proposed [4, 5, 14, 20]. These methodologies leverage on the estimation of the Tonic Stretch Reflex Threshold (TSRT) for the evaluation of neural-related motor impairment. Furthermore, the application of EPH to stroke patient rehabilitation was proven to produce promising results in terms of an increase in muscular recruitment [7]. Under this regard, it is worth mentioning the *Grimaldi's method* [15]. This therapeutic manoeuvre specifically acts on TSRT of a selected muscular group, which is shortened in a passive way by applying a series of fast accelerations, while the limb, bound by a mechanical system, is being moved by forces applied in the opposite direction.

It is hence clear that the correct estimation of the TSRT is of paramount importance not only for diagnostic but also for therapeutic purposes. This estimation usually relies on the usage of a mechatronic device—thus minimizing the dependence of the evaluation outcomes on the manual capabilities of the operator. The device applies a controllable torque that is co-axial to the joint under investigation, and record joint angular position and surface ElectroMyoGraphyc (EMG) patterns, e.g. at the elbow level in the adults [18, 19], and the ankle level in children with muscular spasticity associated to cerebral palsy [14].

However, the development of this kind of systems for the upper limb is not a trivial problem, since it comes with very strict requirements, which are mainly related to the need for: (i) a portable design, to enable the usage in clinical environments; (ii) different regulation and control possibilities to accommodate various pathological conditions, with special focus on shoulder abduction and flexion, and wrist flexion-extension; (iii) a reduced active interaction between patients and therapists during the evaluation task.

In this chapter, we present the design and characterization of a new mechatronic device, depicted in Fig. 7.1, which targets the estimation of the TSRT in adult subjects at the elbow level, where spasticity is preeminent and present in the 79% of stroke cases [24]. This system was thought to correctly satisfy the aforementioned requirements and apply torque-velocity stimuli, which are coherent with the dedicated literature. It is also worth noticing that the fulfilment of the latter ones for the

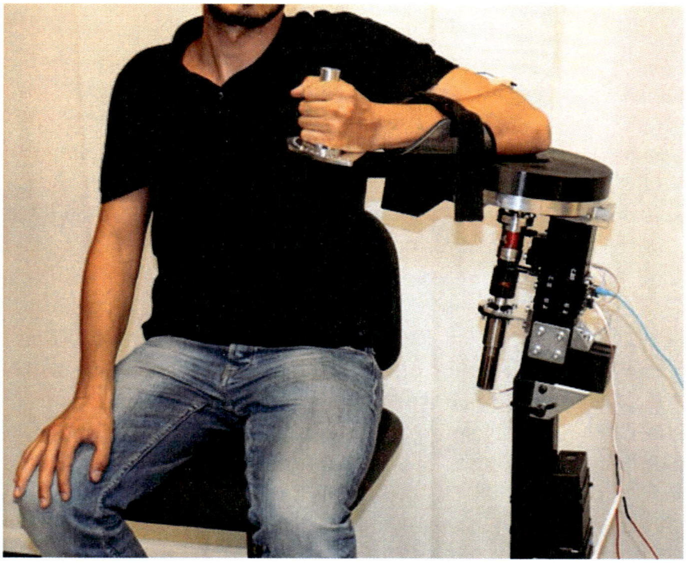

Fig. 7.1 Overview of the proposed device while a subject is performing the task

TSRT estimation phase also enables the possible usage of this device as an active therapeutic tool based on [15]. To conclude, although a clinical validation still lacks and is needed, the system described in this preliminary work potentially represents the first technology that allows for both the measurement and rehabilitation of TSRT in patients affected by muscular spasticity. The content of this chapter is adapted from [1].

7.2 Spasticity and Equilibrium Point Hypothesis

Introduced by A. Feldman, the equilibrium point hypothesis (EPH) is an interesting yet debated theory in motor control [10–12]. In a nutshell, the idea is that the CNS does not directly control the activation of all body muscles, but rather is able to modify the stretch-reflex (SR) threshold, defined as the spatial value of muscle length at which a given muscle begins to be activated, if stretched at a given stretch velocity.

For a given coordinated set of thresholds, the system results in a stable equilibrium between internal visco-elastic forces and external loads, and movements are generated through a time-dependent coherent modification of these equilibrium points. In physiological conditions, the SR threshold can be arbitrarily "set" by the CNS to any value within and beyond the range of angular variation in the joint considered. Therefore, the central adjustment range of the control variables (muscle SR thresholds) exceeds the biomechanical range of angular variation in a joint.

On the contrary, it has been suggested that some CNS injuries may result in a reduction of the admitted range of variability of the SR threshold, which can be also observed considering the torque/angle plot for the analyzed joint, i.e. the maximum torque that the joint is able to actively resist at a given angular value [19]. This behavior can justify the occurrence of muscular spasticity (see Fig. 7.2).

This concept is supported by several experimental findings (see e.g. [16, 18, 19]), in which authors reported on changes of the SR thresholds for both flexor and extensor muscular groups at the elbow level, in subjects affected by stroke or cerebral palsy. In these cases, threshold values are reduced within the biomechanical joint nominal range. Similar results were also discussed at the ankle level in adult patients after stroke [2], and in children with cerebral palsy [14]. The SR thresholds discussed so far are intended in static conditions and typically named with the symbol λ - also referred to as *Tonic* Stretch Reflex Threshold (TSRT). It is, however, worth mentioning that the real behavior of SR depends on the stretch velocity (v), and is described as *Dynamic* Stretch Reflex Threshold (DSRT) $\lambda^*(v)$, whose relation with the TSRT can be linearly approximated as $\lambda^*(v) = \lambda - \mu v$. Due to the extreme difficulty in directly measuring the TSRT, its value is then typically estimated from a series of observation of the DSRT. The joint is passively stretched using bell-shaped velocity profiles with different peak values and, for each trial velocity peak $\hat{v}$, the examiner records the joint angular value $\lambda^*(\hat{v})$ at which the muscle begins to be activated. Hereinafter we refer to this stimulation to as mode A (see Sect. 7.4). A linear regression is then used to estimate the static value of λ (See Fig. 7.3).

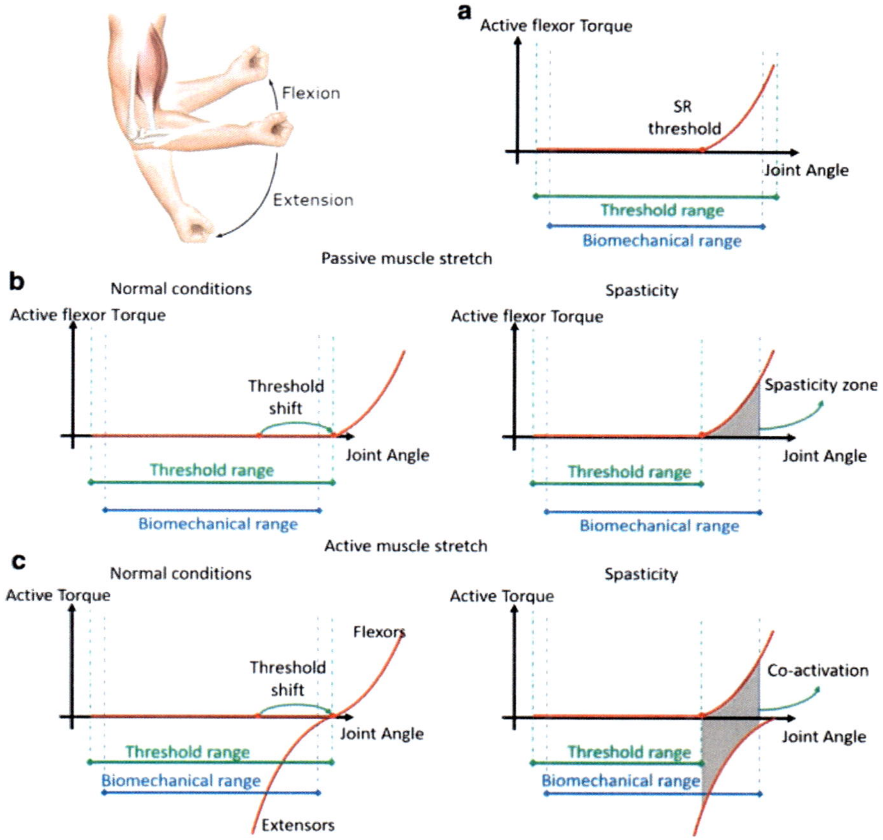

Fig. 7.2 Schematics of elbow movements in passive and active conditions. **a** Nominal conditions in steady state; **b** Passive extension of the elbow in normal and pathological conditions. The latter case shows the torque actively generated by the flexor muscle over joint angles, during spasticity of the flexor muscle; **c** Active extension of the elbow in normal and pathological conditions. The latter case shows a co-activation pattern, required to overcome the resistance to motion of flexor muscles due to spasticity

To quantify pathological conditions based on EHP, we then propose a device that can (i) precisely implement mode A stimulation and apply a torque co-axially with the examined joint (in our case the elbow flexion-extension) for retrieving the torque/angle characteristics (mode B) in agreement with [19], (ii) enable synchronous software recording of the actual joint angles, torques and the surface EMG patterns of specific muscles (flexors and extensors), and (iii) potentially acts as an active therapeutic tool implementing the manouver in [15].

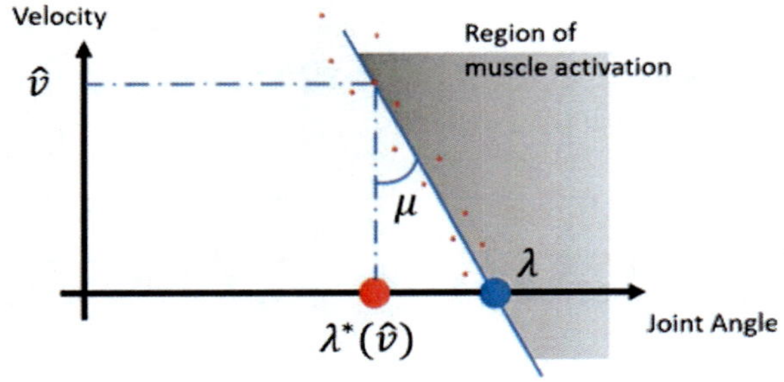

Fig. 7.3 Linear regression of multiple DSRT measurements used to estimate the TSRT. The angular value $\lambda^*(v)$ at which the muscle begins to be activated is measured considering different peak velocities $\hat{v}$ (red dots). A linear regression (blue line, with μ coefficient of the linear approximation) is then used to estimate the value of λ in static conditions (i.e. $v \approx 0$)

7.3 Mechanical Design

The proposed device consists of two rigid segments, one supporting the upper arm and the second supporting the forearm (see Fig. 7.5e, f for a full view). These are mounted on a support platform that grounds the whole structure on the floor and whose height can be adapted for different subjects' heights (from 145 to 210 cm) by maneuvering a passive prismatic joint (see Fig. 7.4b). The forearm support is

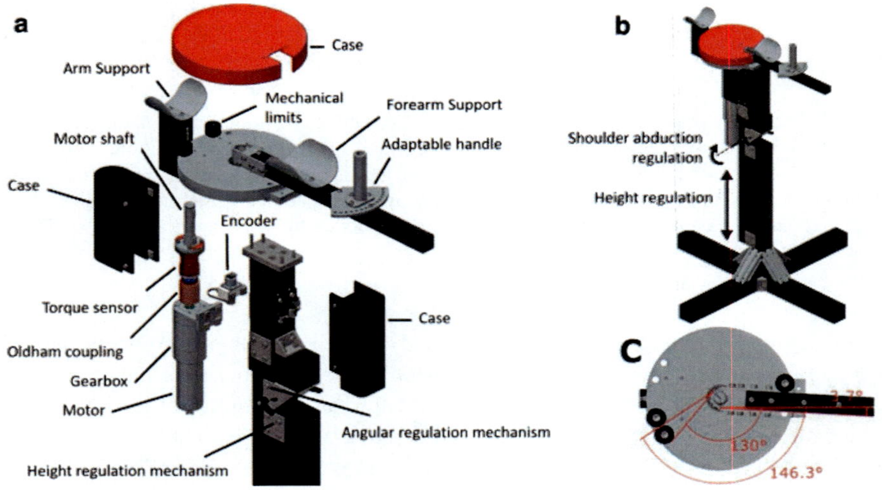

Fig. 7.4 Mechanical design of the proposed device. **a** Exploded view; **b** Full view, with details on possible regulations to adapt the device to the user; **c** Top view, with details on mechanical limit bumpers and joint range-of-motion. The device can be mounted as to be used with both arms, in **c** the components are mounted targeting the usage with the left arm

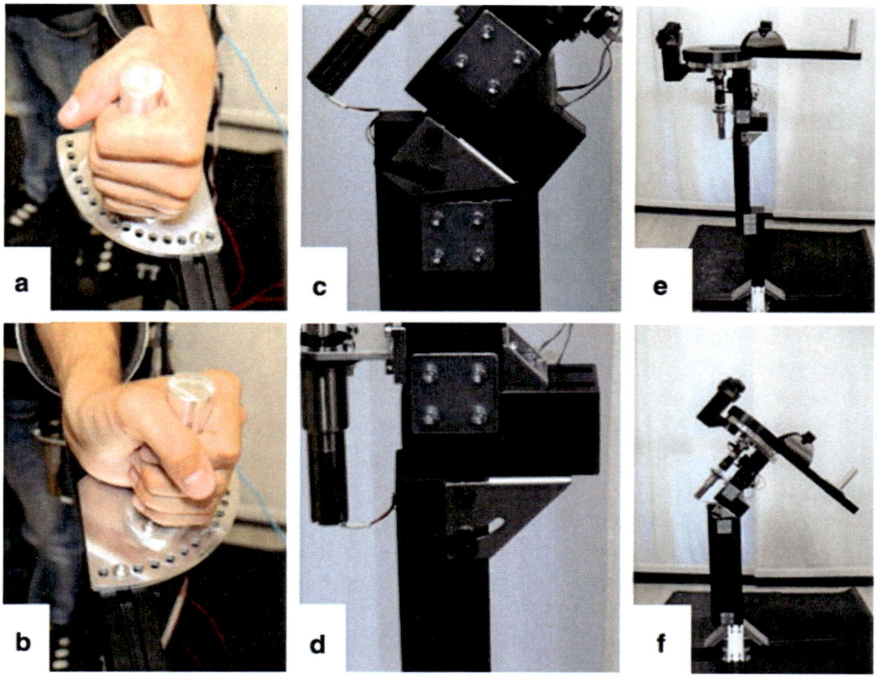

Fig. 7.5 Details of the proposed device. In Subfigs **a**, **b** regulations of the wrist flexion-extension. In Subfigs **c**, **d** linear and angular regulations at the shoulder level. In subfigs **e**, **f** full view of the proposed design in two different configurations

actuated trough a DC motor, mounted with the axis along the main axis of the elbow flexion-extension DoF and coupled with a torque sensor. An encoder is also included to measure joint angular values, together with torque and surface EMG sensors. The whole system is intended to be used by a subject seating on an adjustable chair, to accommodate for different subject body dimensions and pathological conditions. The total device weight is less than 20 Kg. A detailed exploded view is reported in Fig. 7.4a.

The elbow can rotate (actively or passively) around the motor axis in the range [34°, 176°], considering a safety margin of 4° w.r.t. the nominal elbow range [9] (see Fig. 7.4c). The elbow total movement can be also mechanically limited to the range [50°, 176°], through the usage of mechanic bumpers, for patients in which the wrist is completely flexed due to spasticity (see Fig. 7.4c for details). Shoulder abduction can be manually regulated through a passive revolute joint mechanism in the range [65°, 90°] (see Figs. 7.5c, d and 7.4b), while hand support inclination w.r.t. the forearm (around the flexion-extension axis) can be mounted in multiple configurations (regulation step of 5° in the range [0°, 60°]) through the adaptable handle shown in Fig. 7.5a, b. The handle can also slide along the forearm main axis to account for different forearm lengths, with a distance between handle and motor axis ranging between 29 and 42 cm. All the regulation ranges were designed considering average biomechanical dimensions (see [9])

To identify the minimum motor characteristics, we considered the requirements of the two experimental conditions: TSRT estimation and torque/angle diagram estimation [19]. During the first experimental condition, the subject is asked to keep the arm relaxed while a velocity profile is applied. Assuming that we can neglect the dynamics after the SR (not relevant for the analysis), we considered a required maximum peak velocity equal to 300°/s, to be applied for 1.2 s (see [23], equivalent to 50 rpm). Considering the second experimental condition, instead, the maximum velocity required is 8°/s, to be applied for less than 30 s. The maximum torque considered is derived by the experiments discussed in [19] (40 Nm) and over-respected with a design limit set to 60 Nm.

To satisfy all these requirements, we selected a 24V DC brushed motor (Maxon Motors, mod RE50), coupled with a gearbox with a reduction factor equal to 71 : 1 (Maxon Motors, mod GP62). This system is able to perform a maximum peak velocity equal to 63 rpm and a maximum torque equal to 63.9 Nm, thus satisfying the design specifics. Motor and gearbox are coupled though a Oldham coupling (mod MOM-38K from NBM, maximum torque 110 Nm, maximum velocity 2000 rpm). The torque sensor used is the TFF-350-FSH04063 by Futek, with full-scale equal to 150 Nm, maximum measure error equal to 0.45 Nm and dimensions comparable with those of the motor. The encoder used is an Austrian Microsystems AS5045 (12 bit, resolution 0.0875°). The electronic board, together with the software and libraries used in the proposed device are extracted from Natural Machine Motion Initiative[1] (NMMI) platform [8]. The microcontroller CY8C3246PVI-147 48-SSOP from PSoC is embedded in the control board, connected to a PC via USB. MATLAB/Simulink libraries are used to control the motor, record force, velocity data and EMG signals (in our case we consider the 4 electrodes by Ottobock. Note, however, that this architecture is modular) a 1 KHz loop.

7.3.1 *Evaluation of Mechanical Strength*

Due to the high forces involved during the usage of the device, we performed mechanical strength evaluation of all the components included in the design. Aluminium bars used to support arm and forearm were analysed using a beam model, while the other connection components were analysed using a Finite Element Modeling (FEM) analysis using the software PTC Creo Engineer.

[1] https://www.naturalmachinemotioninitiative.com/.

7.3.1.1 Frame

The frame is made by aluminium bars, with a yielding stress of 205 Mpa. Considering the forearm support, the critical section is at the connection between the bar and the pulley of the crankshaft. The equivalent Tresca stress is 41.7 Mpa, then with a safety coefficient $\eta = 4.9$. Regarding the upper arm support bar, the critical section is at the coupling with the base support, with a critical condition for maximum extension of the bar. In this case, the equivalent Tresca stress is 47.9 Mpa, then with a safety coefficient $\eta = 4.3$.

7.3.1.2 Crankshaft

The crankshaft is made in aluminium 7075-T6 (UNI EN 573-3), supporting up to 434 Mpa and with a Young module of 71 Gpa. In its critical section (higher bearing) the equivalent Tresca stress is 114.7 Mpa, safety coefficient $\eta = 3.75$. For deformation evaluation, we used the Mohr method and verified that at the joint level the angular and radial deformations are negligible w.r.t. to the margin admitted by the Oldham coupling.

7.3.1.3 Bearings

Radial loads on the higher and lower bearings are 2.79 kN and 2.37 kN respectively, while axial loads are negligible. Radial static load coefficient is 4.4 kN, then the safety coefficient is equal to $\eta = 1.58$.

7.4 Control

The two evaluation procedures discussed above require two different control strategies: a position control for the estimation of the TSRT (mode A), and a torque regulation for the estimation of the torque/angle plot (mode B). Both the controllers were developed in Matlab Simulink to enable a fast prototyping and easy tuning of parameters. The input considered is the tension applied to the motor (PWM), while the output is the position of the motor (mode A) or the applied torque (mode B).

7.4.1 Mode A: Position Control

To precisely regulate the angular position of the motor, we implemented a PID controller $C(s) = K_c(1 + \frac{1}{\tau_i s} + \tau_d s)\frac{1}{1+\tau_p s}$. Parameters were tuned following a the closed-loop Ziegler-Nichols rule with no overshoot. Tuned parameters of the controller were $K_c = 0.64$, $\tau_i = 0.19$, $\tau_d = 0.125$ and $\tau_p = \frac{1}{300}$.

7.4.2 *Mode B: Torque Control*

To control the torque we used a PI regulator $C(s) = K_c(1 + \frac{1}{\tau_i s})$, tuned through the Frequency-Response Based Tuning available in Matlab. Design requirements included a step response lower than 0.3s with no overshoot, a bandpass of 10°/s and a phase margin of 90°. Tuned parameters were $K_c = 0.3$ and $\tau_i = 0.025$.

7.5 Preliminary Experiments with Healthy Subject

To test the performance of our device, and its control, during usage in both the experimental conditions (mode A and B), we performed a set of experiments with one healthy participant (right-handed, M. 27).

Preliminary tests in mode A (see Fig. 7.7) were performed by commanding trajectories with bell shaped velocity profiles (peak velocity ranging between 10 and 110°/s) (the upper limit is chosen for safety reasons) with a load on the arm and forearm supports equal to 1.7 and 3 Kg respectively, coherently with [9]. Maximum tracking error recorded was lower than 1°. Then, a simulation of the experiment for the estimation of the TSRT was performed. A naive subject was asked to hold the handle—whose configuration was carefully adapted to his specific body dimensions - and to keep upper limb muscles relaxed. Then, bell-shaped velocity profiles were applied to the motor considering the peak values in the range from 10 to 110°/s. Subject was asked to contract the muscles in the middle of the trajectory (close to the

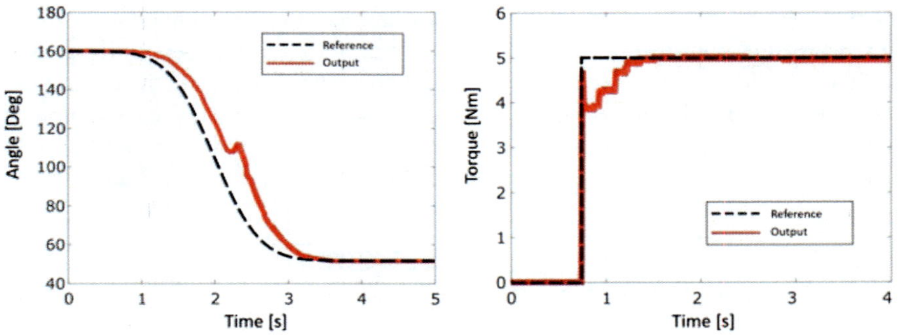

Fig. 7.6 On the left, reference and output during a test with position control and an external disturbance due to muscle contraction. Peak velocity was equal to 90°/s. On the right, reference and output during a test in torque control for the estimation of torque/angle plot

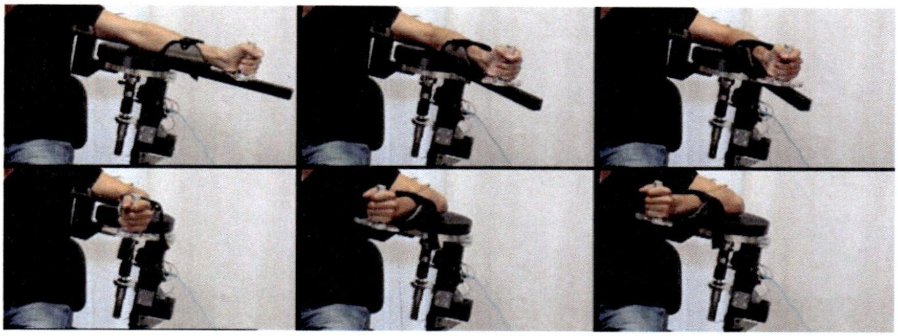

Fig. 7.7 Snapshot of a test simulating the estimation of the TSRT (mode A), performed using the device presented in this chapter. Subject was asked to grasp the handle and keep the arm muscles relaxed. Then, a bell-shaped velocity profile was applied to the motor imposing a passive motion to the subject forearm

velocity peak, to simulate the stretch reflex). Performance of the controller during these tests were collected and analysed by recording commanded and real position of the forearm support. In Fig. 7.6-left we show the outcome of these experiments for the case of 90°/s. It is possible to observe a prompt and smooth response of the controller to the external disturbance caused by the subject muscular contraction.

Tests in mode B were first performed with the axis of the motor parallel to the floor and with a weight at the extremity of the forearm support equal to 5 Kg (i.e. a pendulum configuration). Experiments were conducted by evaluating the step response, considering input reference torques ranging between 1 and 10 Nm. Maximum steady state error recorded was always lower than 0.3Nm. Then, a simulation of the experiment for the estimation of the torque/angle plot was performed. The subject was asked to slowly move the arm in flexion (average velocity 10°/s) multiple time. The first repetition was performed in free motion, then at each iteration the opposite torque was increased of 1 Nm, up to the threshold of 10 Nm. The measured and commanded torque value for the case of opposite torque equal to 5 Nm is reported in Fig. 7.6-right. Note that the time-to-rise of the step response in mainly related to the muscle response time.

7.6 Conclusions

In this chapter we presented a new mechatronic device to be used for the diagnosis and the evaluation of spasticity phenomena at the elbow level. Moving from current state-of-the-art devices and clinical specifications, we proposed a new implementation that is portable and enable different regulations and control possibilities, opening the possibility to minimize the active interaction between patients and therapists during the evaluation task, thus the operator-dependent effects. It is also worth mentioning that our design could be potentially used as an active therapeutic tool based on the manoeuvre presented in [15]. Preliminary experiments with one healthy participant demonstrated that our device is able to replicate both the experimental procedures envisioned (mode A and B) with good performance in terms of tracking errors. A careful and extended experimental campaing with healty and stroke-affected subjects is currently under evaluation and is left for future developments of this work.

References

1. Averta, G., Abbinante, M., Orsini, P., Felici, F., Lippi, P., Bicchi, A., Catalano, M.G., Bianchi, M.: A novel mechatronic system for evaluating elbow muscular spasticity relying on tonic stretch reflex threshold estimation. In: Proceedings of the Engineering in Medicine and Biology Society, 2020. EMBC'20. 42th Annual International Conference of the IEEE (2020)
2. Blanchette, A.K., Mullick, A.A., Moïn-Darbari, K., Levin, M.F.: Tonic stretch reflex threshold as a measure of ankle plantar-flexor spasticity after stroke. Phys. Ther. **96**(5), 687–695 (2016)
3. Bourbonnais, D., Noven, S.V.: Weakness in patients with hemiparesis. Am. J. Occup. Ther. **43**(5), 313–319 (1989)
4. Calota, A., Feldman, A.G., Levin, M.F.: Spasticity measurement based on tonic stretch reflex threshold in stroke using a portable device. Clin. Neurophysiol. **119**(10), 2329–2337 (2008)
5. Calota, A., Levin, M.F.: Tonic stretch reflex threshold as a measure of spasticity: implications for clinical practice. Top. Stroke Rehabil. **16**(3), 177–188 (2009)
6. Charalambous, C.P.: Interrater reliability of a modified ashworth scale of muscle spasticity. In: Classic Papers in Orthopaedics, pp. 415–417. Springer (2014)
7. Crippa, A., Cardini, R., Pellegatta, D., Manzoni, S., Cattaneo, D., Marazzini, F.: Effects of sudden, passive muscle shortening according to grimaldi's method on patients suffering from multiple sclerosis: a randomized controlled trial. Neurorehabil. Neural Repair **18**(1), 47–52 (2004)
8. Della Santina, C., Piazza, C., Gasparri, G.M., Bonilla, M., Catalano, M.G., Grioli, G., Garabini, M., Bicchi, A.: The quest for natural machine motion: an open platform to fast-prototyping articulated soft robots. IEEE Robot. Autom. Mag. **24**(1), 48–56 (2017)
9. Dempster, W.T., Gaughran, G.R.: Properties of body segments based on size and weight. Am. J. Anat. **120**(1), 33–54 (1967)
10. Feldman, A.G.: Functional tuning of the nervous system with control of movement or maintenance of a steady posture-ii. controllable parameters of the muscle. Biofizika **11**, 565–578 (1966)
11. Feldman, A.G.: Once more on the equilibrium-point hypothesis (λ model) for motor control. J. Mot. Behav. **18**(1), 17–54 (1986)
12. Feldman, A.G., et al.: Referent control of action and perception. In: Challenging Conventional Theories in Behavioral Neuroscience. Springer, New York (2015). https://doi.org/10.1007/978-1-4939-2736-4
13. Feldman, R.G., Young, R.R., Koella, W.P.: Spasticity: Disordered Motor Control. Symposia Specialists, Incorporated (1980)
14. Germanotta, M., Taborri, J., Rossi, S., Frascarelli, F., Palermo, E., Cappa, P., Castelli, E., Petrarca, M.: Spasticity measurement based on tonic stretch reflex threshold in children with cerebral palsy using the pedianklebot. Front. Hum. Neurosci. **11**, 277 (2017)
15. Grimaldi, L., Lippi, P., Marri, P., Fantozzi, M., Bracci, R.: Revoking of absent motor components in CNS lesions. Giardini **1–3** (1986–1988)
16. Jobin, A., Levin, M.F.: Regulation of stretch reflex threshold in elbow flexors in children with cerebral palsy: a new measure of spasticity. Dev. Med. Child Neurol. **42**(8), 531–540 (2000)

17. Levin, M.F.: Interjoint coordination during pointing movements is disrupted in spastic hemiparesis. Brain **119**(1), 281–293 (1996)
18. Levin, M.F., Feldman, A.G.: The role of stretch reflex threshold regulation in normal and impaired motor control. Brain Res. **657**(1–2), 23–30 (1994)
19. Levin, M.F., Selles, R.W., Verheul, M.H., Meijer, O.G.: Deficits in the coordination of agonist and antagonist muscles in stroke patients: implications for normal motor control. Brain Res. **853**(2), 352–369 (2000)
20. Marques, I.A., Silva, M.B., Silva, A.N., Luiz, L.M.D., Soares, A.B., Naves, E.L.M.: Measurement of post-stroke spasticity based on tonic stretch reflex threshold: implications of stretch velocity for clinical practice. Disabil. Rehabil. **41**(2), 219–225 (2019)
21. Santello, M., Lang, C.E.: Are movement disorders and sensorimotor injuries pathologic synergies? when normal multi-joint movement synergies become pathologic. Front. Hum. Neurosci. **8**, 1050 (2015)
22. Sommerfeld, D.K., Eek, E.U.B., Svensson, A.K., Holmqvist, L.W., von Arbin, M.H.: Spasticity after stroke: its occurrence and association with motor impairments and activity limitations. Stroke **35**(1), 134–139 (2004)
23. Turpin, N.A., Feldman, A.G., Levin, M.F.: Stretch-reflex threshold modulation during active elbow movements in post-stroke survivors with spasticity. Clin. Neurophysiol. **128**(10), 1891–1897 (2017)
24. Wissel, J., Schelosky, L.D., Scott, J., Christe, W., Faiss, J.H., Mueller, J.: Early development of spasticity following stroke: a prospective, observational trial. J. Neurol. **257**(7), 1067–1072 (2010)

Part III
Transferring Human Principles to Cobots and Autonomous Robots

Chapter 8
Natural Motion: Embedding Human-Likeliness in Robot Movements

Human-likeliness of robot movements is a key component to enable a safe and effective human-robot interaction, resulting in an increase of acceptance and motion predictability of robots that have to closely interact with people. This is of particular interest for example for assistive or rehabilitative devices, but also for companion technologies. Several parameters have been used to quantify how much a robot behaves like a human, which encompass aspects related to both the robot appearance and motion. The latter point is fundamental to allow the operator to interpret robotic actions, and plan a meaningful reactions. While different approaches have been presented in literature, which aim at devising bio-aware control guidelines, a direct implementation of human actions for robot planning is not straightforward, still representing an open issue in robotics. In this chapter, we report on the possibility to encode human likeliness in a synergistic representation of human motion, which in turn can be used to generate arbitrary movements. To do this, we recorded human upper-limb motions during daily living activities. We used functional Principal Component Analysis (fPCA) to extract principal motion patterns. We then formulated the planning problem by optimizing the weights of a reduced set of these components. For free-motions, our planning method results into a closed form solution which uses only one principal component. In case of obstacles, a numerical routine is proposed, incrementally enrolling principal components until the problem is solved with a suitable precision. Results of fPCA show that more than 80% of the observed variance can be explained by only three functional components. The application of our method to different meaningful movements, with and without obstacles, show that our approach is able to generate complex motions with a very reduced number of functional components. We show that the first primitive alone accounts for the 96% of cost reduction and that three components are able to achieve a satisfactory motion reconstruction in all the considered cases. Then, we present a technique to directly embed human upper limb principal motion modes

G. Averta, *Human-Aware Robotics: Modeling Human Motor Skills for the Design, Planning and Control of a New Generation of Robotic Devices*, Springer Tracts in Advanced Robotics 145, https://doi.org/10.1007/978-3-030-92521-5_8

computed through functional analysis in the robot trajectory optimization. We report on the implementation with manipulators with redundant anthropomorphic kinematic architectures—although dissimilar with respect to the human model used for functional mode extraction—via Cartesian impedance control. In our experiments, we show how human trajectories mapped onto a robotic manipulator still exhibit the main characteristics of human-likeness, namely low jerk values and bell-shaped velocity profiles. We discuss the results with respect to the state of the art, and their implications for advanced human-robot interaction in industrial co-botics and for human assistance.

8.1 A New Method to Generate Human-Like Movements for Anthropomorphic Manipulators

8.1.1 Introduction

There are many examples in literature that have highlighted the importance of human-likeness (HL) to ensure a safe and effective Human-Robot Interaction (HRI) [6, 15]. This aspect is particularly relevant for the control of devices that interact with users for rehabilitation purposes. Indeed, human-inspired movement profiles may be preferred references for the active control of exoskeletons. Moreover, having a robot that moves embedding the same underlying characteristics of the user is proved to increase the predictability and acceptability of the device [6, 38]. This play a key role for the usability and the effectiveness of assistive and/or companion robots [10, 11, 33].

Motion planning for highly articulated manipulators is a very well-studied topic in robotics research, and several solutions relying on sampling-based algorithms have been proposed [12], such as Rapidly-exploring Random Trees or Probabilistic Road-maps [23, 26], also modified to include constraints [41] and improve efficiency [18]. Among the others, some interesting solutions to this problem are discussed in [40] and in [47], leading to efficient algorithms for motion planning of manipulators. However, these approaches were not originally designed for the generation of human like motions, as discussed in [47].

HL has been quantified through a network of parameters [46], including movement indexes. The latter are critical to allow the human user to intuitively interpret robot motion and prepare a suitable response. Usually, HL is achieved leveraging on a vast neuroscientific literature to devise cost functions, whose optimization results in human-like characteristics of the motion. In his seminal work [30], Morasso analyzed the principal characteristics of point-to-point human arm motions, identifying in a single peak of the velocity profile a salient characterizing feature. In [17], Flash et al. theorized that human generate motions minimizing the jerk. Taking inspiration from this finding, in [36] human-like artificial motions were generated through jerk minimization.

Other indexes have been also identified as the ones whose minimization leads to anthropomorphic movements, and applied to robotic planning such as minimum joint torques [7] and Virtual Spring-Damper Hypothesis [1]. Methods for anthropomorphic motion generation are also widely and effectively adopted for character animation. However, most of these techniques focus on locomotion [32], and/or strongly rely on the usage of neural network to find suitable solutions [31, 34]. Direct human movements observations have been also used for getting human-like motions, e.g. to train Neural Networks [45], or to compute a set of parameters defining non-linear dynamical systems [14]. However, optimal control-based methods usually come with hypotheses on motion generation that may limit the variability of the planned movement and, sometimes also lacks of experimental support [29]. On the other hand, learning methods typically requires a large dataset whose dimensionality dramatically increases with task complexity.

To the best of authors' knowledge, a direct exploitation of human observations for robotic motion generation has not been applied yet. This approach would come with several advantages, since human-likeness would be intrinsically guaranteed. However, a mere copy-cutting from nature would be unfeasible and clearly a daunting task. What we propose instead is to use neuroscientific data and model them with a mathematical language, that can be understood by an artificial body, and thus implemented in an effective manner. This approach has been investigated in literature for robotic hands (see hand postural synergies defined in [39]), for the design and control of robotic hands or for grasp planning [8, 13]. In this chapter, we propose to directly embed human upper limb principal motion modes in the planning of anthropomorphic manipulators. To this end, this Chapter builds upon the experimental framework of Chap. 2 and extend the results with a dataset of 33 right-handed able-bodied subjects performing a selection of 30 Activities of Daily Living (ADLs). We then applied functional Principal Component Analysis (fPCA) to extract a low number of base functions, which explain, for each joint, most of the trajectory variability. As reported later, we demonstrate that a weighted sum of only three functional components takes into account more than 80% of the variance. Capitalizing on these results, we formulate the planning problem, for a given anthropomorphic manipulator, as an optimization problem. In particular, the final motion is obtained as the optimal weighted combination of a reduced set of functional principal components. For free-motions, our method results into a closed form solution, which uses only one principal component.

This methodology comes with a significant perspective shift: from the search for optimal paths to the identification of a reduced number of scalars weighting the functional components. This could enable a fast convergence toward a solution for the planner, which is intrinsically human-like. To further increase the cost-effectiveness of our method, we propose an incremental enrollment of the functional components, as suggested in [2]. In this way, the number of primitives needed to perform the task is tailored on its complexity, avoiding the useless inclusion of higher order primitives.

We demonstrate in simulations that our techniques can generate human-like motion using a reduced number of functional component. Human likeliness of generated movements is evaluated according to main indexes reported in [46] observing velocity profiles and jerk values [16, 30]. In the final part of the Chapter these results are generalized to different kinematic structures and complemented with experiments using a 7DoFs manipulator.

To conclude, this chapter contributes with: (i) An extensive study of upper limb motion primitives in humans, applying functional analysis and extending our previous work [4] substantially increasing the number of participants involved in the study; (ii) A new methodology for human like motion generation, applied in the case of point-to-point motions with obstacle avoidance, which intrinsically embed the main principal modes of human upper limb motion; (iii) Simulation results to show the effectiveness of our method; (iv) a control-based approach to map HL movements generated for the human kinematics onto a generic redundant manipulator. The content of this chapter is adapted from [3, 5].

A companion video is available at the following link: https://www.youtube.com/watch?v=Grg4L4fljdg.

8.1.2 Functional Principal Components of Upper Limb Motion

In Chap. 2 we discussed a functional characterization of upper-limb movements during Activities of Daily Living (ADLs), showing that a reduced-dimensionality set of principal functions account for a large part of upper-limb movements variability [4].

To expand and confirm the validity of our analysis, in this chapter we report on a completely new dataset, enrolling 33 subjects (17 women, 26.6 years old on average). The experimental protocol comprises 30 activities of daily living involving a predominant component of upper limb movements. Each activity is repeated three times (i.e., a grand total of 90 tasks per person). All experimental procedures were approved by the local ethical committee. The movements are grouped into three categories, characterized by a different kind of interaction with the object. Please refer to Fig. 8.1 for a graphical representation of these actions. Note that each subject was instructed on the task to accomplish and not on the specific motion to perform for achieving the goal.

8.1.2.1 Motion Identification

In the following, we will recall some concepts useful for the development of this chapter, additional details are provided in Chap. 2. Movements were recorded using a 3D motion tracking system with active markers (Phase Space). Ten stereo-cameras worked 100 Hz tracking 3D position of active markers linked to rigid supports and attached to upper limb links (see Fig. 8.2). To map the recorded movements on an analytic kinematic model, we developed a two-phases procedure: a preliminary

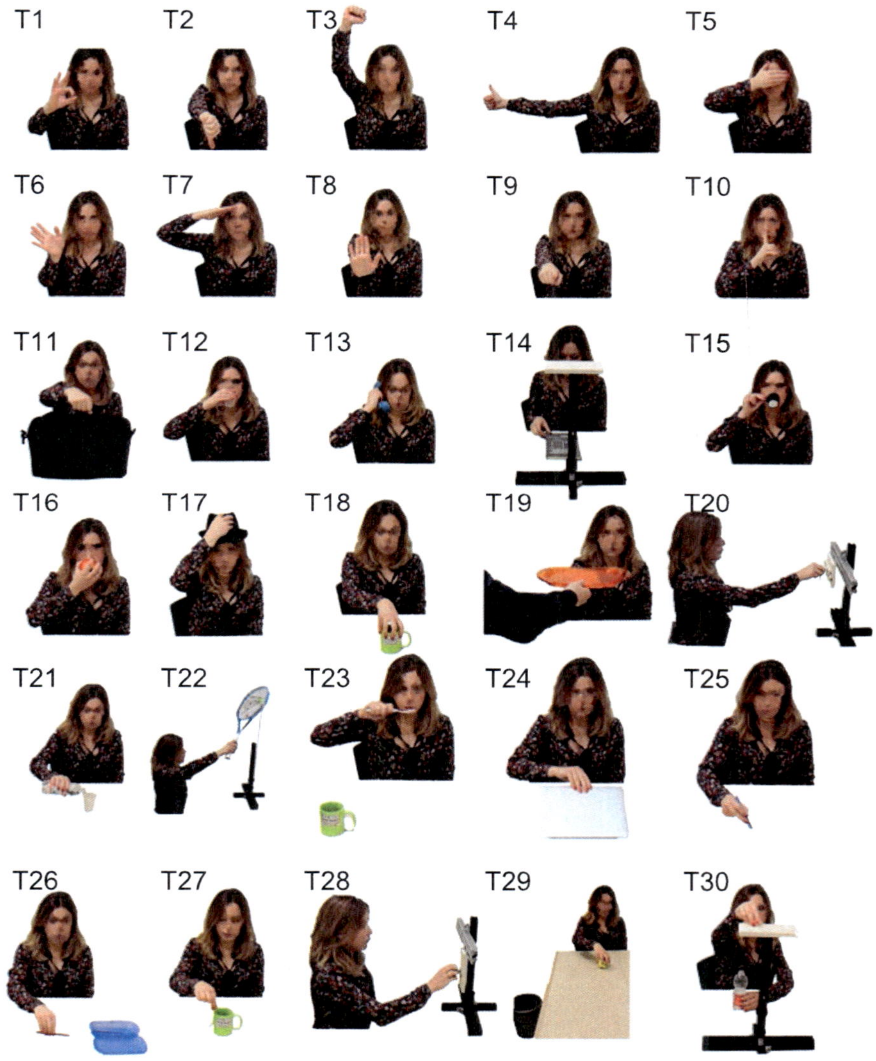

Fig. 8.1 From top to bottom, and from left to right, pictures of the task considered in this study

calibration of the model followed by a Kalman-based identification procedure. In this chapter, we used the same kinematic model discussed in [4], i.e. a serial manipulator with three rigid links connected by seven revolute joint (see Fig. 8.3). This description is completely defined using 14 parameters: 2 parameters for bones lenght and 12 parameters to identify the position of the markers w.r.t. the kinematic chain (more technical details may be found in [4]). This set of parameters (p_G) is calibrated, for each subject, by solving the following optimization problem:

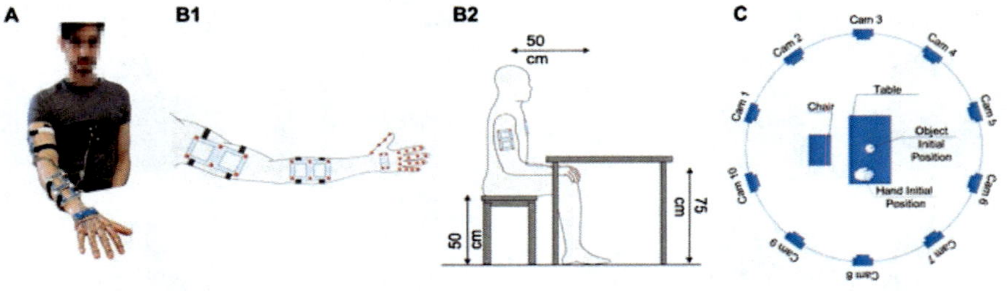

Fig. 8.2 Schematics of the experimental setup. **a** shows a picture of a subject wearing markers on the upper limb. **b** and **c** detail the experimental setup. Arm and forearm are tracked using six markers, while the hand is tracked using four markers. Four additional marker are placed on the subject's chest to track the reference. Markers are redundant and fastened on rigid supports. 10 stereo-cameras are placed around the scene to minimize markers occlusion

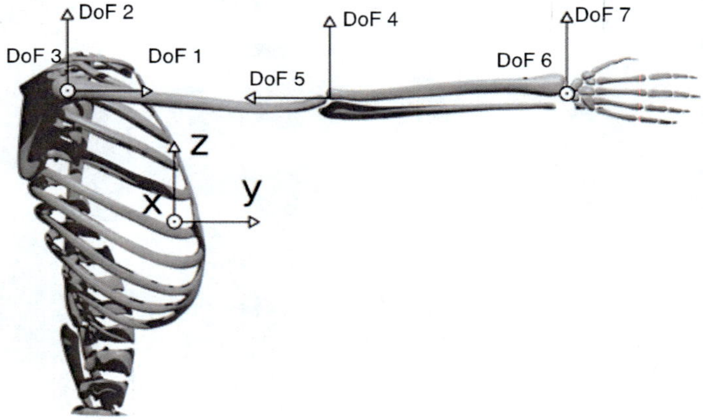

Fig. 8.3 Kinematic parametrization, the labels DoF 1, ..., DoF 7 refer to the joint angles of the model

$$(q^*, p_G^*) = \arg \min_{q_k \in D_q, p_G \in D_p} \frac{1}{2} \sum_{k=1}^{N_p} r_k^T r_k,$$

where r_k is the residual function

$$r_k(q_k, p_G) = y_k - f(q_k, p_G),$$

calculated as difference between the measured position of markers y_k and their estimation via Forward Kinematics (FK) $f(q_k, p_G)$. The FK is function of q_k, the current estimation of joint angles, and p_G, the parameters of the kinematic model. The optimization problem is constrained by D_q, the joints range of motion, and D_p, the maximum variation of parameters around a preliminary manual estimation. To improve the effectiveness of this procedure, it is possible to extend the vector of measures y_k and the vector of estimations $f(q_k, p_G)$ by concatenating the output at different time frames. In this case, we selected 10 equally time-spaced samples. Given the calibrated vector p_G for the specific subject, the upper limb pose is completely described by 7 joints angles $[q_1, \ldots, q_7]^T$.

The calibrated model is then used as a component for a Kalman-based identification procedure. More specifically, if we consider, at time frame k, the joints angle vector q_k as the state of a stochastic process with w_k and v_k process and observation zero mean Gaussian noises, with covariance Q_k and R_k respectively, we have that the system can be written as

$$\begin{cases} q_k = q_{k-1} + w_k \\ y_k = f(q_k) + v_k \end{cases} \tag{8.1}$$

Recursively, the state q_k can be estimated from q_{k-1} with the following implementation of a Kalman filter [42]: *prediction* of the future state $\hat{q}_{k|k-1} = \hat{q}_{k-1}$; *update* of the state estimation as $\hat{q}_{k|k} = \hat{q}_{k|k-1} + K_k\tilde{r}_k$. The prediction correction is the product between the residual values vector $\tilde{r}_k = y_k - f(\hat{q}_{k|k-1})$ and the Kalman Gain K_k, defined as product between the covariance matrix estimation of the predicted state $P_{k|k-1}$, the jacobian matrix $H_k = \frac{\partial(f(q))}{\partial q}$ and the inverse of the residual covariance.

Finally, to effectively jointly process different acquisitions with different temporal lengths, a time normalization is required. To this end, we used Dynamic Time Warping. Given a reference signal extracted from the dataset $q_{ref}(t)$, all the trajectories are normalized in time w.r.t. the reference by solving the following minimization problem

$$(S_i, T_i) = \arg \min_{S_i > 0, T_i} (||q_{ref}(t) - q_i(S_i t - T_i)||) \tag{8.2}$$

8.1.2.2 Functional Analysis

In the following, we will briefly summarize the main concepts behind functional component analysis and how we apply these techniques to upper limb motion description.

Let us consider a generic upper-limb movement

$$q(t) : \mathbb{R} \rightarrow \mathbb{R}^7$$

where $t \in [0, 1]$ is the normalized time, and 7 is the number of upper limb degrees of freedom for the considered model. What we are interested to investigate here is if a suitable reduced functional representation exists, able to closely approximate $q(t)$, which represents joint trajectories.

Functional Principal Component Analysis (fPCA) can be regarded as a functional extension of Principal Component Analysis. Indeed, as Principal Components identify joint couplings that account for most kinematic pose variability, functional PCs (fPCs) are functions that allow to describe most of the movement variability. Using fPCA framework, a generic upper limb motion $q(t)$ can be described as weighted sum of the basis functions $S_i(t)$, or functional Principal Components

$$q(t) \simeq \bar{q}(t) + S_0(t) + \sum_{i=1}^{s_{\max}} \alpha_i \circ S_i(t), \tag{8.3}$$

where $\alpha_i \in \mathbb{R}^{\mathrm{k}}$ is a vector of weights, $S_i(t) \in \mathbb{R}^{\mathrm{n}}$—in our case n equals to 7—is the ith basis element or primitive and $s_{\max}$ is the number of basis elements. The operator $\circ$ is the Hadamard product, i.e. an element-wise product. $\bar{q} \in \mathbb{R}^7$ is the average posture of q evaluated as

$$\bar{q} = \int_0^1 q(\tau)\mathrm{d}\tau, \tag{8.4}$$

while $S_0 : \mathbb{R} \to \mathbb{R}^7$ is the *zero-order* primitive. We obtain it as average trajectory across all the trajectories q in the dataset, for all the tasks and subjects. fPCA is used to identify a basis of functions $\{S_1, \ldots, S_{s_{\max}}\}$ that maximizes the explained variance of the movements included in the dataset. The first fPC $S_1(t)$ is the function that solves the following problem

$$\begin{aligned} \min_{S_1} \quad & \sum_{j=1}^{R} \left(\int S_1(t) x_j(t) dt \right)^2 \\ \text{subject to} \quad & ||S_1(t)||_2^2 = \int_0^1 S_1^2(t)dt = 1. \end{aligned} \tag{8.5}$$

Subsequent fPCs $S_i(t)$ are the functions that solves the following:

$$\begin{aligned} \min_{S_i} \quad & \sum_{j=1}^{R} \left(\int S_1(t) x_j(t) dt \right)^2 \\ \text{subject to} \quad & ||S_i(t)||_2^2 = 1 \\ & \int_0^1 S_i(t) S_k(t) dt = 0, \ \forall k \in \{1, \ldots, i-1\}. \end{aligned} \tag{8.6}$$

In this way, it is possible to build an ordered set of functions that—given a number of basis elements—maximizes the explained variability of the dataset.

In Fig. 8.4, we report, for each joint, the principal components over the time. As shown in Fig. 8.5, the first four fPCs together (i.e. a weighted sum of them) account for around the 85% of the total variance of the dataset. What is also noticeable is that lower order primitives (i.e. fPC1 and fPC2) introduce low-frequency contributions to the mean function, while higher order primitives include higher-frequency terms to the signal. In other terms, this behavior suggests that lower order primitives are mainly devoted to a gross definition of the whole trajectory, while higher order components

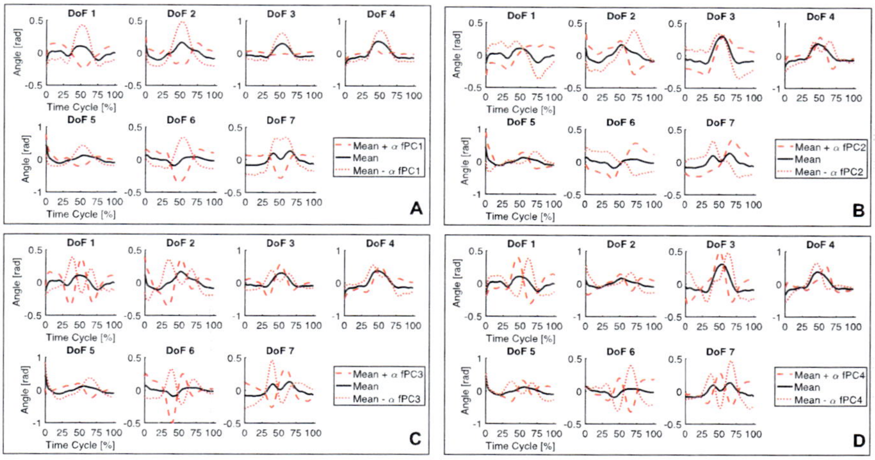

Fig. 8.4 From figure A to D, mean trajectory (in black) and the contribution (signed) of principal functions, from fPC1 to fPC4 respectively. Components ordering follows the indexing of Fig. 8.3

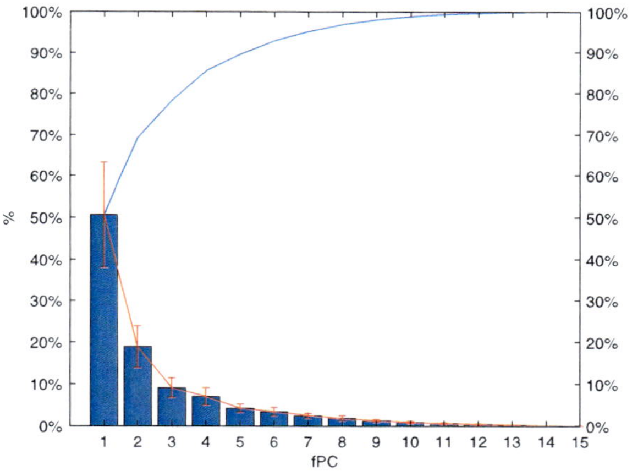

Fig. 8.5 Average variance explained by each fPC (blue bars) ± standard deviation (in red). The blue line on top reports the incremental sum of mean values

deal with the fine tuning of the specific movement. Note that these findings are totally coherent and, hence, further confirm the outcomes of [4].

8.1.3 Proposed Strategy

As discussed in the introduction, typical approaches used in the literature to achieve human likeness [36] rely on the strong assumption that human movements are generated by optimizing a known cost function $J_{\mathrm{hl}}(q) : \mathrm{C}^1_7[0, t_{\mathrm{fin}}) \rightarrow \mathbb{R}^+$. This function is used to produce artificial natural motions by solving the problem

$$\min_{q \in \mathrm{C}_7^1[0,t_{\mathrm{fin}})} \quad J_{\mathrm{hl}}(q). \tag{8.7}$$

How to choose J_{hl} is not obvious, and it is indeed a very debated topic in literature. Moreover, assuring human-likeness is not sufficient by itself, since the artificial agent must at the same time execute at least one task. Suppose we can describe it through the minimization of another cost function $J_{\mathrm{task}} : \mathrm{C}_7^1 \rightarrow \mathbb{R}^+$.

As soon as the need of minimizing J_{task} is introduced, (8.7) becomes a multi--objective optimization which is not clear even how to formulate apart from very simple cases [36].

We propose here an approach allowing to by-pass this issue. Instead of deriving a $J_{\mathrm{hl}}(\cdot)$ from data, and then explicitly optimize it, we propose to directly embed human likeness into the choice of the functional subspace in which the optimization occurs. More specifically, we move from a generic infinite dimensional space $\mathrm{C}_7^1[0, t_{\mathrm{fin}})$, to the finite dimensional family of functions

$$q(t) = \bar{q} + S_0\left(\frac{t}{t_{\mathrm{fin}}}\right) + \sum_{i=1}^{M} \alpha_i \circ S_i\left(\frac{t}{t_{\mathrm{fin}}}\right) \tag{8.8}$$

with $\bar{q}$, S_i, α_i defined as in previous section. $M \leq s_{\mathrm{max}}$ is the amount of primitives considered in the optimization (with s_{max} as in (8.3)). According to the results presented in Sect. 8.1.2, it is plausible to expect that a low number of primitives shall be sufficient to implement most of the human-like motions that the human arm can present. The following general optimization problem results:

$$\begin{aligned} \min_{\bar{q},\alpha_1,\ldots,\alpha_M} \quad & J_{\mathrm{task}}(q) \\ \text{subject to} \quad & q(t) = \bar{q} + S_0\left(\frac{t}{t_{\mathrm{fin}}}\right) + \sum_{i=1}^{M} \alpha_i \circ S_i\left(\frac{t}{t_{\mathrm{fin}}}\right), \end{aligned} \tag{8.9}$$

Thus, narrowing the search space serves to the twofold purpose of assuring human likeness, and strongly simplifying the control problem (the search space is now of dimension $M + 1$). In the following subsections we present the application of this approach to J_{task} describing either simple point-to-point free movements and more complex motions with obstacle avoidance.

8.1.3.1 Point-to-Point Free Motions

We propose here to generate a human-like point-to-point motion by solving the following optimization problem

$$\begin{aligned} \min_{\bar{q},\alpha_1,\ldots,\alpha_M} \quad & ||q(0) - q_0||_2^2 + ||q(t_{\text{fin}}) - q_{\text{fin}}||_2^2 \\ \text{subject to} \quad & q(t) = \bar{q} + S_0\left(\frac{t}{t_{\text{fin}}}\right) + \sum_{i=1}^{M} \alpha_i \circ S_i\left(\frac{t}{t_{\text{fin}}}\right), \end{aligned} \tag{8.10}$$

where $q(0)$ and $q(t_{fin})$ are initial and final poses of the calculated trajectory, while q_0 and q_{fin} are the desired initial and final poses.

In this very simple case a single primitive (i.e. $M = 1$) is already sufficient to solve (8.10) with zero error.

We start by imposing the zero error conditions $q_0 = q(0)$, $q_{\text{fin}} = q(t_{\text{fin}})$, which can be rewritten by using (8.8)

$$\begin{aligned} q_0 &= \bar{q} + S_0(0) + \alpha_1 \circ S_1(0) \\ q_{\text{fin}} &= \bar{q} + S_0(1) + \alpha_1 \circ S_1(1). \end{aligned} \tag{8.11}$$

We subtract the second equation from the first, and exploit the associativeness of Hadamard product, to obtain

$$\begin{aligned} q_{\text{fin}} - q_0 = {} & S_0(1) - S_0(0) + \alpha_1 \circ (S_1(1) - S_1(0)) \\ \Rightarrow \alpha_1 = {} & [(q_{\text{fin}} - q_0) - (S_0(1) - S_0(0))] \circ (S_1(1) - S_1(0))^{\circ -1} \end{aligned} \tag{8.12}$$

where $^{\circ-1}$ is the Hadamard inverse, as defined in [37]. Substituting (8.12) back into (8.11) yields

$$\begin{aligned} \bar{q} &= q_0 - \alpha_1 \circ S_1(0) \\ &= q_0 - [[(q_{\text{fin}} - q_0) - (S_0(1) - S_0(0))] \\ &\circ (S_1(1) - S_1(0))^{\circ -1}] \circ S_1(0), \end{aligned} \tag{8.13}$$

which concludes the analysis, fully specifying the trajectory.

8.1.3.2 Obstacle Avoidance

Let us consider the case in which we also desire to avoid one or more obstacles along the trajectory.

We thus generalize the generation of the human-like point-to-point motion as

$$\begin{aligned} \min_{\bar{q},\alpha_1,\ldots,\alpha_M} \quad & \left|\left|\begin{bmatrix} q(0) - q_0 \\ q(t_{\text{fin}}) - q_{\text{fin}} \end{bmatrix}\right|\right|_2^2 + \rho P(q, P_O) \\ \text{subject to} \quad & q(t) = \bar{q} + S_0\left(\frac{t}{t_{\text{fin}}}\right) + \sum_{i=1}^{M} \alpha_i \circ S_i\left(\frac{t}{t_{\text{fin}}}\right), \end{aligned} \tag{8.14}$$

in which we distinguish two terms. The first contribution guarantees that the desired initial and final poses are achieved, as for the free motion case (8.10). The second term takes into account the distance w.r.t. obstacles. For the sake of conciseness, and without any loss of generality, we assume here N_{O} spherical obstacles. We call $P_{\text{O}} = \{P_{\text{O}_1}, \ldots, P_{\text{O}_{N_{\text{O}}}}\}$ the set containing the Cartesian coordinates of all the centers of these obstacles.

$P(q, P_O)$ is a potential-based function [21] that sums up, for each obstacle, a term inversely proportional to the minimum distance between the obstacle and the closest joint trajectory, i.e.

$$P(q, P_O) = \sum_{i=1}^{N_O} \frac{1}{m_i(q([0, 1]), P_{\text{O}_i})^2} \tag{8.15}$$

where m_i is the distance between the arm and the $i-th$ obstacle, defined as

$m_i(q([0, 1]), P_{\text{O}_i}) = \min_k\{d(h_k(q([0, 1])), P_{\text{O}_i})\}$.

The distance between the $k-th$ point of contact with forward kinematics h_k, and the $i-th$ sphere is

$$d(h_k(q([0, 1])), P_{\text{O}_i}) = \max\left\{\min_{x \in h_k(q([0,1]))} ||P_{\text{O}_i} - x||_2, R_{\text{O}_i}\right\}, \tag{8.16}$$

with R_{O_i} radius of the sphere.

8.1.3.3 Incremental Optimization Procedure

The problem of motion generation with obstacle avoidance does not have a closed-form solution, hence the optimal trajectory is calculated via numerical optimization. To do this, we took inspiration by the ordering of fPCs basis, according to a descending amount of the associated explained variance, and implemented an incremental procedure reported in Algorithm 1.

The proposed approach, for each step, calculates the optimal trajectory that minimizes the error in starting and final position while maximizing the distance from the obstacles. We consider as initial condition the one specified by (8.12) and (8.13). If the corresponding solution is sufficiently far from the obstacles, this choice already defines the globally optimal solution. If the obstacles are nearer but not too close to the said trajectory, then solving (8.14) with $M = 1$ would fine tune the initial guess, achieving good results.

In case of obstacles close of even intercepting the free-motion trajectory, at least one more fPC should be enrolled to suitably solve the problem. The more are the basis elements enrolled, the more complex final trajectories can be implemented. However, a higher cardinality of enrolled elements produce larger computational cost and dramatically increase the number of local minima, thus increasing the complexity of the problem.

Algorithm 1 Procedure of Motion Generation

procedure INITIALIZATION
  Load fPCs basis set;
  Load Obstacles set;
  Set initial and final pose;
  $M \leftarrow 1$;
  $r_0 \leftarrow \infty$;
  $i \leftarrow 0$;
  Initialize α_{in} as in (8.12);
  Initialize $\bar{q}_{\text{in}}$ as in (8.13);
procedure OPTIMIZATION
  $i \leftarrow i + 1$;
  $[\alpha_{\text{opt}}, \bar{q}_{\text{opt}}] \leftarrow \text{InteriorPoint}(\alpha_{\text{in}}, \bar{q}_{\text{in}})$;
  Calculate residual cost r_i via Eq. (8.14);
  if $r_i - r_{i-1} \leq R_{\text{tol}}$ **or** $M = S_{\max}$ **then**
    Break;
  else
    $\alpha_{\text{in}} \leftarrow \alpha_{\text{opt}}$;
    $\bar{q}_{\text{in}} \leftarrow \bar{q}_{\text{opt}}$;
    $M \leftarrow M + 1$;
    Go to Step 10;
  end
end

8.1.4 *Simulations*

To prove the effectiveness of our approach, we performed simulations of both free-motion and obstacle-avoidance cases. We used a robotic arm simulator with the same kinematic description used for the experiments with human subjects. We report four representative motions that span the upper limb movement workspace: (T.1) is a movement from right to left side of the trunk, a typical action used to maneuver objects on a table; (T.2) represents a movement towards a position in front of the subject. (T.3) is a movement of arm elevation towards a position upon the head; (T.4) is a reaching movement toward the subject's own face, typical action for self-care tasks, e.g. self-feeding etc. Numerical values of desired initial and final conditions for each task are reported in Table 8.1.

In terms of obstacle distribution, we consider three scenarios for each task: zero obstacles (i.e. free motion), one obstacle, and two obstacles, placed along the optimal trajectories calculated in the free-motion case so to force the algorithm to modify the initial guess. We report he the case of a 90 mm radius sphere for T.1, a 50 mm radius sphere for T.2, a 70 mm radius sphere for T.3 and a 40 mm radius sphere for T.4.

8.1.4.1 Results

Figures 8.6, 8.7, 8.8, and 8.9 present graphical representations of the human-like trajectories generated by the proposed method, from T.1 to T.4. The first row of

Table 8.1 Initial (q_0) and final (q_{fin}) configurations for the four tasks considered. Angles are expressed in radians

Task		dof 1	dof 2	dof 3	dof 4	dof 5	dof 6	dof 7
T.1	q_0	−0.31	−0.58	−0.8	0.47	−1.23	0.22	0.65
	q_{fin}	−0.33	0.34	−0.39	0.44	−0.69	0.36	0.15
T.2	q_0	0.09	−0.79	−0.62	0.65	−1.50	0.06	0.76
	q_{fin}	0.65	−0.40	0.07	0.79	−1.32	0.87	0.63
T.3	q_0	−0.48	−0.07	−0.71	0.03	−0.86	0.52	0.47
	q_{fin}	−1.67	−0.68	−0.33	−0.21	−0.09	−0.26	0.52
T.4	q_0	−0.28	−0.86	−0.92	0.86	−1.52	−0.12	0.77
	q_{fin}	0.65	−0.78	−0.12	2.18	−1.52	0.57	0.59

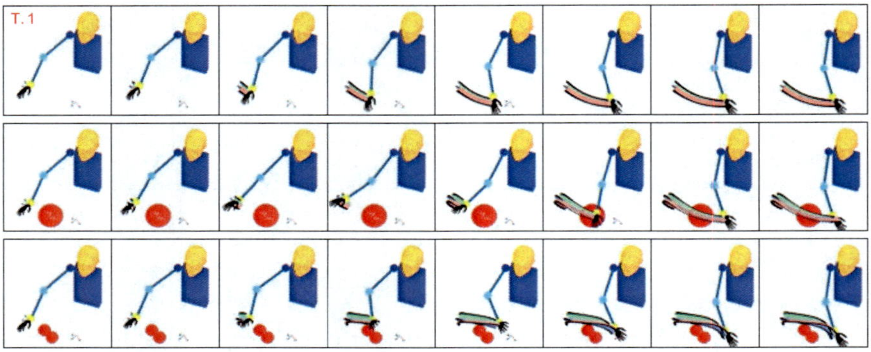

Fig. 8.6 From top to bottom, the reconstruction for task 1 is reported for the free motion case, the single and the double obstacle avoidance cases respectively. For each row, motion execution is intended from left to right, with time-frames evenly spaced. Reference system at wrist level is finally plotted along the whole trajectory

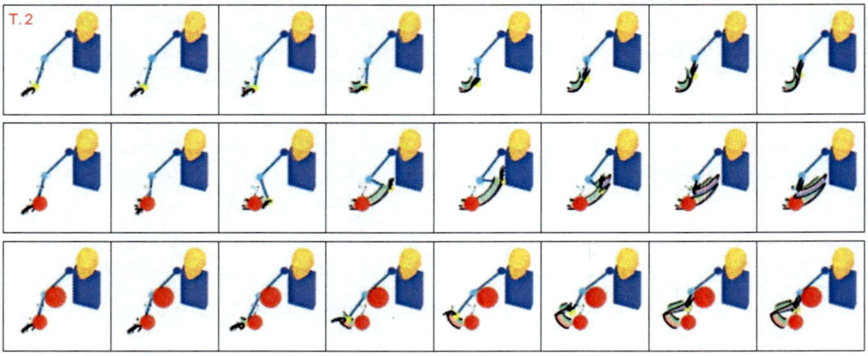

Fig. 8.7 From top to bottom, the reconstruction for task 2 is reported for the free motion case, the single and the double obstacle avoidance cases respectively. For each row, motion execution is intended from left to right, with time-frames evenly spaced. Reference system at wrist level is finally plotted along the whole trajectory

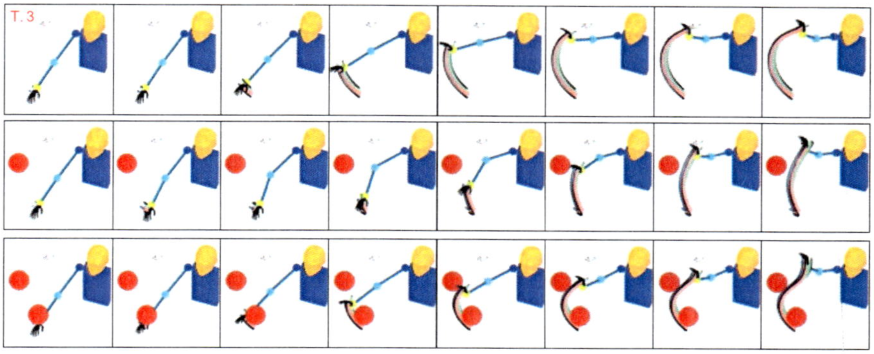

Fig. 8.8 From top to bottom, the reconstruction for task 3 is reported for the free motion case, the single and the double obstacle avoidance cases respectively. For each row, motion execution is intended from left to right, with time-frames evenly spaced. Reference system at wrist level is finally plotted along the whole trajectory

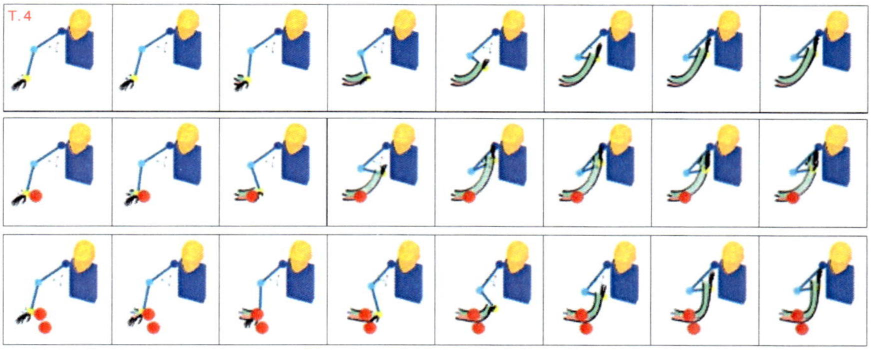

Fig. 8.9 From top to bottom, the reconstruction for task 4 is reported for the free motion case, the single and the double obstacle avoidance cases respectively. For each row, motion execution is intended from left to right, with time-frames evenly spaced. Reference system at wrist level is finally plotted along the whole trajectory

each figure depicts the free motion case, the second row the one-obstacle case, and the third the two-obstacles one. In all the considered cases the result is a smooth trajectory connecting the two postures without interacting with the obstacles. This is also evident from Fig. 8.11, where we presents the profiles of angles and angular velocities in time.

Table 8.2 reports, for each task and for each environment, the amount of functional Principal Components retrieved by algorithm 1. As expected, in the zero-obstacle case all tasks are executed with only one primitive (cfr. subsection 8.1.3.1). When obstacles are introduced, the number M of primitives used increases. However, the complex strategies that we can observe in Figs. 8.6, 8.7, 8.8, and 8.9 are produced with a maximum of three primitives required for T.1. In the considered simulations,

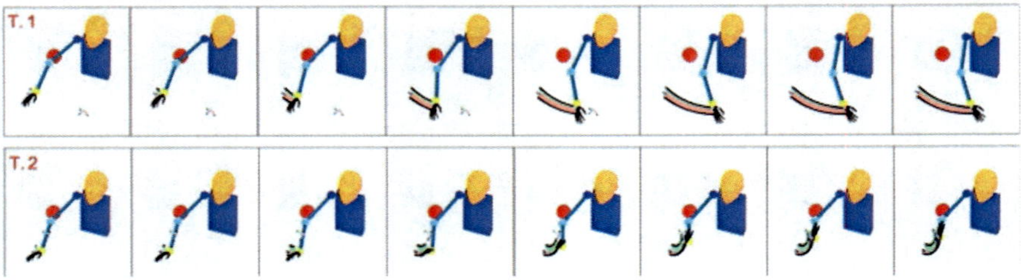

Fig. 8.10 From top to bottom, the reconstruction for task 1 and 2 are reported in case of obstacles along the elbow trajectory. For each row, motion execution is intended from left to right, with time-frames evenly spaced. Reference system at wrist level is finally plotted along the whole trajectory

Table 8.2 # of primitives (M) enrolled by Algorithm 1

Number of obstacles	Task 1	Task 2	Task 3	Task 4
0	1	1	1	1
1	2	2	2	2
2	3	2	2	2

the first primitive alone accounts for the 96% of the cost reduction, while with only the first three primitives the total reduction is achieved. This is in accordance with Table 8.2, confirming the effectiveness of the proposed strategies for selecting the number of primitives to be used. Additional experiments have been performed to show how the algorithm works in case of obstacles close to the elbow trajectory. We reported in Fig. 8.10 results for tasks 1 and 2. The interested reader is invited to refer to the attached video for different points of view.

8.1.4.2 Human-Likeliness of Trajectories

The method proposed in this chapter embeds human-like profiles for upper-limb movements without explicitly constraint this behavior in the optimization problem. Human movements are commonly characterized by the following principal characteristics: (i) bell-shape velocity profiles [46]; (ii) minimization of jerk values [30, 46]. For this reason, in this chapter we leverage on these two characteristics to verify the human likeliness of our results. Regarding the velocity profiles, in Fig. 8.11, we report the angular profiles and the corresponding velocities, for all the considered simulations. The bell-shaped distributions are effectively embedded in the considered evolutions, not only in the case of point-to-point free motion but also while avoiding contact with one or more obstacles. Moreover, the norm of the trajectories jerk is always lower than $5\ 10^{-7}$rad ($7.92\ 10^{-8} \pm 8.53\ 10^{-8}$rad), in line with the observations of [17]. Finally, all the evolutions of angular velocity present a marked

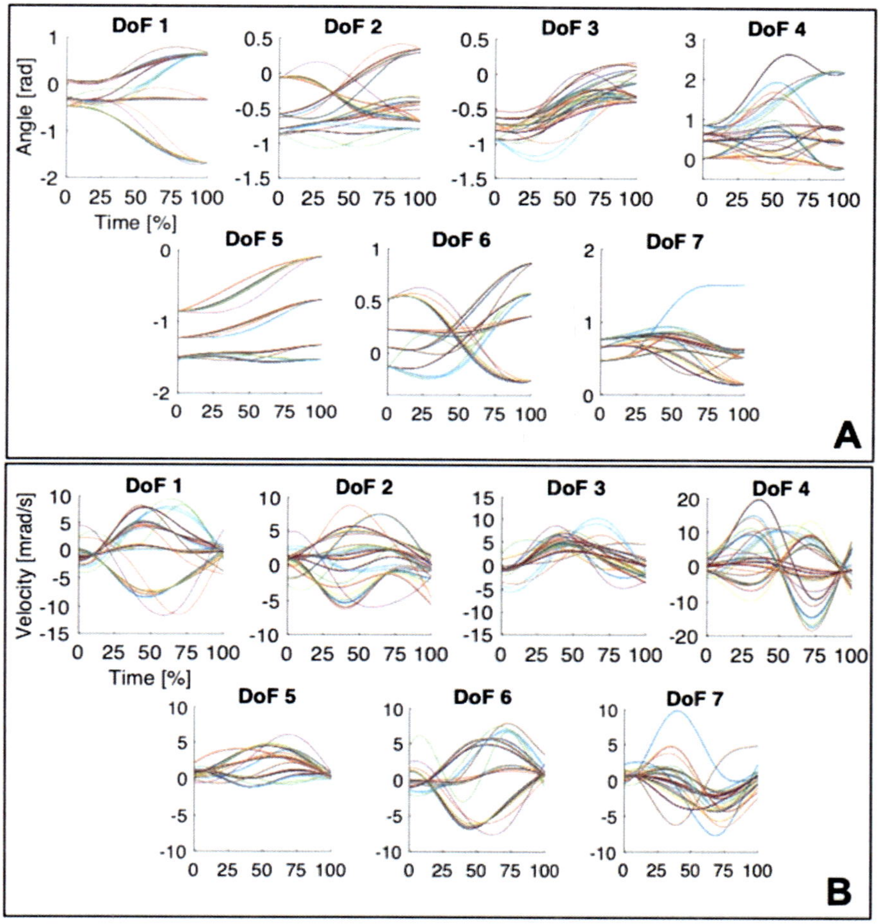

Fig. 8.11 Angular positions and velocities for a set of simulations performed on movements T.1, T.2, T.3 and T.4 with different obstacles. On subplot **a**, we report the angular positions for the seven DoFs, arranged following the order of Fig. 8.4, while on subplot **b** we show the corresponding angular velocities

bell-shaped profile, as observed in [30]. Few exceptions can be evidenced mostly for the elbow joint (dof 4), which however can be seen as a sequence of two bell shaped trajectories, one for the opening and one for the closing.

8.1.5 Discussion and Conclusions

In this section, we presented a novel method to endow robotic manipulators with human-like movements by directly using human principal motion modes identified through functional analysis. In this way, human-likeness is intrinsically guaranteed.

This leaves room to optimize other aspects related to motion generation, such as obstacle avoidance—as done in this chapter—energetic consumption, joint torques, time efficiency and so on. It is worth noticing that the aim of our work is not to solve planning problems but instead to propose a new method for intrinsically embedding human-likeness in motion generation. Future work will encompass the evaluation with sophisticated planning problems and state of the art planning benchmarks.

In the general case, the resulting optimization problem deals with a non-linear cost function which may present local minima. The current implementation relies on the `MATLAB` optimization routine `fmincon`, which implements a gradient-based local optimization. This does not necessarily guarantee the convergence to the global optimum. The use of global optimization algorithms will be considered as future work. Moreover, any comparison with other existing algorithms in terms of execution time is beyond the scope of this chapter. Nevertheless, we do believe that the proposed method could also enable a strong reduction of the computational costs w.r.t. state of the art algorithms, since the optimization space is constrained to a reduced set of parameters (i.e. the principal functions weights) instead of the whole evolution of the robot pose. Future work will focus on more efficient implementation of the proposed algorithm.

Note that the method is specifically tailored for systems having the same kinematic structure of human upper limb. However, the outcomes of this work can be generalized to different kinematic descriptions—and this mapping problem is discussed in the next section.

The methodology presented in this chapter is designed to easily and uniquely generate human-like trajectories. We do believe that this work may have a major impact for the control of active devices during the interaction with humans. This is particularly relevant for the applications in which the robot is used as an instrument for human assistance or rehabilitation. Indeed, human-like robotic platforms that interact with humans, for example in self-feed or elderly care tasks, may benefit in acceptability and predictability of the movements. Applications for rehabilitation devices are also envisioned. Our strategy could be used to suggest preferred movements to rehabilitation exoskeletons, in conjunction with an analogous modulation of the robot impedance. Indeed, our future works will investigate the extension of our approach also to the modulation of impedance parameters.

8.2 A Control-Based Approach to Motion Mapping

8.2.1 Introduction

The method introduced in this Chapter faces important translational changes, when it comes to be applied to robotic anthropomorphic manipulators, whose kinematic structure is dissimilar with respect to the model used for extracting human principal modes (Fig. 8.12).

In this Section, we tackle the problem of mapping these modes on a generic anthropomorphic robotic manipulator. This problem is not trivial, since it is not possible to rely on joint-based one-by-one mapping and, at the same time, it is necessary to guarantee that the mapped trajectories still maintain the human-likeliness. Classic mapping algorithms implement optimization-based methods to maximize the similarity between the human and robot kinematic chain motions, see e.g. [19, 27]. These methods can map human movements onto robotic structures, but these can be hardly used on-line. Here we propose a control-based approach that allows to implement onto a redundant robotic arm with a given kinematic architecture the trajectories obtained using the functional approach reported in [5], overcoming the limitations of the optimization methods described above. This approach relies on two Cartesian impedance controllers, which also allow to manage the redundancy problem:

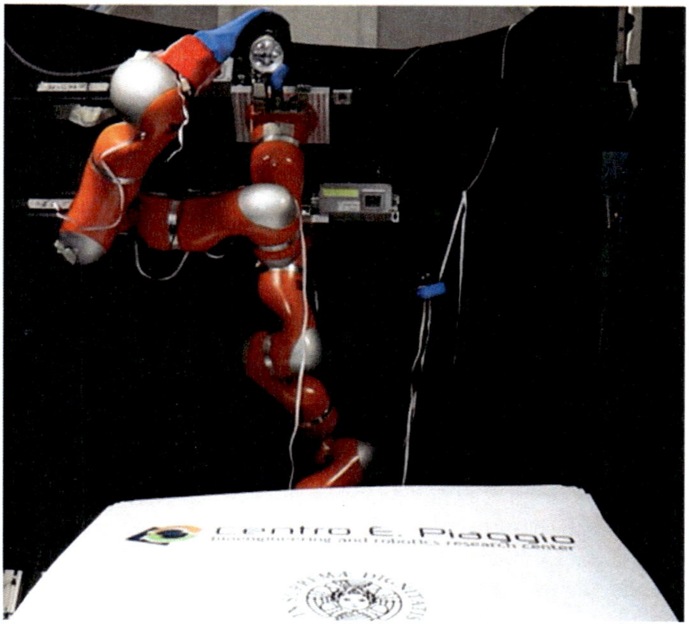

Fig. 8.12 A Kuka LWR+ 7 DoFs robot "drinking" water from a bottle

the first controller ensures that the robot end effector trajectories follow the ones of the human end-effector, while the second controller implemented at the elbow level projects in the null-space of the end-effector Jacobian. This implementation enables to keep the robot elbow as close as possible to the reference-human elbow trajectory—which is crucial for increasing HL [28, 44]—with no influence on the trajectory tracking at the end-effector level—controlled by the first control law. We report the details of this approach and show the implementation on an anthropomorphic robotic manipulator KUKA LWR+ platform, whose kinematic architecture is different from the kinematic model used to extract human functional modes. We also discuss the experimental outcomes in terms of HL as well as the implications for collaborative robotics.

A companion video is available at the following link: https://www.youtube.com/watch?v=FHOsVVXBV74.

8.2.2 *Mapping Between Kinematics via Impedance Control*

The framework introduced in the previous section enables the efficient and effective generation of human-inspired movements. However, the generated trajectory is defined in the joint space and, thus, is strictly related to the specific manipulator kinematic definition. In other terms, it can work only for robots with the same kinematic architecture as the one used for describing human motions. In the following, we will refer to this architecture as to the *paradigmatic human arm*. To overcome this limitation and generalize to any kinematic chain the approach reported in Algorithm 1, a mapping problem needs to be solved. This has been usually achieved in literature via optimization, minimizing the distance between the robot and the paradigmatic human arm: i.e. given the definition of an end effector (EE), a shoulder and an elbow—which can be easily identified in an anthropomorphic architecture [27], the goal is to minimize a cost function proportional to the sum of the area of the triangles between these points (for additional details the interested reader may refer to [27]). Another approach, namely Virtual Object Method introduced in [19], maps postures of hands with dissimilar kinematics, considering reference points on the two kinematic structures, and then mapping the movement of these reference points in the velocity domain. However, to the best of our knowledge, all the methods presented so far rely on time consuming optimization procedures, which need to be executed offline. To overcome this issue, we proposed to map the movements using a different control-based approach. Since the mapping must preserve the main HL characteristics of the original planned movement, i.e. bell shaped velocity profiles, low jerk values etc., the control law needs to be transparent w.r.t. the motion dynamic characteristics. To this aim, we used a Cartesian Impedance Control (IC) [20]. More specifically, given the trajectory of the planned motion as it results from Alg. 1 and computed via forward-kinematics at the end-effector level, the following control law is adopted:

$$\tau = M(q)\ddot{q} + c(q, \dot{q}) + J_1(q)^T (K_1 e_1 + D_1 \dot{e}_1), \tag{8.17}$$

where τ represents the joint torques, $M(q)$ is the inertia matrix of the manipulator, $c(q, \dot{q})$ takes into account the Coriolis term and the gravity of the manipulator, $e_1 = x^o - x$ is the Cartesian pose error at the EE, $\dot{e}_1$ is the error time derivative, and $J_1(q)$ is the Jacobian for the EE task. K_1 and D_1 are positive definite gain matrices.

In addition, since the majority of the anthropomorphic manipulators is redundant, i.e. the number of controllable joints is higher than the dimensionality of the end-effector space, an additional Cartesian IC at the elbow level is introduced.

Implementing a control at the elbow is important not only to solve the redundacy problem [9, 24, 25] but also for increasing motion functional anthropomorphis. Indeed, it was proven that making the pose of the elbow of the slave robot as close as possible to the pose of the paradigmatic human arm enhances the HL of the commanded trajectories. Another aspect that increases functional anthropomorphism is the alignment of the robot end-effector with the paradigmatic human arm EE [27]. Note that, since the references of the end-effector and the elbow controllers are different, this may result in steady state error at the EE w.r.t. the reference trajectory. To have an exact mapping, i.e. zero tracking error, a null-space projection of torques produced by the second impedance controller is hence needed [9, 24, 25]. Our implementation of the second Cartesian IC can be achieved trough the following control law:

$$\nu = J_2(q)^T \big(K_2 e_2 + D_2 \dot{e}_2\big), \tag{8.18}$$

where $J_2(q)$ is the elbow Jacobian, e_2 is the Cartesian position error (no orientation is considered) for the elbow, $\dot{e}_2$ its time derivative and K_2, D_2 positive definite gain matrices. This control action projects in the null space of the robotic manipulator EE. This is achieved with the projector $P_1(q) = I - J_1(q)^T J_1(q)$.

The total control strategy can be hence written as:

$$\tau = M(q) + c(q, \dot{q}) + J_1(q)^T (K_1 e_1 + D_1 \dot{e}_1) + P_1(q) J_2(q)^T \nu. \tag{8.19}$$

8.2.3 Experiments

To prove the effectiveness of our planning and mapping framework, we carried out a set of experiments on a real platform, i.e. a 7° of Freedom (DoFs) Kuka LWR+. The main difference between the paradigmatic human arm (see [4]) and the Kuka kinematics is that, in the latter case, the joint defining the shoulder, elbow and wrist degrees of freedom are not coincident. Note that our implementation can be applied to any robotic manipulator with redundant kinematic architecture (see Fig. 8.13).

With reference to the first part of this Chapter, we identified four different sample movements that span the upper limb motion workspace, referred a first person view: (T.1) is a translation from right to left side of the body in the transverse plane, a typical movement used to manipulate objects on a table; (T.2) is a movement towards a pose in front of the robot, in the sagittal plane; (T.3) is an arm elevation mimicking a human motion towards a point upon the head; (T.4) is a reaching movement reproducing a

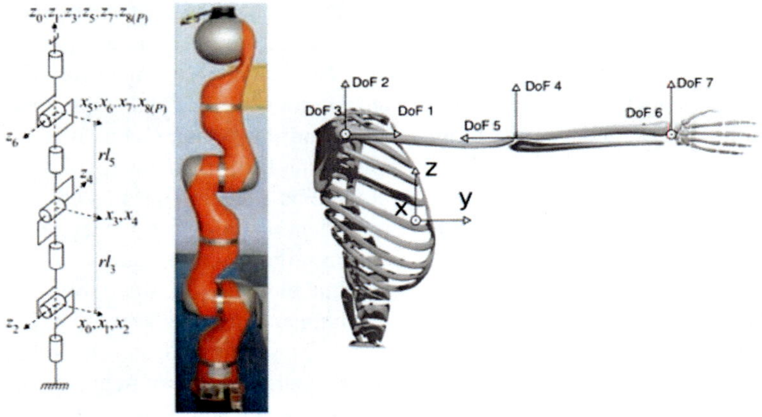

Fig. 8.13 Right plot. Kinematic model of the paradigmatic human arm. Left plot. Kinematic model of the Kuka LWR antrophomorphic manipulator. Figure partially adapted from [5, 22]

human motion towards the face, e.g. for self-feeding tasks. Numerical values of the initial and final joint configurations on the paradigmatic human arm kinematics are listed in Table 8.3.

Each task was planned in different conditions, i.e. with and without the presence of obstacles along the trajectory. The trajectories are then mapped on the Kuka kinematics in two different configurations: (i) anthropomorphic configuration; (ii) industrial configuration (see Fig. 8.14). Indeed, also in the latter case it is important to ensure HL to favour robot action predictability by the human operator [43]. Reference end-effector position defined on the human kinematics is scaled by a factor L_r/L_h, where L_r and L_h are the robot and human arm total arm length, respectively, when all the joints are set to zero, i.e. total arm extension.

Table 8.3 Initial (q_0) and final (q_{fin}) joint posture for the four tasks. Angles are expressed in radiants and defined for the paradigmatic human arm kinematics.

Task		dof 1	dof 2	dof 3	dof 4	dof 5	dof 6	dof 7
T.1	q_0	−0.31	−0.58	−0.8	0.47	−1.23	0.22	0.65
	q_{fin}	−0.33	0.34	−0.39	0.44	−0.69	0.36	0.15
T.2	q_0	0.09	−0.79	−0.62	0.65	−1.50	0.06	0.76
	q_{fin}	0.65	−0.40	0.07	0.79	−1.32	0.87	0.63
T.3	q_0	−0.48	−0.07	−0.71	0.03	−0.86	0.52	0.47
	q_{fin}	−1.67	−0.68	−0.33	−0.21	−0.09	−0.26	0.52
T.4	q_0	−0.28	−0.86	−0.92	0.86	−1.52	−0.12	0.77
	q_{fin}	0.65	−0.78	−0.12	2.18	−1.52	0.57	0.59

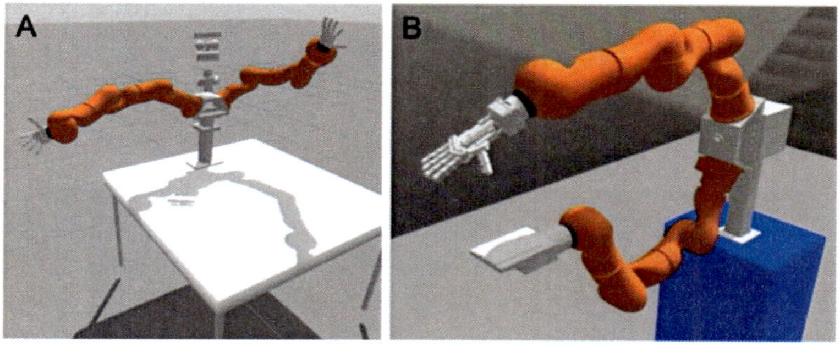

Fig. 8.14 Two different Kuka configurations. **a**, robot in anthropomorphic dual arm configuration. **b**, robot in industrial configuration, typically used to pick and maneuver objects from pallets. Note that in both cases only one arm (the right arm) is used

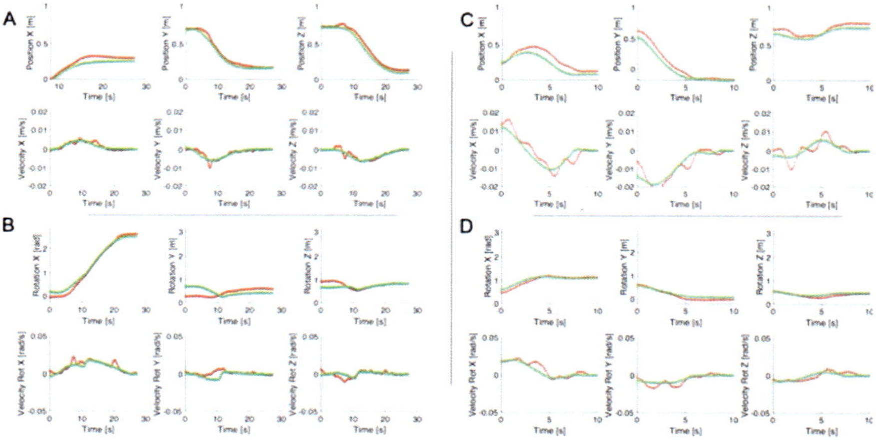

Fig. 8.15 Mapped motions. **Subplot A** Position (first row) and velocity (second row) of the reference (green line) and the real (red line) end-effector position for task T.1 with two obstacles. **Subplot B** Position (first row) and velocity (second row) of the reference (green line) and the real (red line) end-effector orientation for task T.1 with two obstacles. **Subplot C** Position (first row) and velocity (second row) of the reference (green line) and the real (red line) end-effector position for task T.4 with no obstacle. **Subplot D** Position (first row) and velocity (second row) of the reference (green line) and the real (red line) end-effector otientation for task T.4 with no obstacle

In Fig. 8.15 we show the end-effector commanded and real robotic EE pose in two sample cases. More specifically, we report the motion in the transverse plane (T.1) with two obstacles on the table (20×30 cm cylinder—height equal to 20; and $20 \times 30 \times 40$ cm paralleliped—height equal to 40, at a center-to-center distance of approx. 10 cm), and one movement in the medial plane (T.4) for the free-movement case (see also Fig. 8.16). Results show a good approximation of the commanded end-effector motion, with a mean error between the reference and real EE position lower than $5 * 10^{-2}$ m, and an EE average orientation error lower than 10^{-1}. All the other tested

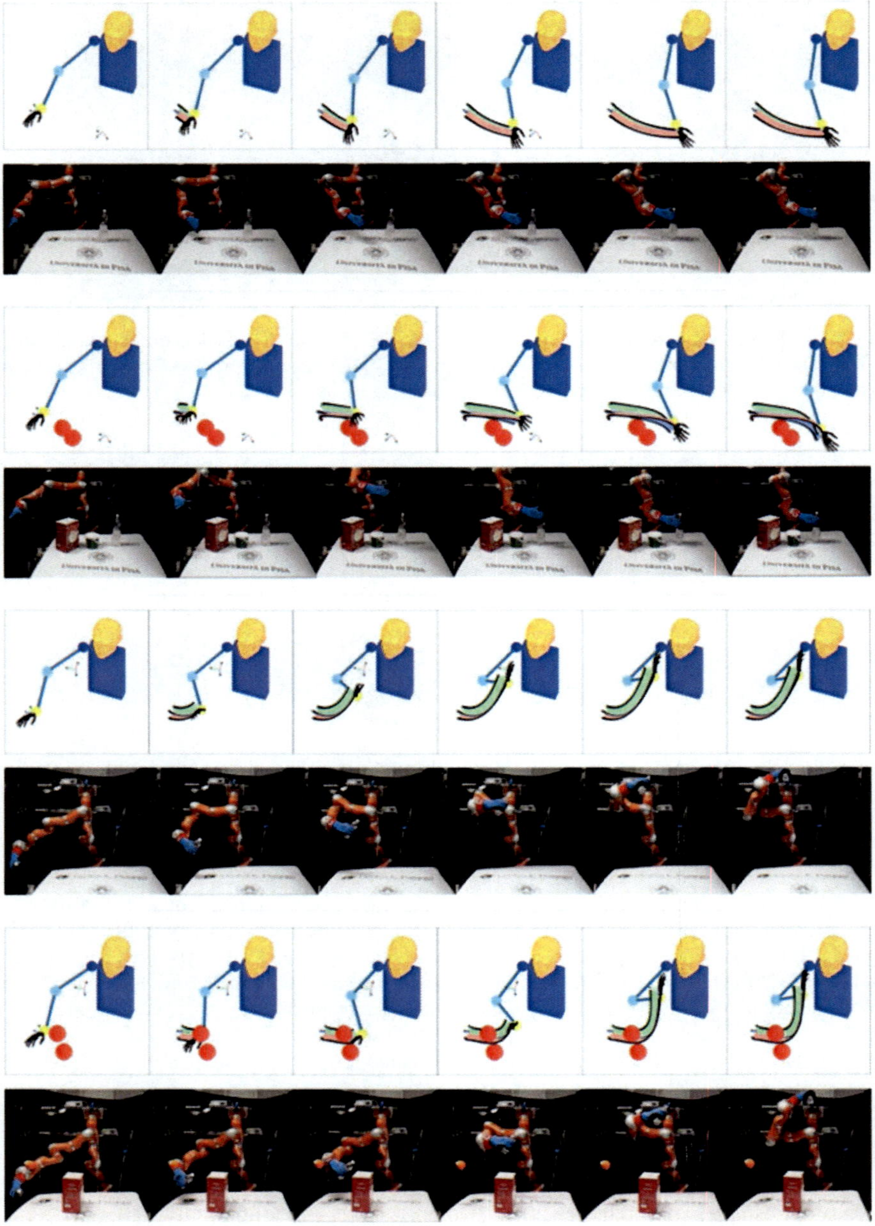

Fig. 8.16 From top to bottom, snapshots of human reference and real robot execution of T.1 free motion, T.1 with two obstacles; T.4 free motion; T.4 with two obstacles. First controller param: $K_x = 100$; $K_y = 100$; $K_z = 100$; $K_{rx} = 1.5$; $K_{ry} = 1.5$; $K_{rz} = 1.5$; $D_x = 2.0$; $D_y = 2.0$; $D_z = 2.0$; $D_{rx} = 1.0$; $D_{ry} = 1.0$; $D_{rz} = 1.0$. Second controller param: $K_x = 5.0$; $K_y = 5.0$; $K_z = 5.0$; $K_{rx} = 0$; $K_{ry} = 0$; $K_{rz} = 0$; $D_x = 0$; $D_y = 0$; $D_z = 0$; $D_{rx} = 0$; $D_{ry} = 0$; $D_{rz} = 0$

movements presented similar performances, details are not reported here for sake of space. The corresponding velocities are depicted in Fig. 8.15. Notwithstanding the presence of oscillations due to vibrations of the manipulator not related to our control law, the robot movement shows bell-shaped velocity profiles with jerk values always lower than 10^{-2}, in line with the observations in [16].

The overall movement is depicted in Fig. 8.16 for two sample actions (T.1 and T.4), i.e. for the free motion and the two-obstacle case.

8.2.4 Discussions and Conclusions

This Chapter contributes with a novel technical framework for human-like motion generation for anthropomorphic robots. The goal is achieved through two main building blocks. First, we discussed a motion generation algorithm based on the optimal combination of human data-driven functional modes [4]. This method was preliminarily theorized in [5] and is here extended for its usability with generic redundant anthropomorphic manipulators, whose kinematic architecture differs from the one used to extract human functional modes. This is achieved though a mapping approach that exploits torque control of the manipulator. More specifically, we implemented a Cartesian impedance control at the end-effector level. Functional anthropomorphism of the overall motion is preserved via a second Cartesian impedance controller at the elbow level. This applies an additional torque in the null-space of the end-effector Cartesian controller to keep the robot elbow as close as possible to the reference of the human elbow. The reference pose commanded is the movement planned on the human kinematics, scaled by a proportionality factor. Our experiments show that such control strategy is able to on-line map the desired trajectory of the paradigmatic human arm at the robot end-effector level. Robot trajectories appear to maintain the basic characteristics of human-likeliness of the commanded poses, i.e. low jerk values, bell-shaped velocity profile. Note that—due to kinematic constraints—the second controller is not meant to achieve zero error, while the end-effector is expected to follow with a good approximation the desired trajectory (position and orientation).

In the future developments, the approach described in this chapter will be tested with other manipulators and configurations and will be integrated in a more general framework together with a vision based target and obstacles identifier. It is important to underline here that the aim of this chapter is not to solve a planning problem per-se, but rather to propose a new method for intrinsically embedding human-likeness in motion generation. Future work will encompass the evaluation with sophisticated planning problems and state of the art planning benchmarks. The outcomes of this chapter could successfully inform the next developments for the control of redundant anthropomorphic co-bots in industrial settings and for human assistance. Indeed, thanks to the human-likeness of the robot trajectories, the robot interaction with the humans in collaborative tasks, e.g. for synergistic action completion, or during robot-assisted actions, e.g. self feeding, will be performed in a predictable, ergonomic and safe manner. This will finally and likely increase the acceptability of robotic technology by the human user. Under this regard, the analysis of the ergonomy of humans interacting with robotic manipulators implementing the control strategies reported in this chapter (see e.g. [35]), and the evaluation of the perceived human-likeliness via experiments with human participants are already envisioned.

References

1. Arimoto, S., Sekimoto, M., Hashiguchi, H., Ozawa, R.: Natural resolution of ill-posedness of inverse kinematics for redundant robots: A challenge to bernstein's degrees-of-freedom problem. Adv. Robot. **19**(4), 401–434 (2005)
2. Averta, G., Angelini, F., Bonilla, M., Bianchi, M., Bicchi, A.: Incrementality and hierarchies in the enrollment of multiple synergies for grasp planning. IEEE Robot. Autom. Lett. **3**(3), 2686–2693 (2018)
3. Averta, G., Caporale, D., Della Santina, C., Bicchi, A., Bianchi, M.: A technical framework for human-like motion generation with autonomous anthropomorphic redundant manipulators. In: 2020 IEEE International Conference on In Robotics and Automation (ICRA), Paris. IEEE (2020)
4. Averta, G., Della Santina, C., Battaglia, E., Felici, F., Bianchi, M., Bicchi, A.: Unvealing the principal modes of human upper limb movements through functional analysis. Front. Robot. AI **4**, 37 (2017)
5. Averta, G., Della Santina, C., Valenza, G., Bianchi, M., Bicchi, A.: Exploiting upper-limb functional synergies for human-like motion generation of anthropomorphic robots. J. Neuro-Engineering Rehabil.
6. Bartneck, C., Kulić, D., Croft, E., Zoghbi, S.: Measurement instruments for the anthropomorphism, animacy, likeability, perceived intelligence, and perceived safety of robots. Int. J. Soc. Robot. **1**(1), 71–81 (2009)
7. Breteler, M.D.K., Gielen, S.C., Meulenbroek, R.G.: End-point constraints in aiming movements: effects of approach angle and speed. Biol. Cybern. **85**(1), 65–75 (2001)
8. Ciocarlie, M.T., Allen, P.K.: Hand posture subspaces for dexterous robotic grasping. Int. J. Robot. Res. **28**(7), 851–867 (2009)
9. Dietrich, A., Ott, C., Albu-Schäffer, A.: An overview of null space projections for redundant, torque-controlled robots. Int. J. Robot. Res. **34**(11), 1385–1400 (2015)
10. Duffy, B.R.: Anthropomorphism and the social robot. Robot. Auton. Syst. **42**(3), 177–190 (2003)
11. Edsinger, A., Kemp, C.C.: Two arms are better than one: A behavior based control system for assistive bimanual manipulation. In: Recent Progress in Robotics: Viable Robotic Service to Human, pp. 345–355. Springer (2007)
12. Elbanhawi, M., Simic, M.: Sampling-based robot motion planning: a review. IEEE Access **2**, 56–77 (2014)
13. Ficuciello, F.: Synergy-based control of underactuated anthropomorphic hands. IEEE Trans. Ind. Inform. (2018)
14. Figueroa Fernandez, N.B., Billard, A.: Modeling compositions of impedance-based primitives via dynamical systems. In: In Proceedings of the Workshop on Cognitive Whole-Body Control for Compliant Robot Manipulation (COWB-COMP), CONF (2018)
15. Fink, J.: Anthropomorphism and human likeness in the design of robots and human-robot interaction. In: International Conference on Social Robotics, pp. 199–208. Springer (2012)
16. Flash, T.: The control of hand equilibrium trajectories in multi-joint arm movements. Biol. Cybern. **57**(4–5), 257–274 (1987)
17. Flash, T., Hogan, N.: The coordination of arm movements: an experimentally confirmed mathematical model. J. Neurosci. **5**(7), 1688–1703 (1985)

18. García, N., Suárez, R., Rosell, J.: Hg-rrt*: Human-guided optimal random trees for motion planning. In: 2015 IEEE 20th Conference on Emerging Technologies & Factory Automation (ETFA), pp. 1–7. IEEE (2015)
19. Gioioso, G., Salvietti, G., Malvezzi, M., Prattichizzo, D.: Mapping synergies from human to robotic hands with dissimilar kinematics: an approach in the object domain. IEEE Trans. Rob. **29**(4), 825–837 (2013)
20. Hogan, N.: Impedance control-an approach to manipulation. i-theory. ii-implementation. iii-applications. ASME Trans. J. Dyn. Syst. Meas. Control. B **107**, 1–24 (1985)
21. Hwang, Y.K., Ahuja, N.: A potential field approach to path planning. IEEE Trans. Robot. Autom. **8**(1), 23–32 (1992)
22. Jubien, A., Gautier, M., Janot, A.: Dynamic identification of the Kuka LWR robot using motor torques and joint torque sensors data. IFAC Proc. Vol. **47**(3), 8391–8396 (2014)
23. Karaman, S., Frazzoli, E.: Sampling-based algorithms for optimal motion planning. Int. J. Robot. Res. **30**(7), 846–894 (2011)
24. Khatib, O.: A unified approach for motion and force control of robot manipulators: the operational space formulation. IEEE J. Robot. Autom. **3**(1), 43–53 (1987)
25. Khatib, O.: Inertial properties in robotic manipulation: an object-level framework. Int. J. Robot. Res. **14**(1), 19–36 (1995)
26. Kuffner, J.J., LaValle, S.M.: Rrt-connect: an efficient approach to single-query path planning. In: Proceedings. ICRA'00. IEEE International Conference on Robotics and Automation, 2000, vol. 2, pp. 995–1001. IEEE (2000)
27. Liarokapis, M., Bechlioulis, C.P., Artemiadis, P.K., Kyriakopoulos, K.J.: Deriving humanlike arm hand system poses. J. Mech. Robot. **9**(1), 011012 (2017)
28. Liarokapis, M.V., Artemiadis, P.K., Kyriakopoulos, K.J.: Functional anthropomorphism for human to robot motion mapping. In: 2012 IEEE RO-MAN: The 21st IEEE International Symposium on Robot and Human Interactive Communication, pp. 31–36. IEEE (2012)
29. Miossec, S., Kheddar, A.: Human motion in cooperative tasks: Moving object case study. In: IEEE International Conference on Robotics and Biomimetics, 2008. ROBIO 2008, pp. 1509–1514. IEEE (2009)
30. Morasso, P.: Spatial control of arm movements. Exp. Brain Res. **42**(2), 223–227 (1981)
31. Mordatch, I., Lowrey, K., Andrew, G., Popovic, Z., Todorov, E.V.: Interactive control of diverse complex characters with neural networks. In: Advances in Neural Information Processing Systems, pp. 3132–3140 (2015)
32. Mordatch, I., Wang, J.M., Todorov, E., Koltun, V.: Animating human lower limbs using contact-invariant optimization. ACM Trans. Graph. (TOG) **32**(6), 203 (2013)
33. Mukai, T., Hirano, S., Nakashima, H., Sakaida, Y., Guo, S.: Realization and safety measures of patient transfer by nursing-care assistant robot Riba with tactile sensors. J. Robot. Mechatron. **23**(3), 360–369 (2011)
34. Peng, X.B., Berseth, G., Van de Panne, M.: Terrain-adaptive locomotion skills using deep reinforcement learning. ACM Trans. Graph. (TOG) **35**(4), 81 (2016)
35. Peternel, L., Fang, C., Tsagarakis, N., Ajoudani, A.: A selective muscle fatigue management approach to ergonomic human-robot co-manipulation. Robot. Comput.-Integr. Manuf. **58**, 69–79 (2019)
36. Piazzi, A., Visioli, A.: Global minimum-jerk trajectory planning of robot manipulators. IEEE Trans. Ind. Electron. **47**(1), 140–149 (2000)
37. Reams, R.: Hadamard inverses, square roots and products of almost semidefinite matrices. Linear Algebr. Appl. **288**, 35–43 (1999)
38. Riek, L.D., Rabinowitch, T.C., Chakrabarti, B., Robinson, P.: How anthropomorphism affects empathy toward robots. In: Proceedings of the 4th ACM/IEEE International Conference on Human Robot Interaction, pp. 245–246. ACM (2009)
39. Santello, M., Flanders, M., Soechting, J.F.: Postural hand synergies for tool use. J. Neurosci. **18**(23), 10105–10115 (1998)
40. Schulman, J., Ho, J., Lee, A.X., Awwal, I., Bradlow, H., Abbeel, P.: Finding locally optimal, collision-free trajectories with sequential convex optimization. In: Robotics: Science and Systems, vol. 9, pp. 1–10. Citeseer (2013)

41. Stilman, M.: Global manipulation planning in robot joint space with task constraints. IEEE Trans. Rob. **26**(3), 576–584 (2010)
42. Welch, G., Bishop, G., et al.: An introduction to the Kalman filter (1995)
43. Zanchettin, A.M., Bascetta, L., Rocco, P.: Acceptability of robotic manipulators in shared working environments through human-like redundancy resolution. Appl. Ergon. **44**(6), 982–989 (2013)
44. Zanchettin, A.M., Bascetta, L., Rocco, P.: Achieving humanlike motion: resolving redundancy for anthropomorphic industrial manipulators. IEEE Robot. Autom. Mag. **20**(4), 131–138 (2013)
45. Zhang, Z., Li, Z., Zhang, Y., Luo, Y., Li, Y.: Neural-dynamic-method-based dual-arm CMG scheme with time-varying constraints applied to humanoid robots. IEEE Trans. Neural Netw. Learning Syst. **26**(12), 3251–3262 (2015)
46. von Zitzewitz, J., Boesch, P.M., Wolf, P., Riener, R.: Quantifying the human likeness of a humanoid robot. Int. J. Soc. Robot. **5**(2), 263–276 (2013)
47. Zucker, M., Ratliff, N., Dragan, A.D., Pivtoraiko, M., Klingensmith, M., Dellin, C.M., Bagnell, J.A., Srinivasa, S.S.: Chomp: covariant hamiltonian optimization for motion planning. Int. J. Robot. Res. **32**(9–10), 1164–1193 (2013)

Chapter 9
A Focus on Motion Dynamics: Planning Impedance Behaviors in Physical Interaction

Physical interaction of robots with their environment is a challenging problem because of the exchanged forces. Hybrid position/force control schemes often exhibit problems during the contact phase, whereas impedance control appears to be more simple and reliable, especially when impedance is shaped to be energetically passive. Even if recent technologies enable shaping the impedance of a robot, how best to plan impedance parameters for task execution remains an open question. In this chapter, we present an optimization-based approach to plan not only the robot motion but also its desired end-effector mechanical impedance. We show how our methodology is able to take into account the transition from free motion to a contact condition, typical of physical interaction tasks. Results are presented for planar and three-dimensional open-chain manipulator arms. The compositionality of mechanical impedance is exploited to deal with kinematic redundancy and multi-arm manipulation.

9.1 Introduction

Recent frontiers of robotics research have encountered a major obstacle in physical human-robot interaction. Contact with the environment requires fine control not only of robot position but also of the wrench exerted. Classical approaches such as position control typically fail to modulate force effectively, while force control is effective only during contact. For this reason, several studies focused on methodologies able to control in both domains. Proposed solutions to this problem can be categorized loosely into two broad classes: (i) Active interaction control; (ii) Passive interaction control. The first group is essentially a 'hybrid' combination of position and force control [36]. The reader may refer to [33, 38, 39] (not a complete list). These approaches typically require end-effector force sensors for their implementa-

G. Averta, *Human-Aware Robotics: Modeling Human Motor Skills for the Design, Planning and Control of a New Generation of Robotic Devices*, Springer Tracts in Advanced Robotics 145, https://doi.org/10.1007/978-3-030-92521-5_9

tion, even if some techniques are available for estimating force [23, 35]. Moreover, force control may result in unstable behavior during contact with external objects [3, 12, 30, 37]. The second group, instead, implement energetically passive control of force which is not affected by typical force-control problems. Among the second group, one of the simplest effective strategies is impedance control [19]. The idea behind the impedance control, originally presented as a general control framework for manipulators in free or constrained motion, with and without dynamic interaction with the environment, is conceptually simple. The goal of the controller is to produce and maintain a desired impedance, i.e. force as a function of the error in position and its derivatives. This concept appears to be effective since it manages contact with external objects in the environment without explicitly programming the transition between free motion and contact, yet preserves tracking of a reference trajectory during free movements. Several applications have benefited from impedance control; the interested reader may refer for example to [1, 2, 8, 18, 20, 28, 34].

One characteristic of the impedance control implementation is an increase in the number of parameters to be specified in the controller. That is, in addition to the reference trajectory, the designer is required to tune the robot stiffness, damping and (ideally) parameters related to higher-order derivatives (e.g. inertia). How best to modulate impedance parameters is still an open question. Typical implementations pre-tune these parameters depending on the specific task. It has been observed that high impedance values are desirable for fine movements, while lower values should be used for large/rapid movements [11, 21, 27]. In [13], the authors propose a strategy to adapt robot impedance depending on the human's behavior during a physical human-robot interaction task. Adaptive impedance control has been used to account for differences between model and manipulator parameters [7, 24]. In addition, the ability to adapt controller parameters may be used to improve system performance and has been proposed to optimize task efficiency [5, 15, 16]. However, to the best of authors' knowledge, all the work on impedance planning presented to date proposed an adaptation of robot stiffness, typically at the joint level, but did not shape the whole impedance at the end-effector level.

In this chapter, we tackle the problem of planning the whole robot behavior in terms of both motion and impedance profiles; and propose an optimization-based approach to modulate the Cartesian impedance of a generic manipulator. Our solution is able to cope with different kinematic structures and exploit the properties of impedance compositionality to manage robot redundancy. To demonstrate our approach, we present a series of simulation experiments, first with a simple planar structure, then generalizing to a multi-arm redundant three-dimensional robot. To test the effectiveness of our methodology, all tasks include a transition from free motion to contact with a compliant environment. The content of this chapter is adapted from [4].

9.2 Background

The theory of impedance control is well established in the literature. This section reviews the main concepts necessary for the rest of this chapter. As mentioned in the introduction, the general idea is to control the relationship between the generalized configuration of a manipulator and its derivative, and the generalized forces. Define the robot generalized configuration x, its derivatives $\dot{x}$, $\ddot{x}$, ..., and the output force/torque F. A common implementation of mechanical impedance is

$$F = Z(x, x_{ref}, \dot{x}, \dot{x}_{ref}, \ddot{x}, \ddot{x}_{ref}, \ldots), \tag{9.1}$$

where Z is an arbitrary non-linear operator and $x_{ref}, \dot{x}_{ref}, \ddot{x}_{ref}, \ldots$ define the motions of the robot that yield zero force/torque, $F = 0$. One typical implementation is to assume a linear operator and truncate the derivatives at the first order, in such a way that the external perceived behavior is the effect of a spring and a damper. This results in the following linear controller [19]

$$F_I = -k(x - x_{ref}) - b(\dot{x} - \dot{x}_{ref}). \tag{9.2}$$

It is important to recognize that all forces/torques—whatever their origin—superimpose their effects on the inertial dynamics of a robot. As a result, multiple controllers defined in the form of Eq. 9.2 may be combined linearly to increase the complexity of the behavior. That is, mechanical impedance has the property of compositionality.

So far we defined x as a generalized configuration variable but Eq. 9.2 can be defined at different levels. For physical interaction, one may want to shape the impedance perceived at the robot end-effector. In that case, without any loss of generality, consider a manipulator with n revolute joints $q = [q_1, \ldots, q_n]^T$ and an end-effector moving in an m-dimensional Cartesian space which has $\text{fk}(q)$ forward kinematics and a Jacobian matirx $J(q) \in \mathbb{R}^{m,n}$ so that Eq. 9.2 may be rewritten as

$$\tau_I = J^T[-K_{EE}(\text{fk}(q) - x_{ref}) - B_{EE}(J\dot{q})], \tag{9.3}$$

where $K_{EE} \in \mathbb{R}^{m,m}$ and $B_{EE} \in \mathbb{R}^{m,m}$ are the endpoint stiffness and damping matrices respectively.

This formulation enables the control of robot trajectory directly in Cartesian space, with a desired impedance also specified in Cartesian space. Note that no kinematic inversion is required. Note also that it is easy to demonstrate the direct map from endpoint impedance to the joint impedance matrices, i.e.

$$K_q = J^T K_{EE} J \tag{9.4}$$

$$B_q = J^T B_{EE} J \tag{9.5}$$

under the hypothesis that the endpoint stiffness is calculated at zero endpoint force. For additional details on this topic the interested reader is referred to [19, 36].

9.2.1 Redundant Robots

In the case of robots with a more degrees of freedom than the dimension of Cartesian space, the impedance control implemented as in Eq. 9.3 has a nullspace cardinality higher than zero. In other words, a desired Cartesian impedance can be achieved via a manifold of robot configurations. This may be an undesired effect, since a small perturbation may produce an arbitrarily large movement in the nullspace. To address this problem, one solution is to include a second impedance control which defines a configuration-space impedance with respect to a desired rest position. In this chapter we implemented this concept at the joint level as:

$$\tau_J = -K_J(q - q_{rest}) - B_J\dot{q}, \tag{9.6}$$

where τ_J are the joint torques required to control the robot configuration toward the rest position q_{rest} with stiffness and damping matrices $K_J \in \mathbb{R}^{n,n}$ and $B_J \in \mathbb{R}^{n,n}$ respectively.

Note that, since the rest configuration defined by the two controllers is different in general, their superimposition may produce a steady-state error at the end-effector with respect to the rest position of the Cartesian impedance control. This may be arbitrarily reduced by design if $K_J << K_{EE}$. Alternatively, zero error may be achieved via null-space projection [10, 25, 26]. Yet another possibility is that the optimization (presented below) could adjust x_{ref} to achieve zero steady-state error.

9.3 Impedance Optimization

In Sect. 9.2 we discussed the stiffness and damping matrices as coefficients of the controller. Typical implementations use fixed parameters selected through heuristic strategies. However, as mentioned in the introduction, having the possibility to modulate the controller parameters during task execution may be a major advantage for robot physical interaction, yet how to shape time-dependent impedance profiles is still an open question. In the following we propose a solution to this problem by presenting a novel approach to jointly optimize the impedance parameters and reference profiles for a given (arbitrary) robot interacting with its environment.

9.3.1 Interaction Modeling

Modeling physical contact with an external object is a complex topic. However, in many applications it is sufficient to schematize it as a force perceived at the contact point. A typical approach to simulating contact with a compliant object is to calculate such a force/torque as the reaction of a spring pushed by the robot against a fixed reference. We will refer to this model in the following as k_E. Note that the particular model of the external environment that we assumed is neither mandatory nor a limitation of this work, since it is only used to simulate a sample task and any other law would produce similar computational cost. The general experiment considered in this chapter is depicted in Fig. 9.1.

9.3.2 Robot Definition

To show the theory behind our implementation clearly, we build this chapter starting from a simple basic building block, a planar $4R$ manipulator (see Fig. 9.2). This keeps the initial complexity low, yet preserves the general goal and the essential challenges as well as the consequences of robot redundancy. Later in this chapter, the complexity of the model will be increased, moving toward bi-manual 4R structures and fully humanoid structures.

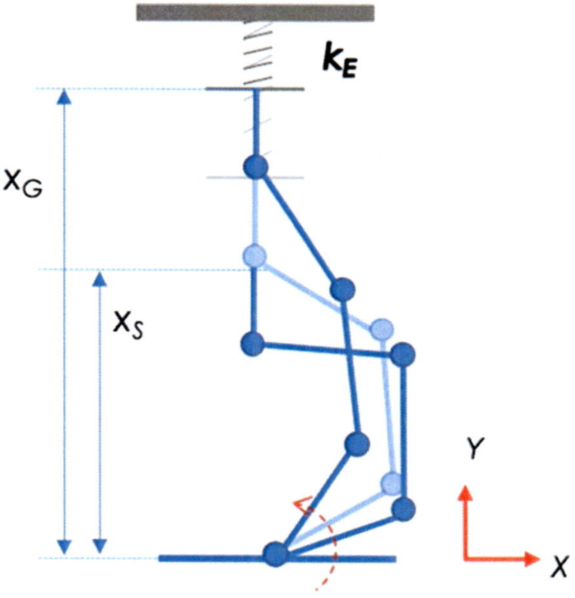

Fig. 9.1 Sample task for a single arm. The robot starts from an initial configuration x_S and moves to push against the spring to reach the goal position x_S. This task includes a transition from a free movement condition to an environment-interaction condition

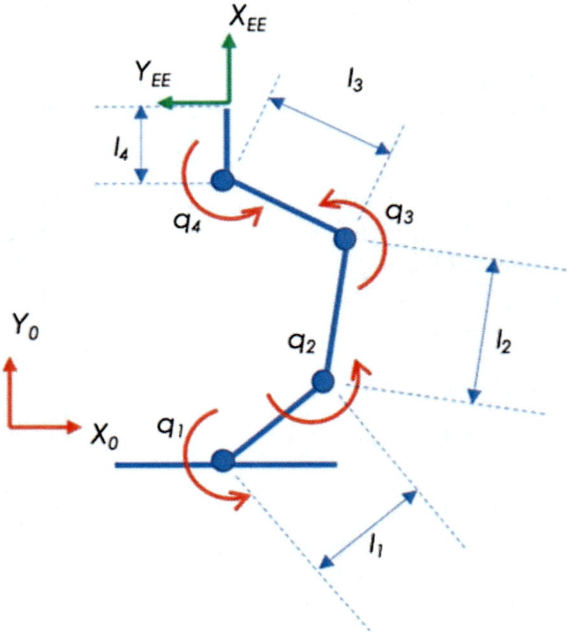

Fig. 9.2 A planar robot with 4 revolute joints

9.3.3 Cost Function

Since mechanical impedance is a relation between force and motion domains, it is reasonable to include in the objective function we aim to minimize at least one term related to the force domain and one related to robot motion at the same point. The reason to do this is also practical. A penalty on end-effector position is needed to follow a desired path, while minimizing end-effector force is useful for several practical purposes, including increased safety while interacting with humans or the environment, lower power consumption, etc. [17]. The general form of the objective function can be written as:

$$\zeta = \alpha_1 ||x - x_{goal}|| + \alpha_2 ||x^{t_{fin}} - x_{goal}^{t_{fin}}|| + \alpha_3 ||W_{ext}||, \tag{9.7}$$

where x is the time-varying configuration of the end-effector, x_{goal} is a desired reference trajectory, $x^{t_{fin}}$ and $x_{goal}^{t_{fin}}$ are the end-effector position and target at the end of the action, and W_{ext} is the external wrench exerted on the end-effector. α_1, α_2 and α_3 are constants $\in \mathbb{R}^+$.

The optimization problem can be formalized as:

$$\min_{K_{EE}, B_{EE}, x_{ref}} \quad \zeta$$
$$\text{subject to} \tag{9.8}$$
$$\zeta = \alpha_1 ||x - x_{goal}|| + \alpha_2 ||x^{t_{fin}} - x_{goal}^{t_{fin}}|| + \alpha_3 ||W_{ext}||,$$

where K_{EE}^*, B_{EE}^* and x_{ref}^* are the optimal parameters for the impedance control of Eq. 9.3.

The problem of Eq. 9.8 is in general non-convex. However a good approximation can be achieved with a reasonable selection of an initial guess. In our case we selected constant values for K_{EE} and B_{EE} while the initial guess for x_{ref} was a minimum-jerk trajectory between the initial and final desired positions. This choice of trajectory was inspired by human movements (see e.g. [14, 29, 31]), and is typically used to generate continuous, smooth movements in robotics [5, 40]. Let x_1 and x_N be the initial and final goal positions of the robot end-effector; the minimum jerk trajectory between x_1 and x_N is calculated as

$$x(t) = x_1 + (x_N - x_1) * (10(t/t_N)^3 - 15(t/t_N)^4 + 6 * (t/t_N)^6) \tag{9.9}$$

9.3.4 Problem Complexity and Generalization to Parallel Robots

The optimization problem formulated in Eq. 9.8 is written in a general form and is valid both for planar and for 3D structures. The main difference between these two conditions is the dimensionality of K_{EE}^*, B_{EE}^* and x_{ref}^*. Indeed, for the planar case the end-effector is defined in $m = 3$ dimensions, while for the general case $m = 6$. This results in an increased computational time, but the formulation of the problem is still valid. For this reason, we will present results on a planar robot first, and on a 3D humanoid robot later. To optimize the whole robot motion, each optimization variable is defined for a series of time frames, i.e. $K_{EE} \in \mathbb{R}^{m \times m \times T}$, $B_{EE} \in \mathbb{R}^{m \times m \times T}$ and $x_{ref} \in \mathbb{R}^{m \times T}$, where T is the number of time-knots considered. Note that the greatest contribution to computational complexity does not arise from the dimensionality of the robot's configuration space or its end-effector workspace but rather from the number of time frames considered in the optimization.

To reduce complexity I defined a set of six evenly spaced knots and optimized the controller parameters at these points. The impedance and reference profiles were then interpolated using a piece-wise cubic polynomial. Specifically, between two adjacent knots the function was approximated by a cubic Hermite polynomial with a constraint on the derivatives at the knots. The resulting profile is guaranteed to be continuous with a continuous first derivative, while the interpolation error at the knots is zero. Interpolation via piece-wise cubic Hermite polynomials is shape preserving. This results in interpolated stiffness and reference profiles that preserve the shape of the optimized data, respect monotonicity and maintain local minima/maxima. Note that, even if the approximation of motion and impedance profiles by piece-wise cubic Hermite polynomials produces continuous variation, this does not imply continuous dynamics; interaction with the external environment may introduce discontinuities in the robot's actual motion. For additional details on the implementation we refer to [6, 22].

9.4 Results

The impedance optimization algorithm discussed in Sect. 9.3 was implemented and tested on several kinematic structures. To show the effectiveness of our approach we present a set of simulations performed on a 4R planar robot (see Fig. 9.2). In this implementation, the objective function introduced in Eq. 9.8 was minimized via an Interior Point Algorithm. For each iteration, the algorithm simulated the robot motion using an updated set of parameters. The end-effector goal trajectory of Eq. 9.7 was a minimum jerk trajectory between an initial configuration in free space and a final configuration that was within the simulated object, thus in a configuration where $W_{ext} > 0$ (See Fig. 9.1). The initial guess for the optimization parameters were constant stiffness and damping and a minimum-jerk trajectory between initial and final configuration (the same as the goal trajectory). Results are depicted in Figs. 9.3 and 9.4.

The algorithm was able to identify a solution that locally minimized ζ in low number of iterations (lower than 40 for the 4R case). For each iteration, the time-varying profiles for the impedance control parameters were updated, until the minimum was approximated with suitable precision (step size lower than 10^{-15}). It is interesting that, without any constraint on the shapes of the impedance profiles and reference trajectory, the calculated optimal solution found impedance variations that would increase the stiffness in proximity of the impact, while maintaining good tracking of the reference trajectory and avoiding overshoot. In addition, since the task space has three dimensions while the robot has four revolute joints, we can expect undesired movements of the manipulator in the nullspace of the controller. As dis-

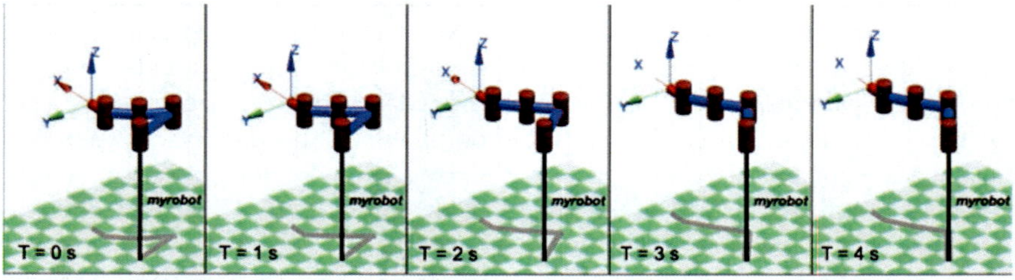

Fig. 9.3 Snapshots of the optimal robot behavior detailed in Fig. 9.4. Visualization was performed using the Robotic Toolbox [9]

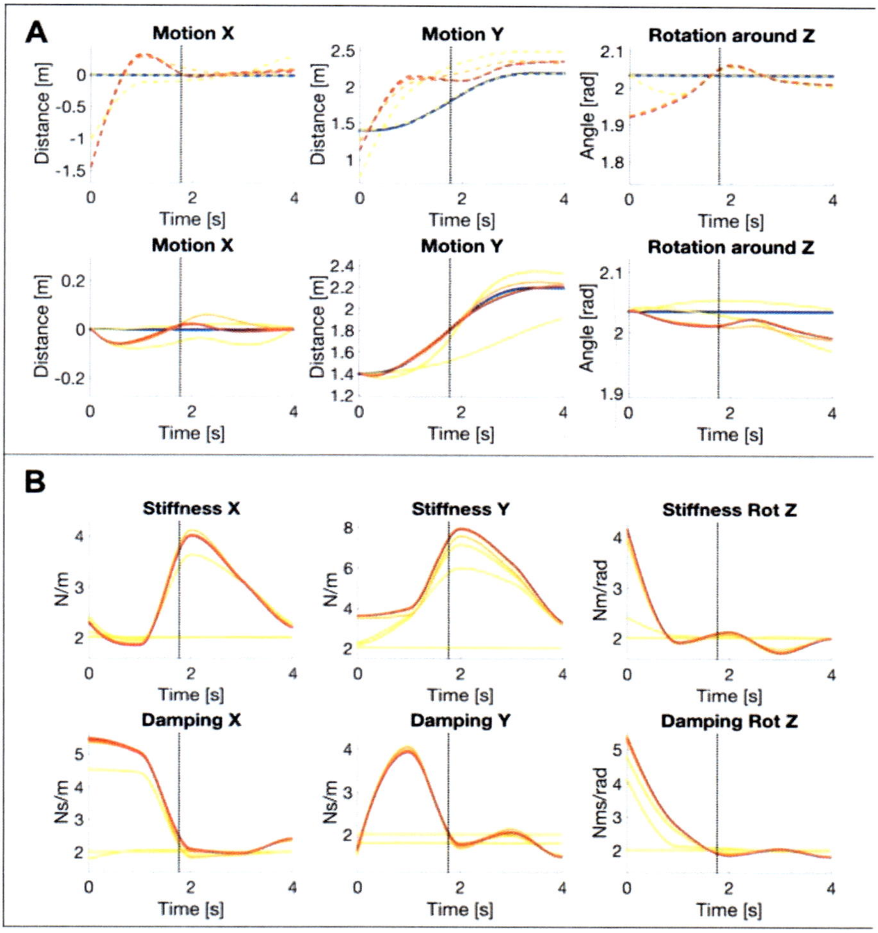

Fig. 9.4 Optimization of impedance control parameters. Sub-figure A. On the first row: in blue the reference trajectory, from the yellow to the red dashed line (along the color gradient) the commanded trajectory given to the impedance control, while increasing number of iterations. Red plots are the final optimal values. On the second row: in blue the reference trajectory, from the yellow to the red filled line (along the color gradient) the actual robot movement. Red plots are the final optimal values. Vertical black dotted lines identify the time at which contact occurs for the optimal case. Sub-figure B. On the first row: From the yellow to the red filled line (along the color gradient), stiffness profile along X and Y axis and around Z axis (rotation), while increasing number of iterations. Red plots are the final optimal values. On the second row: From the yellow to the red filled line (along the color gradient), damping profile along X and Y axis and around Z axis (rotation), while increasing number of iterations. Red plots are the final optimal values. Vertical black dotted lines identify the time at which contact occurs for the optimal case

cussed in Sect. 9.2.1, we solve this problem using a secondary joint-space impedance controller. The parameters of this second controller were not optimized and its reference position was a constant rest configuration. Note that, even though the two controllers have different target behaviors and hence compete, the overall behavior is not affected. The optimization is able to compensate for the effect of the second controller on the end-effector behavior while preserving its effect at the joint level.

9.4.1 Extension to Parallel Robots

Since the procedure presented in this chapter is focused on the end-effector behavior, it is, from a theoretical point of view, straightforward to extend it to more complex structures. In this section, we generalize to multi-arm manipulation. Let us consider two identical robots, the first placed in a configuration resembling a right arm and the second in a left-arm configuration. The choice to use two identical manipulators was only made for convenience and is not a limitation of this work. The task in this case is depicted in Fig. 9.6.

The problem formalized in Sect. 9.3 can be easily extended here with the same computational complexity. Indeed, since the optimization is designed to shape the end-effector behavior, the control of each of the two manipulators, in this specific case, can be mapped as a translation of the optimization result. The final robot behavior is depicted in Fig. 9.5. As expected, the algorithm was able to find a solution that satisfied all of the task requirements.

9.5 Generalization to 3D Robots

In the previous sections, we discussed the application of impedance optimization to planar robots. The extension to non-planar manipulators comes with no additional theoretical complexity, and all of the framework developed in Sects. 9.2 and 9.3 is still valid. Nevertheless, from a computational point of view, the application to 3D robots is not negligible and may require excessive computation time. However, leveraging on the compositionality of impedance, our implementation was still able to provide a solution in a reasonable time. We report here a set of experiments carried out on a model of a 7 degrees of freedom Baxter Robot. In this case, the task was the same as Fig. 9.6 with two 7 degrees of freedom arms moving in a 6 degrees of freedom Cartesian space. The optimal robot behavior is reported in Fig. 9.8, while the resulting impedance profile and reference trajectory are in Fig. 9.7. As before, the optimization was able to provide a desired behavior at the end-effector while the kinematic redundancy was handled through additional joint-space impedance controllers.

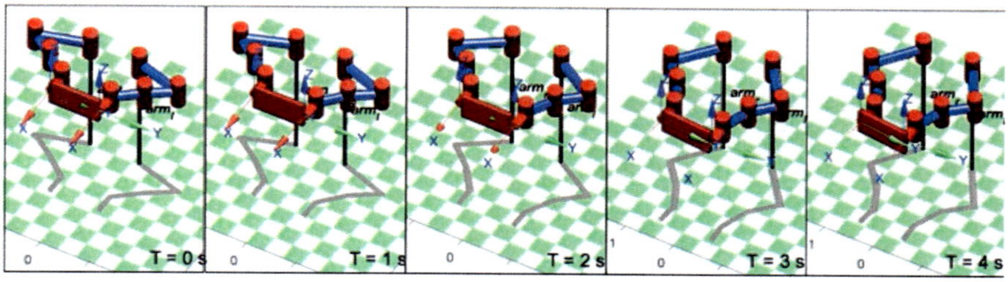

Fig. 9.5 Snapshots of the optimal bi-manual robot behavior. Visualization was performed using the Robotic Toolbox [9]

Fig. 9.6 Sample task for the dual-arm robot. The robot started from the initial configuration x_S and pushed against the spring to reach the goal position x_S. The dual-arm robot end-effector was defined as the mean between the single arms end-effectors. This task included a transition from a free movement condition to an environment-interaction condition

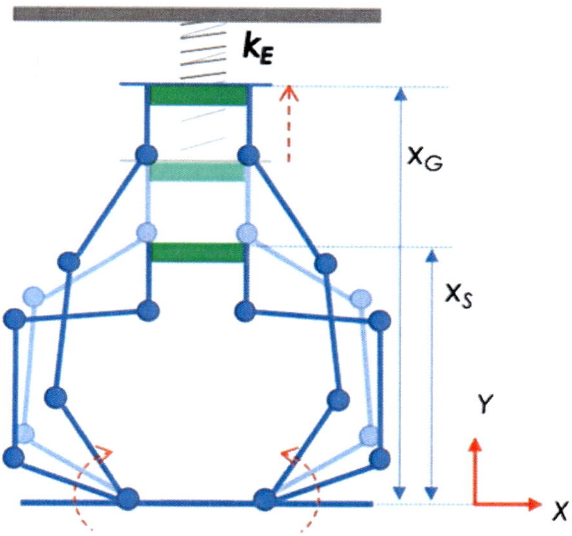

9.6 Discussions and Conclusions

This chapter presented an optimization-based algorithm that identified an optimal End-Effector impedance profile for a generic task of interaction with an external compliant object. This implementation overcomes the limitations of most previous works on impedance planning. While in previous work the planning of the impedance profile was limited to stiffness at the joint level, we also optimized the damping parameters. Another relevant contribution is the implementation of planning in Cartesian space, i.e. calculating the optimal robot behavior at the end-effector level. This task-oriented approach comes with several advantages: it enables a clearer interpretation of the results and generalization to different kinematic structures. The results presented for different kinematics and considering both single- and multi-robot manipulation, show that this approach is effective and reliable.

The results presented here demonstrate that the method is able to generalize across different kinematic structures. Future efforts will apply the outcomes of this algo-

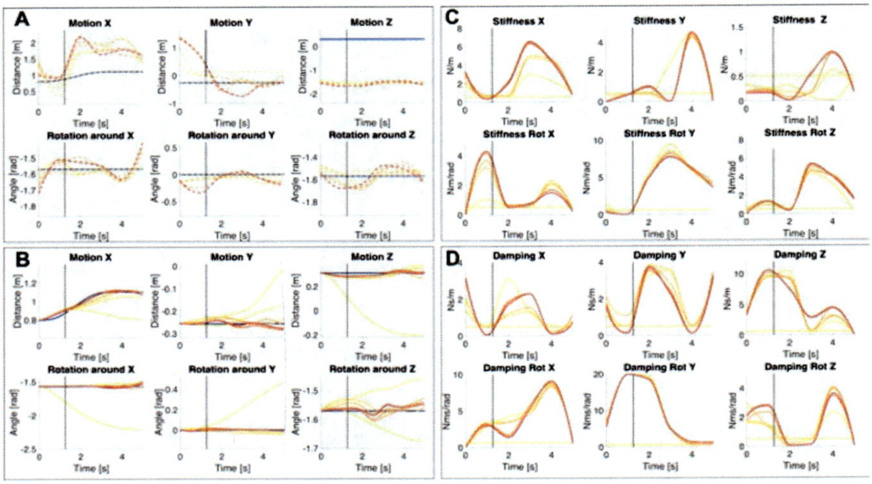

Fig. 9.7 Optimization of impedance control parameters for the 7 degrees of freedom case. Sub-figure A. In blue the reference trajectory, from the yellow to the red dashed line (along the color gradient) the commanded trajectory given to the impedance control, while increasing number of iterations. Red plots are the final optimal values. Vertical black dotted lines identify the time at which contact occurs for the optimal case. Sub-figure B. In blue the reference trajectory, from the yellow to the red filled line (along the color gradient) the actual robot movement, while increasing number of iterations. Red plots are the final optimal values. Vertical black dotted lines identify the time at which contact occurs for the optimal case. Sub-figure C. From the yellow to the red filled line (along the color gradient) stiffness profile, while increasing number of iterations. Red plots are the final optimal values. Vertical black dotted lines identify the time at which contact occurs for the optimal case. Sub-figure D. From the yellow to the red filled line (along the color gradient) damping profile, while increasing number of iterations. Red plots are the final optimal values. Vertical black dotted lines identify the time at which contact occurs for the optimal case

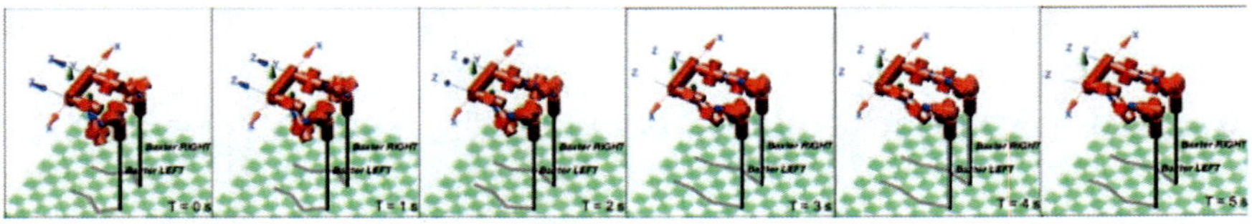

Fig. 9.8 Snapshots of the optimal behavior implemented on a Baxter bi-manual robot. Visualization was performed using the Robotic Toolbox [9]

rithm to real platforms. Applying optimization and reinforcement learning methods directly on real robots presents several practical difficulties—they would require multiple iterations of actions on the real platform, possibly accelerating wear and compromising reliability. From this perspective, the approach presented here offers an advantage: a model of the real robot can be used to perform the optimization. Once an optimal solution has been calculated in simulation, the output can be used for the control of the real manipulator.

The implementation presented in this chapter does not guarantee convergence to the global minimum. However, the particular selection of the optimization method is a free parameter of the problem. Indeed, the same approach can be implemented with global optimization algorithms, such as Genetic Algorithms, Multi-Search approaches or, leveraging on the particular bi-linear formulation of the cost function, with specific optimization algorithms [32]. However, we anticipate that those approaches may incur a significant expansion of time required to converge. For this reason, whether it is worth searching for the global optimum, or instead accepting a local solution that can be identified in a reduced amount of time, remains an open question to be addressed in future developments of this work.

References

1. Ajoudani, A., Tsagarakis, N., Bicchi, A.: Tele-impedance: teleoperation with impedance regulation using a body-machine interface. Int. J. Robot. Res. **31**(13), 1642–1656 (2012)
2. Albu-Schäffer, A., Ott, C., Hirzinger, G.: A unified passivity-based control framework for position, torque and impedance control of flexible joint robots. Int. J. Robot. Res. **26**(1), 23–39 (2007)
3. An, C.H., Hollerbach, J.M.: The role of dynamic models in cartesian force control of manipulators. Int. J. Robot. Res. **8**(4), 51–72 (1989)
4. Averta, G., Hogan, N.: Enhancing robot-environment physical interaction via optimal impedance profiles. In: Proceedings of the 8th IEEE RAS/EMBS International Conference for Biomedical Robotics and Biomechatronics (BioRob), pp. 973–980. IEEE (2020)
5. Buchli, J., Stulp, F., Theodorou, E., Schaal, S.: Learning variable impedance control. Int. J. Robot. Res. **30**(7), 820–833 (2011)
6. Carlson, R.E., Fritsch, F.N.: Monotone piecewise bicubic interpolation. SIAM J. Numer. Anal. **22**(2), 386–400 (1985)
7. Colbaugh, R., Seraji, H., Glass, K.: Direct adaptive impedance control of robot manipulators. J. Robot. Syst. **10**(2), 217–248 (1993)
8. Colgate, J.E., Hogan, N.: Robust control of dynamically interacting systems. Int. J. Control **48**(1), 65–88 (1988)
9. Corke, P.: Robotics, vision and control: fundamental algorithms In MATLAB® second, completely revised, vol. 118. Springer (2017)
10. Dietrich, A., Ott, C., Albu-Schäffer, A.: An overview of null space projections for redundant, torque-controlled robots. Int. J. Robot. Res. **34**(11), 1385–1400 (2015)
11. Duchaine, V., St-Onge, B.M., Gao, D., Gosselin, C.: Stable and intuitive control of an intelligent assist device. IEEE Trans. Haptics **5**(2), 148–159 (2012)
12. Eppinger, S., Seering, W.: Introduction to dynamic models for robot force control. IEEE Control Syst. Mag. **7**(2), 48–52 (1987)
13. Ficuciello, F., Villani, L., Siciliano, B.: Variable impedance control of redundant manipulators for intuitive human-robot physical interaction. IEEE Trans. Robot. **31**(4), 850–863 (2015)
14. Flash, T.: The control of hand equilibrium trajectories in multi-joint arm movements. Biol. Cybern. **57**(4–5), 257–274 (1987)
15. Gabellieri, C., Palleschi, A., Mannucci, A., Pierallini, M., Stefanini, E., Catalano, M.G., Caporale, D., Settimi, A., Stoyanov, T., Magnusson, M., et al.: Towards an autonomous unwrapping system for intralogistics. IEEE Robot. Autom. Lett. (2019)
16. Gasparri, G.M., Fabiani, F., Garabini, M., Pallottino, L., Catalano, M., Grioli, G., Persichin, R., Bicchi, A.: Robust optimization of system compliance for physical interaction in uncertain scenarios. In: 2016 IEEE-RAS 16th International Conference on Humanoid Robots (Humanoids), pp. 911–918. IEEE (2016)

17. Haddadin, S., Albu-Schäffer, A., Hirzinger, G.: Safe physical human-robot interaction: measurements, analysis and new insights. In: Robotics Research, pp. 395–407. Springer (2010)
18. Hoffman, E.M., Laurenzi, A., Muratore, L., Tsagarakis, N.G., Caldwell, D.G.: Multi-priority cartesian impedance control based on quadratic programming optimization. In: 2018 IEEE International Conference on Robotics and Automation (ICRA), pp. 309–315. IEEE (2018)
19. Hogan, N.: Impedance control-an approach to manipulation. I-theory. II-implementation. III-applications. ASME Trans. J. Dyn. Syst. Meas. Control B **107**, 1–24 (1985)
20. Hyun, D.J., Seok, S., Lee, J., Kim, S.: High speed trot-running: implementation of a hierarchical controller using proprioceptive impedance control on the mit cheetah. Int. J. Robot. Res. **33**(11), 1417–1445 (2014)
21. Ikeura, R., Moriguchi, T., Mizutani, K.: Optimal variable impedance control for a robot and its application to lifting an object with a human. In: Proceedings of the 11th IEEE International Workshop on Robot and Human Interactive Communication, pp. 500–505. IEEE (2002)
22. Kahaner, D., Moler, C., Nash, S.: Numerical Methods and Software. Prentice Hall, Englewood Cliffs (1989)
23. Katsura, S., Matsumoto, Y., Ohnishi, K.: Modeling of force sensing and validation of disturbance observer for force control. IEEE Trans. Ind. Electron. **54**(1), 530–538 (2007)
24. Kelly, R., Carelli, R., Amestegui, M., Ortega, R.: On adaptive impedance control of robot manipulators. In: Proceedings of the 1989 International Conference on Robotics and Automation, pp. 572–577. IEEE (1989)
25. Khatib, O.: A unified approach for motion and force control of robot manipulators: the operational space formulation. IEEE J. Robot. Autom. **3**(1), 43–53 (1987)
26. Khatib, O.: Inertial properties in robotic manipulation: an object-level framework. Int. J. Robot. Res. **14**(1), 19–36 (1995)
27. Lecours, A., Mayer-St-Onge, B., Gosselin, C.: Variable admittance control of a four-degree-of-freedom intelligent assist device. In: 2012 IEEE International Conference on Robotics and Automation, pp. 3903–3908. IEEE (2012)
28. Lin, H.C., Smith, J., Babarahmati, K.K., Dehio, N., Mistry, M.: A projected inverse dynamics approach for multi-arm cartesian impedance control. In: 2018 IEEE International Conference on Robotics and Automation (ICRA), pp. 1–5. IEEE (2018)
29. Morasso, P.: Spatial control of arm movements. Exp. Brain Res. **42**(2), 223–227 (1981)
30. Ortenzi, V., Stolkin, R., Kuo, J., Mistry, M.: Hybrid motion/force control: a review. Adv. Robot. **31**(19–20), 1102–1113 (2017)
31. Piazzi, A., Visioli, A.: Global minimum-jerk trajectory planning of robot manipulators. IEEE Trans. Ind. Electron. **47**(1), 140–149 (2000)
32. Quesada, I., Grossmann, I.E.: A global optimization algorithm for linear fractional and bilinear programs. J. Glob. Optim. **6**(1), 39–76 (1995)
33. Raibert, M.H., Craig, J.J., et al.: Hybrid position/force control of manipulators. J. Dyn. Syst. Meas. Control **103**(2), 126–133 (1981)
34. Seraji, H., Colbaugh, R.: Force tracking in impedance control. Int. J. Robot. Res. **16**(1), 97–117 (1997)
35. Stolt, A., Linderoth, M., Robertsson, A., Johansson, R.: Force controlled robotic assembly without a force sensor. In: 2012 IEEE International Conference on Robotics and Automation, pp. 1538–1543. IEEE (2012)
36. Villani, L., De Schutter, J.: Force control. Springer Handbook of Robotics, pp. 161–185 (2008)
37. Volpe, R., Khosla, P.: A theoretical and experimental investigation of impact control for manipulators. Int. J. Robot. Res. **12**(4), 351–365 (1993)
38. Whitney, D.E.: Force feedback control of manipulator fine motions. J. Dyn. Syst. Meas. Control **99**(2), 91–97 (1977)
39. Yoshikawa, T., Zheng, X.Z.: Coordinated dynamic hybrid position/force control for multiple robot manipulators handling one constrained object. Int. J. Robot. Res. **12**(3), 219–230 (1993)
40. Zefran, M., Kumar, V., Croke, C.B.: On the generation of smooth three-dimensional rigid body motions. IEEE Trans. Robot. Autom. **14**(4), 576–589 (1998)

Chapter 10
Learning from Humans How to Grasp: A Reactive-Based Approach

Recently, the avenue of adaptable, soft robotic hands has opened simplified yet effective opportunities to grasp different items; however, the potential of soft end effectors (SEEs) is still largely unexplored, especially in human-robot interaction. In this chapter, we propose, for the first time, a simple touch-based approach to endow a SEE with autonomous grasp sensory-motor primitives, in response to an item passed to the robot by a human (human-to-robot handover). We capitalize upon human inspiration and minimalistic sensing, while hand adaptability is exploited to generalize grasp response to different objects. We consider the Pisa/IIT SoftHand (SH), an under-actuated soft anthropomorphic robotic hand, which is mounted on a robotic arm and equipped with Inertial Measurement Units (IMUs) on the fingertips. These sensors detect the accelerations arisen from contact with external items. In response to a contact, the hand pose and closure are planned for grasping, by executing arm motions with hand closure commands. We generate these motions from human wrist poses acquired from a human maneuvering the SH to grasp an object from a table. We obtained a total success rate of 86%, considering many objects passed to the SH in different manners. We also tested our techniques in preliminary experiments, where the robot moved to autonomously grasp objects from a surface. Results are positive and open interesting perspectives for soft robotic manipulation.

10.1 Introduction

Human-robot (HR) handover represents a well-studied topic in robotics, see e.g. [28, 30, 33, 39], which comes with important challenges related to human-to-robot communication [25], human safety and acceptance [18], human intention prediction [29], human-aware planning and execution [20, 24]. According to [30], HR handover consists of three main phases, i.e. *approach*, *passing*, *retraction*. In this chapter we focus on the *passing* phase, which corresponds to the physical transfer of the

G. Averta, *Human-Aware Robotics: Modeling Human Motor Skills for the Design, Planning and Control of a New Generation of Robotic Devices*, Springer Tracts in Advanced Robotics 145, https://doi.org/10.1007/978-3-030-92521-5_10

object from human to robot. Here, the detection of the interaction is mandatory not only to guarantee operator's safety, but also to execute a successful robot grasp response [1]. In biology, we refer to this behavior as sensory-motor response [38]. Robotic sensing is commonly achieved through the employment of visual and/or tactile sensory systems, see e.g. [1, 23, 31, 34], which acquire the needed information on the object (and/or on human hand gesture). Such an information can be then processed to plan meaningful autonomous response [20]. Focusing on the specific action of object grasping, classic approaches, both analytic [6] and data-driven [7], have usually dealt with grasp planning, grasp adaptation and force control. For real-world grasps, which often come with a certain level of uncertainty, visual and/or tactile sensory information can be combined with learning and statistical techniques to enable robots to autonomously manage novel and uncertain situations, see e.g. [2, 4, 12, 27].

Recently, the introduction of under-actuated and soft robotic hands has given new options respect to the traditional way robotic grasping was usually planned and performed with *rigid* end-effectors. Indeed, the embodied capability of *soft* hands to comply and adapt to different objects and the environment, together with the simplicity in control derived from under-actuation—e.g. following the principled simplification approach inspired by human hand synergies [38]—has led to simple, adaptable yet robust systems. These systems can mold around items and exploit physical environmental constraints as opportunities to guide adaptive grasping of different objects [11, 16]. Thanks to these characteristics, a rough approximation of object geometry and robot hand pose are enough to generate candidate successful grasps despite external uncertainty [8], e.g. due to partial knowledge through point-cloud data. In this case, final grasping pose can be refined using low-cost infrared sensors [26].

The latter result further sustains the evidence that more effective grasps can be obtained when information from short-range or non-ranged sensors is used [19]. Under this regard, touch/contact—based sensing allows for a more direct and less problematic detection of important contact-related aspects than artificial vision [35], as well as for a simplified sensor-based implementation of *reactive* behavior [17, 19, 26]. This aspect is particularly crucial in proximate human-robot interaction (HRI), hence in handover tasks, where a prompt robot reaction is a key factor for a successful implementation of the *sense-plan-act* model [32].

To the best of authors' knowledge, there are no solutions that combine (i) the adaptability of soft robotic hands, (ii) a minimalistic tactile sensing and (iii) a reactive robotic behavior for robotic autonomous grasping in human to robot object passing tasks. In this chapter, we propose a solution that, for the first time, targets points (i), (ii), (iii). The goal is to endow a soft robot hand/arm system with purely touch-based grasp primitives. We refer to grasp primitives as coordinated hand closure and wrist movements of a soft manipulator, triggered by touch-based sensor readings. Our leading idea is to take inspiration from the human example, to implement effective sensory-motor response on the robot side, having in mind the trade-off between performance maximization and resource usage minimization. The content of this chapter is adapted from [5].

A companion video is available at the following link: https://www.youtube.com/watch?v=S_771wvvOdY.

10.2 The Method

Looking at biology, the simplest level of sensory-motor integration consists of primitive reflex actions [9], i.e. nearly instantaneous (and automatic) motor reactions in response to a sensed event. Robotics often takes inspiration from biology [38], and the implementation of primitive-based control in robots is no exception. In this context the term *primitives* (or, sometimes, *reflexes* [26]) is used to define automated movements that are triggered in response to certain sensory inputs experienced by the robot. In the last years, primitive control has become somewhat popular in different fields of robotics. In [40], predefined biologically inspired postural primitives were used with a robotic arm to imitate reflex-like withdrawal behaviors. Similar approaches were used e.g. in [13, 41].

For what concerns grasp, the specific definition of *grasp primitives* can vary slightly, but it usually involves an automated control mechanism triggered by certain conditions. In [3], authors implemented grasp reflexes for robotic manipulation using leaky integrate-and-fire neurons. The work in [22] developed specialized reflexes for an anthropomorphic robotic hand based on force and velocity control. Tactile-based grasp reactive behavior was implemented e.g. in [17, 19, 26], although in these cases a partial knowledge of object pose and location was needed. To the best of our knowledge, our work is the first one where a primitive-based approach is used with a soft manipulator, relying only on contact based information as triggering input.

We use as end effector the Pisa/IIT SoftHand (SH), an under-actuated anthropomorphic soft robotic hand [11], which is mounted on a robotic arm and equipped with six IMUs, five on the back of the fingertips and one on the back of the hand. The fingertip sensors detect the accelerations arisen from contact with external items, while the IMU on the back of the hand is used to keep track of the acceleration generated by the robot movement. In response to a given contact, the hand pose and closure state are planned for grasping, by executing arm motions with hand closure commands, i.e. a primitive. Data used to generate these motions are extracted from an experiment, where a human user controlled the SH through a suitable interface to grasp a tennis ball placed on a pole, considering a set of approaching directions and contact locations at distal phalanxes. Since the SH can adapt to grasp different items with various geometrical properties [11], the crucial part for a successful grasp is the correct determination of the pose of the hand w.r.t. the target. To do this, we correlate the accelerations with the user's wrist pose to devise primitive grasp strategies (see Fig. 10.1), which we evaluated in a simple human to robot object passing task. More specifically, the SH was still and placed over the wrist of a 7 DOF Kuka LWR 4+ arm, and the user passed the object to it. During the experimental evaluation, we considered many objects passed to the SH along different contact directions. Our working hypothesis was that despite the fact that the primitives were obtained for only one

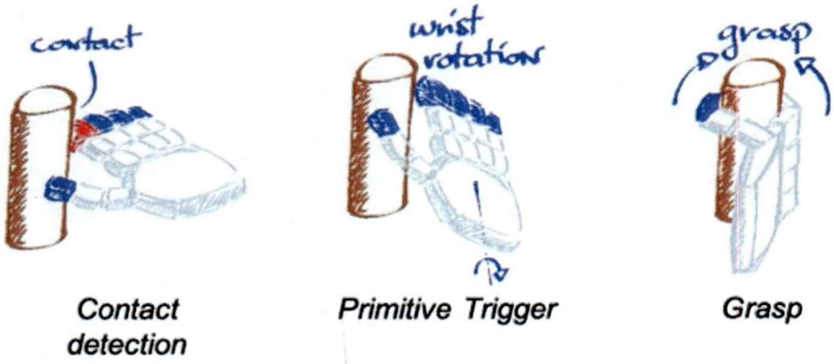

Fig. 10.1 The leading idea of our approach. In blue the sensorized fingertips, in red the contact

object and limited number of contact conditions, the hand should have been able to perform successful grasps with novel situations, thanks to its intrinsic adaptability.

Our choice to use human-inspired motion primitives was motivated by several studies in literature, which proposed human-like robotic actions to improve robot-motion legibility [30, 42]. However, a comparison between different planning methods (including anthropomorphic vs. non-anthropomorphic) is currently out of the scope of this chapter.

Finally, it is worth noticing that we are aware that a human-to-robot handover task would require a combination of both feed-forward and feedback control components, the latter is crucial to assess the correctness of task execution and modify the action if needed. In this chapter, we decided to focus only on the feed-forward part, since our goal was to verify to which extent the combination of reactive primitives and robotic hand softness enables to generalize to different object grasps, under controlled conditions. Future works will be devoted to include also a feedback part, which will allow to take into account unforeseen events related to human behavior and provide a control method to check for the success of the robotic response—and eventually to change it, also to guarantee users safety. Further tests in more unstructured environments are envisioned. The inclusion of additional sensors is also under evaluation (e.g. increasing the number of IMUs in use or adding force sensors to the SH), still keeping in mind the trade-off between resource usage and system performance.

We conducted additional experiments with some objects placed on the table, while the robotic hand was controlled to move and contact them at different locations. Although it is well-known that humans tend to naturally adapt to robot grasping capabilities in handover tasks [15], the objective of these experiments was to preliminarily evaluate the success of our approach, against possible helps from the human operator. This is important, especially for the interesting scenarios that could be opened by the proposed method, e.g. for purely tactile-based autonomous grasps of soft robotic hands.

In the following, we describe the experimental apparatus for primitive extraction and the implementation on the robot side.

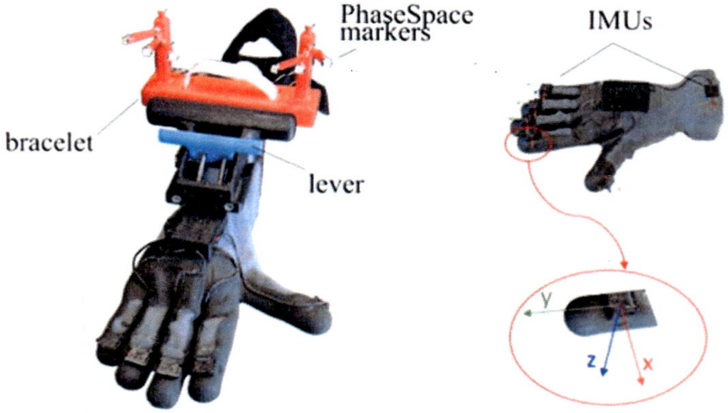

Fig. 10.2 Experimental setup used for primitive identification

10.2.1 The Human-Robot Interface and Motion Capture System

The SH was mounted on a human-robot interface [8] (hereinafter, we will refer to it as the handle), which enabled the user to control SH aperture through a lever.

In the primitive identification experiments we used the PhaseSpace Motion Capture system to record the pose of the wrist of the user. More specifically, the system recorded the x-y-z coordinates of eight active infra-red markers (sampling frequency 480 Hz) attached to a bracelet in ABS fastened on the handle immediately before the SH.

During the experiments, an off-the-shelf working glove with padded rubber surface was placed on the SH to increase contact compliance and grip.

Five IMUs were glued on the glove, in correspondence with the back of the distal phalanx of each finger, roughly speaking where the "nails" of the SH should be (Fig. 10.2). An additional IMU was placed on the back of the hand to be able to compensate both for acceleration caused by movement of the hand and for the contribution due to the gravity. In other terms, the accelerometer mounted close to the wrist of the hand (from now on it will be called reference IMU) is used to reject the common mode noise, as it follows. First, a calibration procedure computes five rotation matrices that describe the orientations of the fingertips IMUs w.r.t. the reference IMU, through a passive complementary filter [37]. Second, during the task, for each step the accelerations read from the fingertips IMUs are expressed in the frame associated to the reference IMU, then the common mode between reference and fingertips IMUs is removed.

We used the InvenSense MPU-9250 IMU, endowed with an on-board Digital Motion Processor (DMP) (capable to process complex 9-axis Motion Fusion algorithms), but relying only on the acceleration signals provided by the embedded 3-axis accelerometer. Data communication was realized via a custom made electronic board through RS485 with a frequency of 60 Hz (for more details on the platform please refer to [14]).

10.3 Grasp Primitives

To build a database of primitives we performed first an acquisition phase, where the poses of the wrist of the human user maneuvering the SH were recorded as well as the acceleration signals, and then an extraction phase, where we defined the primitives (i.e. wrist pose and hand closure) w.r.t. the contact data provided by IMU-measured accelerations.

10.3.1 Data Acquisition

During the acquisition phase, we asked a human user to wear the handle and to control the SH to grasp an object, which was placed on the top of a pole, from different approaching directions. The user was instructed to trigger the grasp (re-)action as soon as he realized the contact. To do it, he relied on both visual information and contact vibration conveyed by the handle.

Since the SH was endowed with the sensing glove and the wrist bracelet, we were able to acquire both the *Detection* (acceleration contact signals) and the *Motion* (wrist motion alongside the hand closure commands). For what concerns the grasped object, we focused on a single spherical object, i.e. a tennis ball, choosing to leverage upon SH adaptability to generalize to different items (see Sect. 10.4). The ball was placed on top of a pole and attached to the center of a small table (50 $\times$ 40 cm) covered with Velcro, at around 120 cm height. This enabled the subject to freely approach the object from several directions: in particular, we chose 13 directions as a good trade-off between exhaustiveness and ease of implementation, as reported in Fig. 10.3. We decided to not consider bottom-up approaching direction, since we associated this to unwanted and undesired contact with external objects in HRI (i.e. false positives, see Sect. 10.4). At the same time, the fact that the object was attached through Velcro at the table required a firm grasp to remove it from the support.

For each approaching direction, the starting configuration was about 20 cm from the object, along that direction. The subject was asked to try to move at constant speed. In addition to that, the subject was asked to initiate contact with the object on the distal phalanges of the SH. The approach and grasp depended on subject's skills: for this reason, preliminary training with the handle was imparted to the subject, until motion with the SH, contact detection and object grasp were as natural as possible.

The experiment protocol to gather relevant data was as follows:

1. We randomly chose an approaching direction, hence a starting configuration (SH palm down, parallel to the ground).

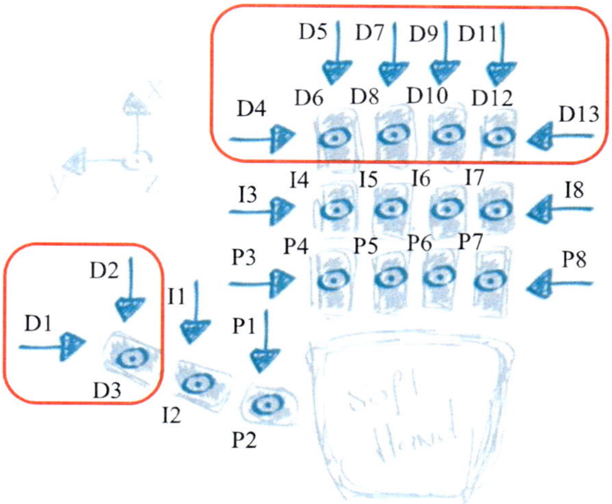

Fig. 10.3 The regions of contact during the acquisition phase and the 13 approaching directions D_i, defined w.r.t. the system of reference in figure. In this phase contacts occur only on the sensorized Distal (D) phalanx (Highlighted in red rectangles), while during the experiments Intermediate (I) and Proximal (P) phalanges were also considered

2. A visual signal was sent to the subject to let him know that the SH was in the starting configuration.
3. The subject moved the SH and initiated a grasp as soon as the hand was in contact with the object.
4. To consider the grasp successful, the subject had to detach the object from the table, lift it up and hold it for 15 s. If this was not verified, he had to repeat the action for the current approaching direction.
5. A visual signal was sent to the subject to let him know that he was allowed to lower the SH and release the object.
6. A 5 min rest followed, after which the cycle started over from (1) for a different approaching direction/starting configuration.

Acceleration data and wrist poses were recorded through the IMUs, and PhaseSpace system, respectively, and synchronized. Each approaching direction was presented three times in the randomized sequence; we used the average values of the sensed signals to perform the primitive extraction phase, for an overall number of 13.

10.3.2 Primitive Identification

The goal was to build a map from IMU measured acceleration to wrist pose evolution (in particular, the final wrist pose used for the grasp action), in order to devise grasping strategies. To do that, we performed the following steps:

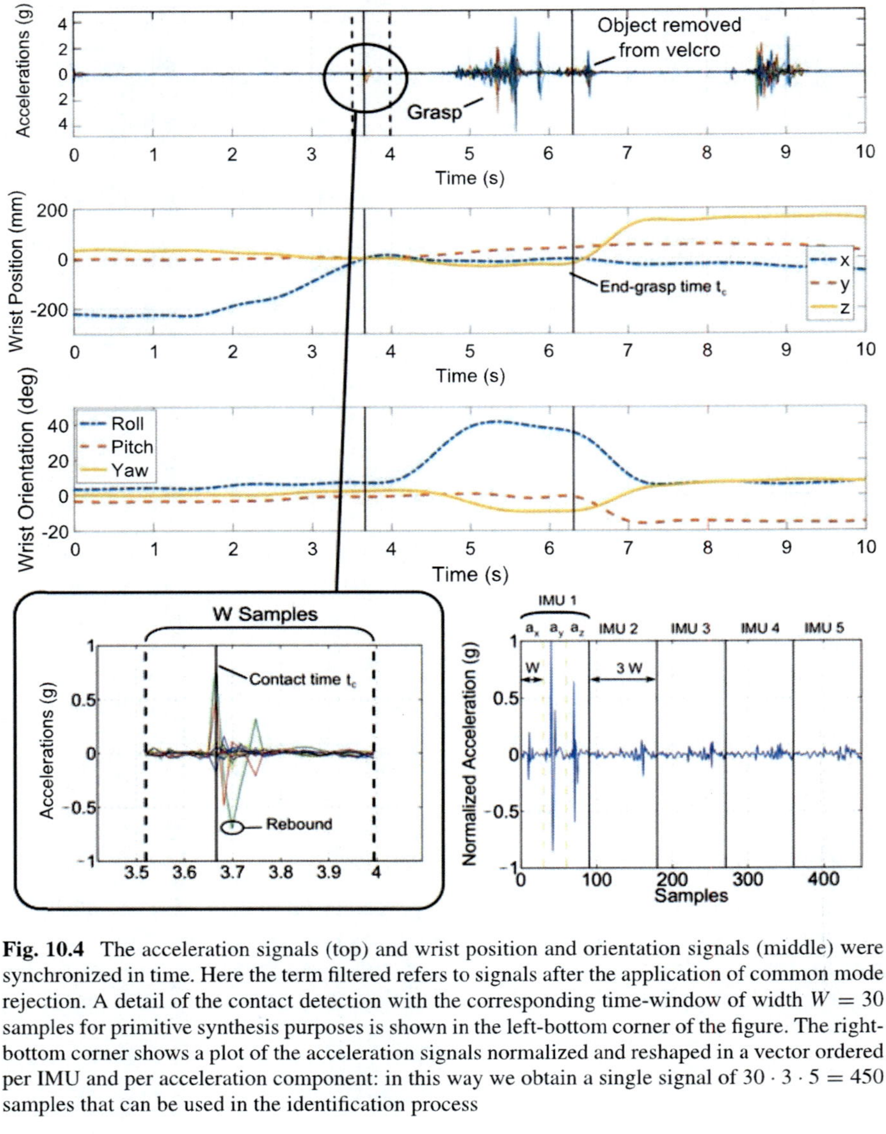

Fig. 10.4 The acceleration signals (top) and wrist position and orientation signals (middle) were synchronized in time. Here the term filtered refers to signals after the application of common mode rejection. A detail of the contact detection with the corresponding time-window of width $W = 30$ samples for primitive synthesis purposes is shown in the left-bottom corner of the figure. The right-bottom corner shows a plot of the acceleration signals normalized and reshaped in a vector ordered per IMU and per acceleration component: in this way we obtain a single signal of $30 \cdot 3 \cdot 5 = 450$ samples that can be used in the identification process

A. **Common-mode rejection**: using the IMU mounted on the back of the hand;
B. **Cropping**: We identified the moment when the contact occurred, and selected a time window where to perform the acceleration processing. In particular, we chose to identify the contact instant t_c as the sample where the first acceleration spike (greater than 0.5 g) occurred, followed by a lower rebound. We heuristically identified this threshold since it guarantees a good signal to noise ratio for contact detection. The window width was 30 samples, starting 10 samples before the contact time (Fig. 10.4);
C. **Reshaping**: Acceleration components in the considered time window were reshaped in a single vector, as detailed at the bottom of Fig. 10.4;
D. **Normalization**: All acceleration values were normalized by the maximum absolute value in the considered time window.

At the end of this procedure we have a vector A of acceleration scalar components in $\mathbb{R}^{3nW}$, where $n = 5$ is the number of fingertip inertial sensors and $W = 30$ is the contact window size.

To complete the primitives extraction, wrist poses need to be associated to the previously considered acceleration values. Wrist poses can in general be represented by six-dimensional vectors, containing position and Roll-Pitch-Yaw orientation; however, since the roll variation was significantly greater than the other two during the grasp phase, as depicted in Fig. 10.4, we chose to simplify the representation of the problem and only consider Roll values for primitives extraction (corresponding to the wrist rotation around the forearm axis). This result is not surprising, since during primitive identification the human subject maneuvered the SH through the handle, which constrained wrist movements mainly along Roll direction. Future works will consider more complex designs, e.g. using the SH in conjunction with an actuated artificial wrist as in [10].

We are thus considering relative translations and Roll values of the final wrist pose measured at the end-grasp time t_{eg}. For a correct association, we need to select the samples of interest: this was done by cropping data considering a window that begins at the contact time t_c and ends at t_{eg} (see Fig. 10.4, middle), which we identified as the time when a variation greater than 5 mm occurs on the vertical (z) component of the human user wrist position. The final wrist pose was then related to the corresponding accelerations in A. At the end of this phase, we got a set of 13 final wrist poses, each associated to acceleration profiles at t_c detected on the distal phalanges of robotic fingers (see Fig. 10.5). It can be interesting to note that the primitives for contacts at thumb are very similar each other: this is caused by the fact that the primitive grasp action generates positive roll rotations, independently from the direction of contact. However, for the moment, we decided to maintain and implement all these primitives.

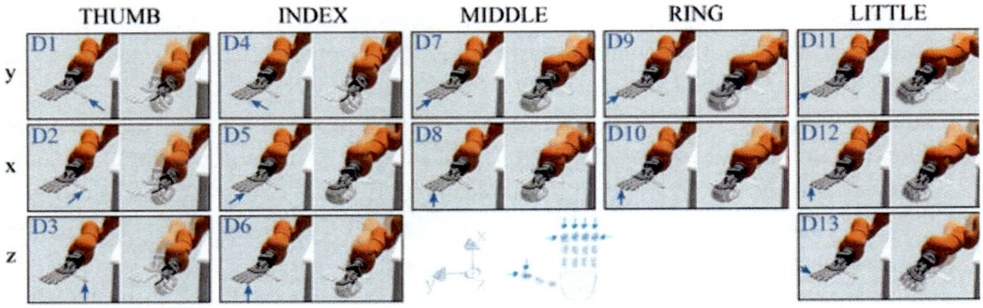

Fig. 10.5 Visual representation of grasp primitives synthesized from the experiments. Contact and sensing areas are indicated with blue arrows, while the corresponding motion primitive are in fade

10.4 Experiments

In this section we first describe how we implemented the primitives for the robotic arm and then how we validated our approach through experiments.

10.4.1 Implementation

In order to implement the 13 grasp primitives previously described we developed a software library composed of two main parts: *Detection* and *Motion*. As the names suggest, the *Detection* implements the sensory system. Here, the raw signals from the IMUs are continuously acquired and quickly processed to infer information about the external environment (contact detection) and to trigger a desired reaction (retrieved from the database of primitives), according to the main steps described in the following lines. The *Motion* part is in charge of the execution of the triggered reaction. During the initialization phase, the robotic system was placed in a starting posture where the proposed perception-action loop was engaged at. After the execution of any primitive, the system came back to the starting posture. The *Motion* and the *Detection* play the server and the client, respectively. The preemption policy was such that only one goal can be active until the action was finished. In other words, it means that once a primitive motion starts, the client becomes "deaf" to any contact until the complete grasp primitive has finished the execution, and the system has came back to the starting posture.

The Detection logic selects the wrist pose, based on the cross-correlation between the acceleration profile recorded by the robot and the 13 acceleration profiles obtained from human demonstration. Of note, each acceleration profile is associated to a given wrist pose, i.e. motion primitive, as described in Sect. 10.3.2. More specifically, this procedure selects the primitive for which the associated acceleration profile presents the highest correlation with the profile observed by the robot.The procedure is structured as follows:

Table 10.1 List of objects

(A) screwdriver	(B) wrench	(C) reel
(D) battery (AA)	(E) pincers	(F) plier
(G) hammer	(H) hot–glue gun	(I) caliper
(J) pen	(K) stapler	(L) bottle
(M) torch	(N) computer mouse	(O) cell phone
(P) eraser	(Q) lighter	(R) table tennis ball
(S) human hand	(T) mug	(U) can
(V) teddy bear		

1. A stream of IMUs data was continuously monitored and common mode rejection applied.
2. Acceleration peaks (threshold 0.5 g) followed by a rebound of the signal revealed the contact occurrences.
3. After a contact was heuristically detected, a time window was selected for the acceleration data stream where the measurements were reshaped and normalized as done for the phase of extraction of primitives (see the bottom right image in Fig. 10.4).
4. Cross correlation of the acceleration between the measured and 13 stored profiles over the time window, in order to select the most similar associated wrist pose.

Steps 1–3 are analogous to the ones reported in Sect. 10.3 for the extraction of grasp primitives.

The method described above is implemented for both the first handover task and the second experiment presented in this section, where the object is grasped from a table.

10.4.2 Experiments with Robotic Arm: Handover Task

For the validation phase the SH was mounted on the Kuka Light Weight Robot (LWR) 4+. The same sensorized glove used for the primitives acquisition was put on the hand.

In the starting configuration, the robotic system was placed with the SH palm facing down, similarly to what was done by the human user in the previous phase. Then, a person, different from the one who performed primitive identification, handed over an object to the SH in a given direction and the acceleration signals due to the contact triggered a grasp primitive, with the directions as shown in Fig. 10.3. The wrist position associated with the triggered primitive is fed as setpoint to a standard motion planning algorithm [36], with the SH closing to grasp the object after the wrist reaches the target posture. The detection process had a duration of about 200 ms, after which the grasp was completed in about 5 s.

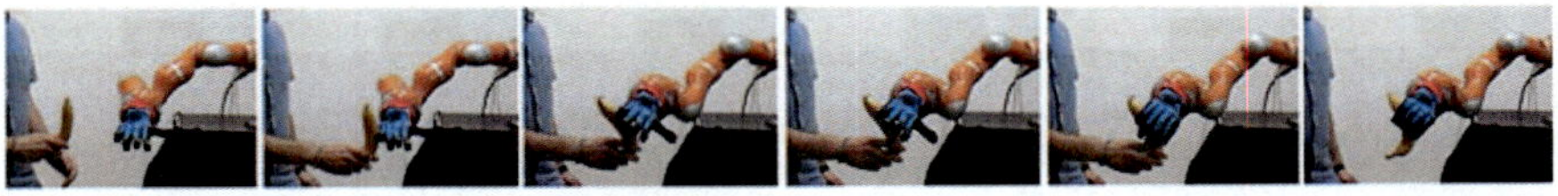

Fig. 10.6 Snapshots of one of the experiments on human-to-robot handover. The contact occurred at the little finger

A grasp was considered successful if the grip held for 15 s, and was robust to external disturbances applied by the experimenter. More specifically, at the end of the grasp, the experimenter hit the SH: in this phase, as soon as any acceleration above 0.5 g was detected from any finger, the SH opened releasing the object and the manipulator returned to the starting configuration, waiting for another input from the user (see Fig. 10.6 for an example. This procedure was repeated three times per object (see Table 10.1), for every direction and target zone on the hand (Fig. 10.3), for a total of 1914 trials. The total average success percentage was around 86%. As previously mentioned in the introduction, the first mandatory requirement for human-robot-interaction is user's safety. To meet this requirement, in our experiments, we considered objects coming in contact with the hand from underneath or from a side. In other terms, we do not want any primitive to be executed for contacts from above the hand, that could be risky for the user or affect task success, e.g. leading to kinematic singularities of the robot arm. Of note, we do not want to claim anything about the safety of our procedures against unforeseen human movements: in these cases, the inclusion of a feedback action is needed as discussed in Sect. 10.2 and investigate in future works. To detect this kind of contacts as false positives, we used the readings from the gyroscope of IMUs to complete accelerometer data: whenever an acceleration peak was detected, if the z-axis reading from the gyroscope was positive and above a threshold of 15 deg/s, then no primitive was implemented. We tested the effectiveness of this approach by hitting the back of the fingers in different zones 10 times, using a wrench: 88% of false positive contacts were correctly rejected. All the code developed in this work is in ROS.

It is worth pointing out that, respect to the grasp primitive identification described in Sect. 10.3, in this experiment we used a considerably higher number of objects (22 against the single tennis ball) and contact approaches (29 against 13). That is, the proposed solution was able to generalize with respect to those variables, also leveraging on the adaptability of the SH.

We can have a better insight in the results by considering the number of successes by object (Table 10.2) and by direction (Table 10.3). For what concerns the former, the lowest success rate (around 68%) was obtained for the bottle, which resulted difficult to grasp for contacts occurred at the distal phalanx of the ring and little finger. Only two other items had a success rate inferior to 80% (the hammer with a 75% success percentage, and the caliper with 78%), on a total of 22 objects. This can be ascribed to the long shape and asymmetric inertial properties, which also affected performance for the bottle that had some water moving inside. In these cases, a feedback action would be required to modify the grasp location on the object, thus counterbalancing

Table 10.2 Results by object

Object	Successes	Failures	Success %
(A)	71	16	81.61
(B)	84	3	96.55
(C)	77	10	88.51
(D)	78	9	89.66
(E)	77	10	88.51
(F)	80	7	91.95
(G)	65	22	74.71
(H)	78	9	89.66
(I)	68	19	78.16
(J)	77	10	88.51
(K)	78	9	89.66
(L)	59	28	67.82
(M)	70	17	80.46
(N)	75	12	86.21
(O)	73	14	83.91
(P)	79	8	90.8
(R)	73	14	83.91
(S)	74	13	85.06
(T)	82	5	94.25
(U)	72	15	82.76
(V)	74	13	85.06
(W)	81	6	93.1

external torques due to long shape and inertial properties of some of the objects. Conclusions that can be drawn are that our method enables to generalize to different objects passed to the hand, thus opening interesting perspectives for human to robot handover, without any claim of exhaustiveness. To properly affirm that our techniques can be applied to real-world HRI tasks, the inclusion of feedback controller and a thorough testing with a larger number of naif human users are needed and will be performed as future works.

10.4.3 Experiments with Robotic Hand: Grasping an Object from a Table

In the previous subsection, we have reported an implementation of the reactive grasp approach we are proposing in a task where a human is handing over an object to the robot. While the success rate in that experiment was fairly high, since a human

Table 10.3 Results by direction

Direction	Successes	Failures	Success %
D1	58	8	87.88
D2	63	3	95.45
D3	61	5	92.42
D4	62	4	93.94
D5	63	3	95.45
D6	65	1	98.48
D7	65	1	98.48
D8	66	0	100
D9	59	7	89.39
D10	63	3	95.45
D11	56	10	84.85
D12	59	7	89.39
D13	53	13	80.3
I1	66	0	100
I2	58	8	87.88
I3	58	8	87.88
I4	63	3	95.45
I5	63	3	95.45
I6	63	3	95.45
I7	52	14	78.79
I8	48	18	72.73
P1	56	10	84.85
P2	54	12	81.82
P3	47	19	71.21
P4	50	16	75.76
P5	45	21	68.18
P6	54	12	81.82
P7	40	26	60.61
P8	35	31	53.03

experimenter was actively participating in the task it is not easy to discriminate between the contribution of the human and the contribution of the robot to the task. This is common in HRI, since humans naturally tend to adapt to robot behavior [15]. To preliminarily evaluate the success of our approach for robotic grasping, against possible helps from the human operator, we also performed a second set of experiments, where the object was autonomously grasped from a table. We placed different objects extracted from our list on a test surface, on which Velcro was attached in order to fix the starting position and to avoid undesired movements before and during the contact. This also required a firm robotic grasp to remove objects from

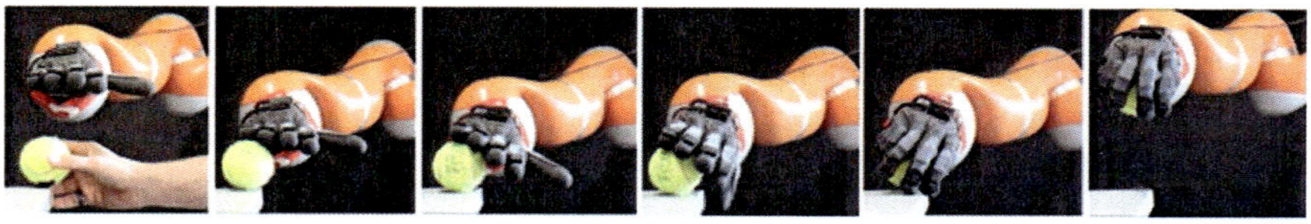

Fig. 10.7 Snapshots of one experiment performed on a table. Contact was on the L/R zone with the hand following traj. 1

the support. The robotic system was controlled to approach the object with three different trajectories:

1. *Vertical trajectory*—approaching direction from the top—contact with the bottom part of the phalanges;
2. *Horizontal trajectory*—approaching direction from the side—contact with the lateral index/little;
3. *Horizontal trajectory*—approaching direction from the front—contact with the frontal fingertips/thumb.

In Fig. 10.7 an example of reactive grasping using the trajectory no.1 is reported. The arm is controlled in order to cyclically follow the selected trajectory. Once the contact is detected from the IMUs, the logic board compute the correct grasp primitive using the approach reported in Sect. 10.4.1. We tested this approach using three objects with different shapes, i.e. a tennis ball, a teddy bear and a water bottle. We arbitrarily changed the contact point with the hand, targeting different regions of the phalanxes and varying robot motion velocity (from 11.5 to 17.5 cm/s). Without any claim of exhaustiveness, since only a reduced set of conditions was tested, we can affirm that the hand was able to grasp the object from the table while approaching on it with the internal phalanges, with the frontal side of the fingertips, with the lateral side of the thumb and the little. In the future, a more thorough validation will be performed. The aim of this second type of experiments was to show a certain degree of robustness of our techniques against possible helps from the users, which is intrinsically unavoidable in HRI. At the same time, the positive outcomes we have obtained, although preliminary, open interesting perspectives for purely tactile based autonomous grasps with soft end effectors.

10.5 Discussions and Conclusions

In this chapter, we have presented a minimalistic approach to endow soft manipulators with human-inspired touch-based sensory-motor grasp primitives in a simple human to robot object passing task. These primitives were then implemented to control the pose of the robotic wrist where the SH was mounted on. We have demonstrated how the adaptability of the SH enables to generalize these primitives for the successful grasp of a wide number of objects, considering a large set of object-to-hand approaching directions and regions of contact on the robotic fingers. We also performed preliminary experiments where the hand was controlled to autonomously grasp objects from a table. The latter results open fascinating perspectives for a new generation of soft robotic manipulators with embedded sensory-motor capabilities. By combining our feed-forward approach with a feedback component, these manipulators could take advantage from these capabilities, and purposefully exploit the contact and the interaction with the environment to increase their autonomy and extend their grasping capabilities, e.g. in purely touch-based environment exploration.

Envisioned applications of the approach reported in this chapter can be also within the framework of shared-control, e.g. in assistive robotics, where robot autonomy is used to help user's input for task accomplishment [21]. The integration of user's intention with the primitives described in this chapter could increase the effectiveness of assistive-rehabilitative systems for motor-impaired people. Indeed, our solution represents a perception-action algorithm that can autonomously generate goals (in our case, human-like object grasps), without the usage of markers on the object, which limit the applicability of shared-control systems only to a set of labeled targets.

Future works will further investigate these methods for a complete handover implementation, addressing all the phases described in [30], together with an implementation of a feedback-feedforward action which could increase safety and generarization. Additional experiments will be suitably designed to evaluate HR communication [18], as well as the effectiveness and users' acceptance of different motion planners. It is important to note that our approach combines the intrinsic adaptability of the SoftHand and tactile-triggered wrist primitive control. To fully highlight the contribution of the latter, future works will aim at providing a quantitative comparison of grasping success/failure results between with and without actively selecting grasp primitives. Finally, we will continue to investigate with more quantitative results the preliminary evidence that demonstrate the success of our approach when the robot grasps the objects from a supporting table by itself—hence without any active role of the human operator.

References

1. Argall, B.D., Billard, A.G.: A survey of tactile human-robot interactions. Robot. Auton. Syst. **58**(10), 1159–1176 (2010)
2. Argall, B.D., Chernova, S., Veloso, M., Browning, B.: A survey of robot learning from demonstration. Robot. Auton. Syst. **57**(5), 469–483 (2009). https://doi.org/10.1016/j.robot.2008.10.024. http://www.sciencedirect.com/science/article/pii/S0921889008001772
3. Bauer, C., Milighetti, G., Yan, W., Mikut, R.: Human-like reflexes for robotic manipulation using leaky integrate-and-fire neurons. In: IEEE/RSJ International Conference on Intelligent Robots and Systems, pp. 2572–2577 (2010)
4. Bekiroglu, Y., Damianou, A., Detry, R., Stork, J.A., Kragic, D., Ek, C.H.: Probabilistic consolidation of grasp experience. In: 2016 IEEE International Conference on Robotics and Automation (ICRA), pp. 193–200. IEEE (2016)

5. Bianchi, M., Averta, G., Battaglia, E., Rosales, C., Bonilla, M., Tondo, A., Poggiani, M., Santaera, G., Ciotti, S., Catalano, M.G., et al.: Touch-based grasp primitives for soft hands: Applications to human-to-robot handover tasks and beyond. In: 2018 IEEE International Conference on Robotics and Automation (ICRA), pp. 7794–7801. IEEE (2018)
6. Bicchi, A., Kumar, V.: Robotic grasping and contact: a review. In: ICRA, vol. 348, p. 353. Citeseer (2000)
7. Bohg, J., Morales, A., Asfour, T., Kragic, D.: Data-driven grasp synthesis–a survey. IEEE Trans. Robot. **30**(2), 289–309 (2014)
8. Bonilla, M., Farnioli, E., Piazza, C., Catalano, M., Grioli, G., Garabini, M., Gabiccini, M., Bicchi, A.: Grasping with soft hands. In: 2014 14th IEEE-RAS International Conference on Humanoid Robots (Humanoids), pp. 581–587. IEEE (2014). https://doi.org/10.1109/HUMANOIDS.2014.7041421
9. Carlson, N.R.: Physiology of Behavior . Allyn & Bacon (1986)
10. Casini, S., Tincani, V., Averta, G., Poggiani, M., Della Santina, C., Battaglia, E., Catalano, M.G., Bianchi, M., Grioli, G., Bicchi, A.: Design of an under-actuated wrist based on adaptive synergies. In: 2017 IEEE International Conference on Robotics and Automation (ICRA), pp. 6679–6686. IEEE (2017)
11. Catalano, M.G., Grioli, G., Farnioli, E., Serio, A., Piazza, C., Bicchi, A.: Adaptive synergies for the design and control of the Pisa/IIT softhand. Int. J. Robot. Res. **33**(5), 768–782 (2014)
12. Chebotar, Y., Hausman, K., Kroemer, O., Sukhatme, G.S., Schaal, S.: Generalizing regrasping with supervised policy learning. In: International Symposium on Experimental Robotics, pp. 622–632. Springer (2016)
13. Dautenhahn, K., Saunders, J.: New Frontiers in Human Robot Interaction, vol. 2. John Benjamins Publishing (2011)
14. Della Santina, C., Piazza, C., Gasparri, G.M., Bonilla, M., Catalano, M.G., Grioli, G., Garabini, M., Bicchi, A.: The quest for natural machine motion: an open platform to fast-prototyping articulated soft robots. IEEE Robot. Autom. Mag. **24**(1), 48–56 (2017)
15. Edsinger, A., Kemp, C.C.: Human-robot interaction for cooperative manipulation: Handing objects to one another. In: The 16th IEEE International Symposium on Robot and Human interactive Communication, 2007. RO-MAN 2007, pp. 1167–1172. IEEE (2007)
16. Eppner, C., Deimel, R., Alvarez-Ruiz, J., Maertens, M., Brock, O.: Exploitation of environmental constraints in human and robotic grasping. Int. J. Robot. Res. **34**(7), 1021–1038 (2015)
17. Felip, J., Morales, A.: Robust sensor-based grasp primitive for a three-finger robot hand. In: 2009 IEEE/RSJ International Conference on Intelligent Robots and Systems, pp. 1811–1816 (2009). https://doi.org/10.1109/IROS.2009.5354760
18. Gharbi, M., Paubel, P.V., Clodic, A., Carreras, O., Alami, R., Cellier, J.M.: Toward a better understanding of the communication cues involved in a human-robot object transfer. In: 2015 24th IEEE International Symposium on Robot and Human Interactive Communication (RO-MAN), pp. 319–324. IEEE (2015)
19. Hsiao, K., Chitta, S., Ciocarlie, M., Jones, E.G.: Contact-reactive grasping of objects with partial shape information. In: 2010 IEEE/RSJ International Conference on Intelligent Robots and Systems (IROS), pp. 1228–1235. IEEE (2010)
20. Huber, M., Rickert, M., Knoll, A., Brandt, T., Glasauer, S.: Human-robot interaction in handing-over tasks. In: The 17th IEEE International Symposium on obot and Human Interactive Communication, 2008. RO-MAN 2008, pp. 107–112. IEEE (2008)
21. Javdani, S., Bagnell, J.A., Srinivasa, S.: Shared Autonomy Via Hindsight Optimization (2015). arXiv:1503.07619
22. Kawasaki, H., Mouri, T., Takai, J., Ito, S.: Grasping of unknown object imitating human grasping reflex. In: CD-ROM of Preprints of the 15th IFAC World Congress (2002)
23. Kehoe, B., Matsukawa, A., Candido, S., Kuffner, J., Goldberg, K.: Cloud-based robot grasping with the google object recognition engine. In: IEEE International Conference on Robotics and Automation, pp. 4263–4270 (2013)
24. Kruse, T., Pandey, A.K., Alami, R., Kirsch, A.: Human-aware robot navigation: a survey. Robot. Auton. Syst. **61**(12), 1726–1743 (2013)

25. Lemaignan, S., Warnier, M., Sisbot, E.A., Clodic, A., Alami, R.: Artificial cognition for social human-robot interaction: an implementation. Artif. Intell. **247**, 45–69 (2017)
26. Luberto, E., Wu, Y., Santaera, G., Gabiccini, M., Bicchi, A.: Enhancing adaptive grasping through a simple sensor-based reflex mechanism. IEEE Robot. Autom. Lett. **2**(3), 1664–1671 (2017)
27. Mahler, J., Pokorny, F.T., Hou, B., Roderick, M., Laskey, M., Aubry, M., Kohlhoff, K., Kröger, T., Kuffner, J., Goldberg, K.: Dex-net 1.0: a cloud-based network of 3d objects for robust grasp planning using a multi-armed bandit model with correlated rewards. In: 2016 IEEE International Conference on Robotics and Automation (ICRA), pp. 1957–1964. IEEE (2016)
28. Mainprice, J., Gharbi, M., Siméon, T., Alami, R.: Sharing effort in planning human-robot handover tasks. In: 2012 IEEE RO-MAN: The 21st IEEE International Symposium on Robot and Human Interactive Communication, pp. 764–770 (2012). https://doi.org/10.1109/ROMAN.2012.6343844
29. Mainprice, J., Hayne, R., Berenson, D.: Predicting human reaching motion in collaborative tasks using inverse optimal control and iterative re-planning. In: 2015 IEEE International Conference on Robotics and Automation (ICRA), pp. 885–892. IEEE (2015)
30. Medina, J.R., Duvallet, F., Karnam, M., Billard, A.: A human-inspired controller for fluid human-robot handovers. In: 2016 IEEE-RAS 16th International Conference on Humanoid Robots (Humanoids), pp. 324–331 (2016). https://doi.org/10.1109/HUMANOIDS.2016.7803296
31. Mittendorfer, P., Yoshida, E., Moulard, T., Cheng, G.: A general tactile approach for grasping unknown objects with a humanoid robot. In: 2013 IEEE/RSJ International Conference on Intelligent Robots and Systems, pp. 4747–4752 (2013). https://doi.org/10.1109/IROS.2013.6697040
32. Murphy, R.: Introduction to AI Robotics. MIT press (2000)
33. Quispe, A.H., Amor, H.B., Stilman, M.: Handover planning for every occasion. In: 2014 14th IEEE-RAS International Conference on Humanoid Robots (Humanoids), pp. 431–436. IEEE (2014)
34. Rautaray, S.S., Agrawal, A.: Vision based hand gesture recognition for human computer interaction: a survey. Artif. Intell. Rev. **43**(1), 1–54 (2015)
35. Romano, J.M., Hsiao, K., Niemeyer, G., Chitta, S., Kuchenbecker, K.J.: Human-inspired robotic grasp control with tactile sensing. IEEE Trans. Robot. **27**(6), 1067–1079 (2011)
36. Sánchez, G., Latombe, J.C.: A Single-Query Bi-Directional Probabilistic Roadmap Planner with Lazy Collision Checking, pp. 403–417. Springer, Berlin, Heidelberg (2003). https://doi.org/10.1007/3-540-36460-9_27. http://dx.doi.org/10.1007/3-540-36460-9_27
37. Santaera, G., Luberto, E., Serio, A., Gabiccini, M., Bicchi, A.: Low-cost, fast and accurate reconstruction of robotic and human postures via imu measurements. In: 2015 IEEE International Conference on Robotics and Automation (ICRA), pp. 2728–2735. IEEE (2015)
38. Santello, M., Bianchi, M., Gabiccini, M., Ricciardi, E., Salvietti, G., Prattichizzo, D., Ernst, M., Moscatelli, A., Jörntell, H., Kappers, A.M., et al.: Hand synergies: integration of robotics and neuroscience for understanding the control of biological and artificial hands. Phys. Life Rev. (2016)
39. Waldhart, J., Gharbi, M., Alami, R.: Planning handovers involving humans and robots in constrained environment. In: 2015 IEEE/RSJ International Conference on Intelligent Robots and Systems (IROS), pp. 6473–6478. IEEE (2015)
40. Williamson, M.M.: Postural primitives: Interactive behavior for a humanoid robot arm. In: Fourth International Conference on Simulation of Adaptive Behavior, pp. 124–131 (1996)
41. Yigit, S., Burghart, C., Woern, H.: Applying reflexes to enhance safe human-robot-co-operation with a humanlike robot arm. In: 35th International Symposium on Robotics (2004)
42. Zacharias, F., Schlette, C., Schmidt, F., Borst, C., Rossmann, J., Hirzinger, G.: Making planned paths look more human-like in humanoid robot manipulation planning. In: 2011 IEEE International Conference on Robotics and Automation (ICRA), pp. 1192–1198. IEEE (2011)

Chapter 11
Learning from Humans How to Grasp: Enhancing the Reaching Strategy

Soft hands are robotic systems that embed compliant elements in their mechanical design. This enables an effective adaptation with the items and the environment, and ultimately, an increase in their grasping performance. These hands come with clear advantages in terms of ease-to-use and robustness if compared with classic rigid hands, when operated by a human. However, their potential for autonomous grasping is still largely unexplored, due to the lack of suitable control strategies. To address this issue, in this letter, we propose an approach to enable soft hands to autonomously grasp objects, starting from the observations of human strategies. A classifier realized through a deep neural network takes as input the visual information on the object to be grasped, and predicts which action a human would perform to achieve the goal. This information is hence used to select one among a set of human-inspired primitives, which define the evolution of the soft hand posture as a combination of anticipatory action and touch-based reactive grasp. The architecture is completed by the hardware component, which consists of an RGB camera to look at the scene, a 7-DoF manipulator, and a soft hand. The latter is equipped with inertial measurement units at the fingernails for detecting contact with the object. We extensively tested the proposed architecture with 20 objects, achieving a success rate of 81.1% over 111 grasps.

11.1 Introduction

The execution of reliable and stable grasping with artificial hands is a main challenge in the robotics field, due to its practical relevance and theoretical complexity. The *classic* approach used to grasp with rigid robotic hands generally favored object-centric analytical solutions. More specifically, a set of available contact points is hypothesized, while their position and contact forces are evaluated from the object knowledge [2]. Although very elegant and theoretically sound, this approach has

G. Averta, *Human-Aware Robotics: Modeling Human Motor Skills for the Design, Planning and Control of a New Generation of Robotic Devices*, Springer Tracts in Advanced Robotics 145, https://doi.org/10.1007/978-3-030-92521-5_11

not yet produced the desired outcomes in practice. To address these limitations, in soft artificial hands part of the control intelligence has been directly embedded in their mechanism, through the purposeful introduction of elastic elements and under-actuation patterns [3, 20].

Thanks to their intrinsic compliance, soft hands can mold around the external items and exploit their environment, thus multiplying their grasping capabilities.

Several papers have shown that soft end-effectors can achieve high-level grasping performance when operated by humans (see e.g. [11, 15]). However, such level of dexterity is still unmatched in autonomous grasp execution. Indeed, *classic* approaches cannot be applied to this kind of hands, which—by their own nature—do not allow fingertips placement with the required precision and relative independence. On the contrary, data driven approaches could be the key to learn from humans how to manage soft hands, towards higher levels of autonomous grasping capabilities. Recently machine learning has become very popular for grasping generation, with positive results [22, 24, 26]. However, so far only few works in literature have applied learning methods in the control of soft hands.

In [14] learning by demonstration is combined with reinforcement learning to transfer grasping capabilities of known objects from a human operator to the robotic system. In [19] autoencoders and generalized regression neural networks are used to learn from examples generated by a human operator, how to manipulate previously unseen thin objects with a soft gripper. While very promising, both works show no generalization capabilities in terms of objects to be grasped. Two works recently attempted to go beyond this limitation. In [1] a library of reactive strategies was collected from a subject operating a soft hand, and successfully translated for robotic grasping of new items, in a human-robot handover scenario. In [5] a 3D convolutional neural network is trained with tens of thousands of labeled images. The network output provides the control input for the hand approaching direction.

These works represent an important step forward, not only in terms of performance, but more fundamentally for recognizing the complementary yet intertwined nature of machine learning methods and soft hands. Indeed, learning based techniques can only achieve solutions that are *close enough* to the desired ones, rather than exact. This uncertainty can be naturally compensated by the ability of soft hands to locally adapt to unknown environments.

The aim of this chapter is to build upon this principle, and fully exploit hardware adaptability to grasp a vast range of very different objects. To this end we propose a human inspired multi-modal, multi-layer architecture (Fig. 11.1) that combines feed-forward components with reactive sensor-triggered actions. We extensively tested the proposed architecture on a set of 20 objects previously unseen by the network. The orientation of the objects placed on a table also varied. We performed three repetitions for each condition, for a total of 111 tests, reaching an overall percentage of grasp success of 81.1%. The content of this chapter is adapted from [7].

A companion video is available at the following link: https://www.dropbox.com/s/00325t5bqi0zikt/SHP_Reactive_DL_ICRA2019_RAL_compressed.mp4?dl=0

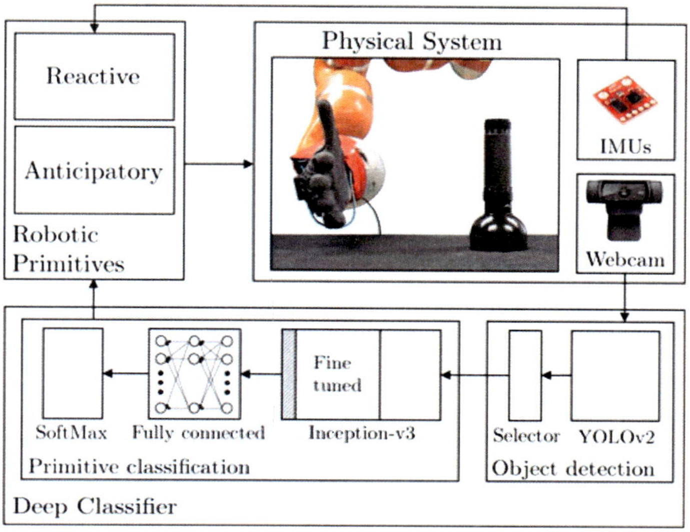

Fig. 11.1 High level organization of the proposed architecture, which combines anticipatory actions and reactive behavior. A deep classifier looks at the scene and predicts the strategy that a human operator would use to grasp the object. This output is employed to select the corresponding robotic primitive. These primitives define the posture of the hand over time, to produce a natural, human-like motion. The IMUs placed on the fingers of the hand detect the contact with the items and triggers a suitable reactive grasp behavior

11.2 Proposed Approach

Humans are able to accomplish very complex grasps by employing a vast range of different strategies [12]. This comes with the challenging problem of finding the right strategy to use for a given scenario. It is commonly suggested that the animal brain addresses this challenge by first constructing representations of the world, which are used to make a decision, and then by computing and executing an action plan [18].

Rather than learning a monolithic end-to-end map, we built the proposed architecture as combination of interpretable basic elements organized as in Fig. 11.1. The *intelligence* is here distributed on three levels of abstractions; (i) high level: a classifier which plans the correct action among all the available ones, (ii) medium level: a set of human-inspired low level strategies implementing both the approaching phase and the sensor-triggered reaction, (iii) low level: a soft hand whose *embodied intelligence* mechanically manages local uncertainties. All the three levels are human-inspired.

We realized the classifier through a deep neural network. This was trained to predict the object-directed grasp action chosen among nine human-labeled strategies, using as input only a first-person RBG image of the scene. These actions were implemented on the robotic side to reproduce the motions observed in the videos. A reactive component was then introduced, following the philosophy of [1]. This component take as input the accelerations coming from six IMUs placed on the soft hand to generate the desired evolution of the hand pose. The lower level of intelligence consists of the soft hand itself, which can take care of local uncertainties relying on its intrinsic compliance. Any robotic hand being soft and anthropomorphic both in its motions and in its kinematics can serve to the scope. Without loss of generality, we use here the Pisa/IIT SoftHand [8]. We report in the next sections the detailed description of these components.

To conclude, the main contributions of this chapter are:

- A deep neural network, which is able to predict with high accuracy the strategy that a human would adopt to grasp an object, using a first-person RGB image of the scene. This result is then used to plan a suitable primitive execution on the robotic side;
- A set of reactive primitives that reproduce human grasps, substantially extending [1];
- The definition and extensive experimental validation of an autonomous grasping system, which combines these two blocks with the adaptability of a soft hand.

11.3 Deep Classifier

The aim of this deep neural network is to associate to an object detected from the scene the correct primitive (i.e. hand pose evolution) humans would perform to grasp it. The deep learning model consists of two stages, depicted in Fig. 11.1: one for detecting the object, and the second one to perform the actual association with the required motion. Before going through the details of these two components, we briefly describe the phases of primitive extraction and labeling from human videos.

Dataset Creation and Human Primitive Labeling

We collected 6336 first person RGB videos (single-object, table-top scenario), from 11 right-handed subjects grasping the 36 objects in Fig. 11.2. The list of objects was chosen to span a wide range of possible grasps, taking inspiration from [10]. During the experiments, subjects were comfortably seated in front of a table, where the

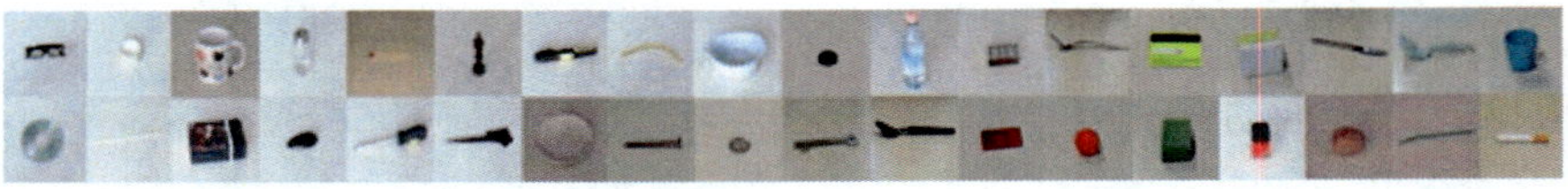

Fig. 11.2 The set of 36 objects used during human videos acquisition. Photos are not in scale. From left to right and from top to bottom we have: a pair of glasses, a tape, a coffee mug, a salt shaker, a match, a piece of chess, a marker, a rubber, a colander, a button, a bottle, a match box, a fork, a credit card, a box, a knife, a spatula, a glass, a CD, a shashlik, a book, a shell, a screwdriver, a comb, a plate, a screw, a coin, a key, a spoon, a game card, a ball, a sponge, a nail polish container, a walnut, a straw and a cigarette

object was placed. They were asked to grasp the object starting from a rest position (hand on the table, palm down). Each task was repeated 4 times from 4 points of view (the four central points of the table edges). To extract and label the strategies, we first visually inspected the video and identified ten main primitives

- Top: the object is approached from the top with the palm down parallel to the table. Object center is approximatively at the level of the middle phalanx. When contact is established, subjects close simultaneously all their fingers, achieving a firm power-like grasp.
- Top left: same for the top grasp, but with the palm rotated clockwise of at least $\pi/9$ radians.
- Top right: as for the top grasp, but with the palm rotated counter-clockwise of at least $\pi/9$ radians.
- Bottom: the object is approached from its right side. The palm is roughly perpendicular to the table, but slightly tilted so that the fingertips are more close to the object than the wrist. When the contact is reached, the hand closes with the thumb opposing the four long fingers. This primitive is used to grasp large and concave objects, e.g. a salad bowl.
- Pinch: same as for the top, but the primitive concludes with a pinch grasp.
- Pinch left: same as for the top left, but the primitive concludes with a pinch grasp.
- Pinch right: same as for the top right, but the primitive concludes with a pinch grasp.
- Slide: the hand is placed on the object from above as to push it toward the surface. Maintaining this hand posture, the object is moved towards the edge of the table until it partially protrudes. A grasp is then achieved by moving the thumb below the object, and opposing it to the long fingers. This strategy is used to grasp objects whose thickness is smaller compared to the other dimensions, such as a book or a compact disk.
- Flip: the thumb is used together with the environment on one side, and the index and/or the middle on the opposite one, to pivot the object. The item rotates of about $\pi/2$ and then it is grasped with a pinch. This strategy is used to grasp small and thin objects, as a coin.
- Lateral: the same as for the top grasp, but the palm is perpendicular to the object during the approaching phase. This strategy is used to grasp tall objects, like a bottle.

The choice of these primitives was done taking inspiration from literature [10, 12], and to provide a representative yet concise description of human behavior, without any claim of exhaustiveness. Note that the selection of the action primitive is not only object-dependent but also configuration dependent. This is clear for the left/right modifier. Consider for example a bottle; if placed on its base it triggers a lateral grasp, while when laying down on its side induces a top grasp.

The first frame of each video showing only the object in the environment was extracted, and elaborated through the object detection part of the network (see next subsection). The cropped image was then labeled with the strategy used by the subject in the remaining part of the video. This is the dataset that we used to train the network.

11.3.1 *Object Detection*

Object detection is implemented using the state of the art detector YOLOv2 [21]. Given the RGB input image, YOLOv2 produces as output a set of labeled bounding boxes containing all the objects in the scene. We first discard all the boxes labeled as person. We assume that the target is localized close to the center of the image. Hence, we select the bounding box closest to the scene center. Once the object of interest is identified, the image is automatically cropped around the bounding box, and resized to 416×416 pixels (size expected by the subsequent layer). The result is fed into the following block to be classified.

11.3.2 *Primitive Classification*

Architecture

Instead of building from scratch a completely new architecture, we follow a transfer learning approach. The idea is to exploit the existing knowledge learned from one environment to solve a new problem, which is different yet related. In this way, a smaller amount of data is sufficient to train the model, and achieve high accuracy with a short training time. We select as starting point Inception-v3 [23], trained on the ImageNet data set to classify objects from images. We keep the early and middle layers and remove the softmax layer. In this way, we have direct access to the highly refined and informative set of neural features that Inception-v3 uses to perform its classification. It is important to note that the object signature is not one-to-one but it aims at extracting high level semantic descriptions that can be applied to objects with similar characteristics. On the top of the original architecture we add two fully connected layers containing 2048 neurons each (with ReLU activation function). These layers operate an adaptive non-linear combination of the high-level features discovered by the convolutional and pooling layers, further refining the information. In this way, the geometric features are implicitly linked each other to serve as the base for the classification. The output of the last fully-connected layer is thus fed into the softmax, which produces a probability distribution over the considered set of motion primitives. We chose the one with maximum probability as output of the network.

Training and Validation

We use the labeled dataset described above to train the network. The parameters of the two fully connected layers at the top of the Inception-v3 architecture are trained from scratch, while the original parameters of the network are fine-tuned. To this end we impose layer-specific learning rates. More specifically, we freeze the weights in the first 172 layers (over the total 249) of the pre-trained network. These layers capture indeed universal features like curves and edges that are also relevant to our problem. We instead use the subsequent 77 layers to capture dataset-specific features. However, we expect the pre-trained weights to be already good if compared to randomly initialized ones. Hence, we avoid to abruptly change them using a relatively small learning rate λ_{ft}. Finally, given that the weights of the two last fully connected layers are trained from scratch, we randomly initialize them and use a higher learning rate λ_{tr} w.r.t. the one we use in previous layers. We further reduce the risk of over-fitting by using dropout ; before presenting a training sample to the network, we randomly disconnect neurons from its structure (actually, this is implemented by masking their activation). Each neuron is removed with probability p_{drop}. In this way, a new topology is produced each time the network is trained, introducing variability and reducing the production of pathological co-adaptation of weights. We use Keras library for network design and training. All the procedures were executed trough an NVIDIA Tesla M40 GPU with 12GB of on-board memory.

To verify the generalization and robustness of primitive classification, we use hold out validation. The goal is to estimate the expected level of model predictive accuracy independently from the data used to train the model. We split our data set in: 70% objects for training, 20% objects for validation and 10% for testing. We maintained a balanced number of objects per class among over the three data sets. We trained 30 different network configurations using the *cross entropy* cost function to adjust the weights by calculating the error between the output of the softmax layer and the label vector of the given sample category. Each configuration was obtained by varying the most relevant model learning hyper-parameters, i.e. learning rates $\lambda_{\text{ft}} \in \{10^{-3}, 10^{-4}, 10^{-5}, 10^{-6}\}$ and $\lambda_{\text{tr}} \in \{10^{-2}, 10^{-3}, 10^{-4}\}$, dropout probability $p_{\text{drop}} \in \{0.4, 0.5, 0.6\}$, number of epochs in $\{10, 20, 30, 40\}$, and the batch size in $\{10, 20, 30, 40\}$. The training time for each network ranged from 1 to 5 h. We selected the configuration that provided the highest $f1$-score accuracy [25] on the validation data set—which is 97%. The selected hyper-parameters are $\lambda_{\text{ft}} = 10^{-5}$, $\lambda_{\text{tr}} = 10^{-3}$, $p_{\text{drop}} = 0.5$, 30 epochs, and batches size 20.

With such parameters, the network is able to classify the primitives in the test set with an accuracy ranging from 86 to 100%, depending on the primitive, and 95% on average. Fig. 11.3 shows the normalized accuracy of the classifier for all ten classes. Visually inspecting the results reveals two main causes behind the occasional failures of the network. The first one is a limitation in the problem formulation itself, which makes intrinsically not possible to achieve 100% classification accuracy. Indeed, it seldom occurs that the same object in the same configuration is grasped in two different ways by two subjects. This happens for example for the coin, which is often grasped through a flip, while sometimes slide primitive is used instead. The second cause is connected to the fact that using only a single RGB image, the network sometimes misinterprets the object size. This could, for example, lead to predict a top grasp rather than a bottom grasp for a bowl, since this object may be interpreted as a ball-like item. In future work we will consider the use of a stereo camera to prevent this issue.

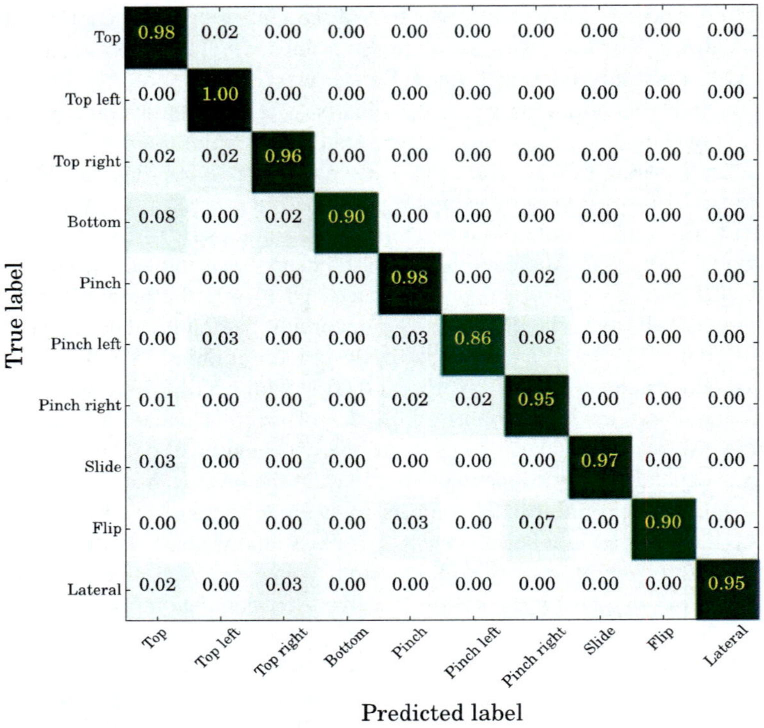

Fig. 11.3 Confusion matrix summarizing the performance of the proposed deep classifier on the test set. Each entry shows the rate at which the primitives identified by the row labels are classified as the ones identified by the column labels. Rate is also color coded, from low rate coded with white to high rate coded with dark green

11.4 Robotic Grasping Primitives

In [16], Johansson and Edin affirm that the Central Nervous System "*monitors specific, more-or-less expected, peripheral sensory events and use these to directly apply control signals that are appropriate for the current task and its phase*". These signals are largely precomputed (i.e. anticipatory, or feedforward). Driven by this observation, we decided to implement the robotic grasping strategies relying mostly on anticipatory actions. To do this, we took inspiration from the visual inspection of the videos described in the previous section, and decided to trigger primitive execution by specific events. The first event is generated by the detection of an object and scene classification. This triggers one primitive among all the available ones. We do not consider here flip, which can not be implemented by the soft hand that we use in this chapter. As a trade-off between performance and complexity, we divide all primitives in two phases: i) approach and ii) reactive grasp. The transition between the first and the second phase is triggered by a contact event, detected as an abrupt acceleration of the fingertips (as read by IMUs).

11.4.1 Experimental Setup

While the proposed techniques are not specifically tailored on this specific setup, it is convenient to introduce it here to simplify the description of the next subsections (see Sect. 11.2). The robotic architecture is composed of two main components: a KUKA LWR-IV arm, and a Pisa/IIT SoftHand [8] as end effector. This anthropomorphic soft hand has 19 degrees of freedom, and only one degree of actuation. The *intelligence embodied* in the hand mechanics is to be considered as an integral part of the control architecture itself, rather than as a simple effector to act onto the environment. A RGB camera is placed on the top of the manipulator to simulate a first-person point-of-view. The robotic hand is equipped with IMUs for contact detection, triggering reactive strategies for grasping. The principal reference frames used in our control framework are depicted in Fig. 11.4.

11.4.2 Approach Phase

During the approach phase, human hand tends to follow straight lines connecting the starting position and the target [13]. We reproduce this behavior through the simple trajectory

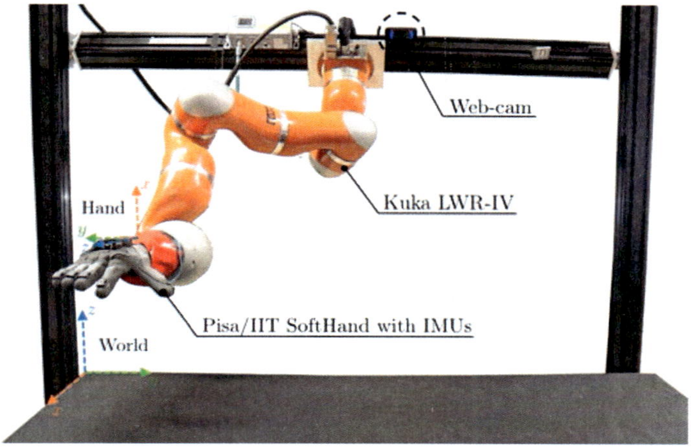

Fig. 11.4 Experimental setup. The Pisa/IIT SoftHand is mounted as end effector of a Kuka LWR. A web-cam records the scene from a first-person point of view. The hand is equipped with inertial measurement units used to detect contacts. Reference frames are reported

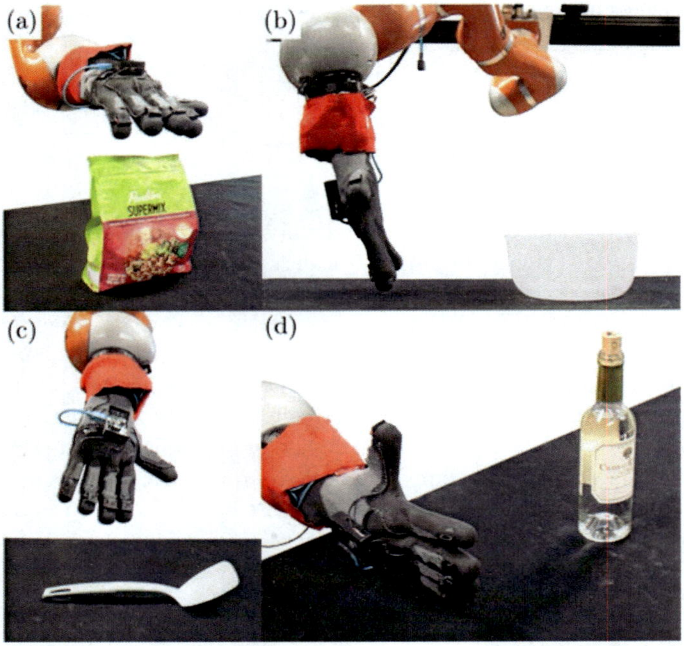

Fig. 11.5 Four significant relative object-hand postures assumed by the hand during the approaching phase. Starting from these initial configurations, the hand translates until a contact is detected by the IMUs. Directions of translation are perpendicular to the table for top and pinch primitives, and parallel to it for lateral and bottom

$$x(t) = x_0 + d\,t\ , \quad Q(t) = Q_0\ , \tag{11.1}$$

where $x \in \mathbb{R}^3$ is the hand base frame position in Cartesian coordinates, and $Q \in \mathbb{R}^4$ its orientation as quaternion, both expressed in global coordinates. $x_0 \in \mathbb{R}^3$ and $Q_0 \in \mathbb{R}^4$ are the initial position and orientation, while $d \in \mathbb{R}^3$ is the direction of approach. All these three quantities are defined by the selected primitive, and dictated by the aim of heuristically reproducing as close as possible the human behavior observed in the videos. Fig. 11.5 shows photos of the hand in $t = 0$ for top, pinch, lateral and bottom grasps. Table 11.1 summarizes directions of approach and initial orientations for all the primitives.

11.4.3 Grasp Phase

The grasp phase is when the grasp actually happens, and thus where the primitives differentiate more from each others. When not differently specified, translations and rotations are here expressed in hand coordinates.

Table 11.1 Initial orientation Q_0 and normalized direction of approach $\hat{d}$ for each primitive

Strategy	Q_0^{T}	$\hat{d}^{\mathrm{T}}$
Top	[0.0 0.711 0.0 0.703]	[0 0 − 1]
Top left	[0.269 0.6570 − 0.2721 0.6496]	[0 0 − 1]
Top right	[0.269 − 0.657 − 0.272 − 0.649]	[0 0 − 1]
Bottom	[0.145 − 0.696 0.701 0.030]	[0 1 0]
Pinch	[0.084 0.816 0.17 0.458]	[0 0 − 1]
Pinch left	[0.116 0.733 0.483 0.463]	[0 0 − 1]
Pinch right	[0.186 0.890 − 0.110 0.400]	[0 0 − 1]
Slide	[0.0 0.711 0.0 0.703]	[0 0 − 1]
Lateral	[0 − 1 0 0]	[0 1 0]

Top and Lateral Grasps

The reactive grasp framework leverages on a dataset of 13 prototypical rearrangements of the hand, extracted from human movements. In [1], a subject was asked to reach and grasp a tennis ball while maneuvering a Pisa/IIT SoftHand. The grasp was repeated 13 times, from different approaching directions. The user was instructed to move the hand until the contact with the object, and then to react by adapting the hand/wrist pose w.r.t. the object. Poses of the hand were recorded through a PhaseSpace motion tracking system. We subtract from the hand evolution recorded between the contact and the grasp (T represents the time between them) the posture of the hand during the contact. The resulting function $\Delta_i : [0, T] \to \mathbb{R}^7$ describes the rearrangement performed by the subject to grasp the object. Acceleration signals $\alpha_1 \ldots \alpha_{13} : [0, T] \to \mathbb{R}^5$ were measured too through the IMUs. To transform these recordings into a grasping strategy, we considered the acceleration patterns as a characteristic feature of the interaction with the object. When the Pisa/IIT SoftHand touches the object, IMUs read an acceleration profile $a : [0, T] \to \mathbb{R}^5$. The triggered sub-strategy is defined by the local rearrangement Δ_j, with

$$j = \arg\max_i \int_0^T a^{\mathrm{T}}(\tau)\alpha_i(\tau)\mathrm{d}\tau \; . \tag{11.2}$$

When this motion is completely executed, the hand starts closing until the object is grasped. This procedure proved its effectiveness in preliminary power grasp experiments on objects approached similarly as specified here by the top primitive [1]. We extend here its use to top left, top right and lateral strategies.

Bottom

To mimic human behavior described in previous section, when a contact is detected we rotate the hand along x of $\pi/3$ and translate 300mm along y. In this way the palm base moves over, and the thumb can enter into the concave part of the object during hand closure.

Pinches

In pinch, left pinch and right pinch strategies the hand just closes without changing its pose.

Slide

To mimic the human behavior we realized an anticipatory routine composed of the following sub-phases, triggered by the initial contact with the object and the environment:

(i) apply a force on the object along x axis to maintain the contact during sliding, by commanding a reference position to the hand 10 mm below the contact position;
(ii) slide the object towards the edge of the table,
(iii) unload the contact to avoid pushing the object out of the table, by translating 10 mm along x,
(iv) rearrange the hand to favor the grasp, by translating 100 mm along x and 50mm along z, and rotating along y of $\pi/12$ radians,
(v) close the hand.

11.4.4 Control

A Jacobian based inverse kinematic algorithm is performed to obtain desired joint positions q_r from the prescribed end effector evolution. A joint-level impedance control is used to realize the motion, with $K = 10^3 \frac{\text{Nm}}{\text{rad}}$ as stiffness and $D = 0.7 \frac{\text{Nms}}{\text{rad}}$ as damping for each joint. The control law is $\tau(t) = Ke(t) + D\dot{e}(t) + \mathbb{D}(q, \dot{q})$, where τ are the applied joint torques, $e = q_r - q$ and $\dot{e} = \dot{q}$ are the error at joint level and its derivate. $\mathbb{D}$ is a compensation of the robot dynamics evaluated by the KUKA embedded controller. All the control and sub-strategies implementation were performed in ROS.

11.5 Experimental Results

We test the effectiveness of the proposed architecture by performing table-top object grasping experiments. A table is placed in front of the system, as depicted in Fig. 11.4. The object is placed by an operator approximatively in the center of the table. RGB information from the web-cam triggers scene classification through the proposed deep neural network, which is followed by primitive execution. The task is repeated three times. The exact position of the object and its orientation vary each time, the first in a circle of radius ~100 mm, the second in the full angle range. All the process is repeated for each of the 20 objects depicted in Fig. 11.6, chosen so as to elicit different grasping strategies. Objects number 5, 6, 7, 8, 9, 10, 16, and 19 are classified with a different strategy depending on their positioning and orientation. We consider three

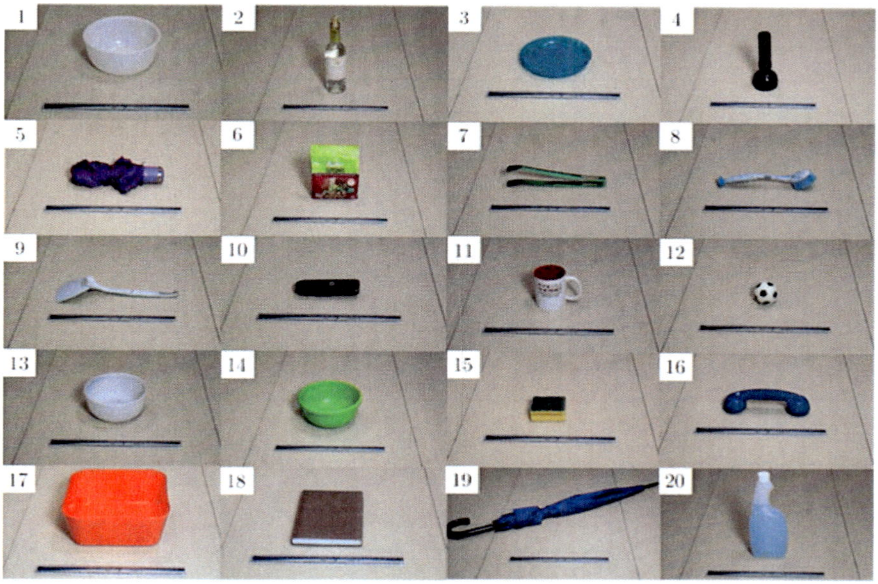

Fig. 11.6 Set of objects used in the experimental validation. None of them was part of the set used during training. A 30cm ruler is present in all the photos to help in qualitatively understanding object sizes

tests for each possible classification. The total amount of grasp tested is 111. None of the selected objects was used during the network training phase.

Table 11.2 summarizes the results in terms of the primitive used, successes, and failures for each object. The overall grasping success rate is 81.1%. A grasp was considered successful if the robot maintained it for 5 s (after which the hand automatically opens). Note that objects 12 and 15 elicit only the top grasp primitive, independently from their orientation. They are indeed both (almost-)rotationally symmetric, so the classifier does not take in account their orientation to select the grasp. Looking instead at primitive-specific success rates we obtain: Top 85.7% (Fig. 11.7a–h), Top left 73.3% (Fig. 11.7i–p), Top right 100% (Fig. 11.7q–x), Bottom 100% (Fig. 11.8), Pinch 55.6% (Fig. 11.9a–d), Pinch left 55.6% (Fig. 11.9e–h), Pinch right 66.7% (Fig. 11.9i-e), Slide 83.3% (Fig. 11.10), Lateral 86.7% (Fig. 11.11).

11.6 Discussion

This chapter represents a substantial improvement w.r.t. [1], where a similar success rate was obtained for human-robot handover, while only exploratory tests were performed on autonomous grasping. It is worth mentioning that this chapter repre-

Table 11.2 Strategy used, successes and failures for each grasp

Object	Strategy	Successes	Failures
1	Bottom	3	0
2	Lateral	2	1
3	Slide	2	1
4	Lateral	3	0
5	Top	3	0
	Top left	2	1
	Top right	3	0
6	Lateral	3	0
	Top	3	0
	Top left	2	1
	Top right	3	0
7	Pinch	3	0
	Pinch left	2	1
	Pinch right	3	0
8	Pinch	2	1
	Pinch left	2	1
	Pinch right	2	1
9	Pinch	0	3
	Pinch left	1	2
	Pinch right	1	2
10	Top	3	0
	Top left	2	1
	Top right	3	0
11	Lateral	3	0
12	Top	2	1
13	Bottom	3	0
14	Bottom	3	0
15	Top	2	1
16	Top	2	1
	Top left	3	0
	Top right	3	0
17	Bottom	3	0
18	Slide	3	0
19	Top	3	0
	Top left	2	1
	Top right	3	0
20	Lateral	2	1
Total	–	90	21

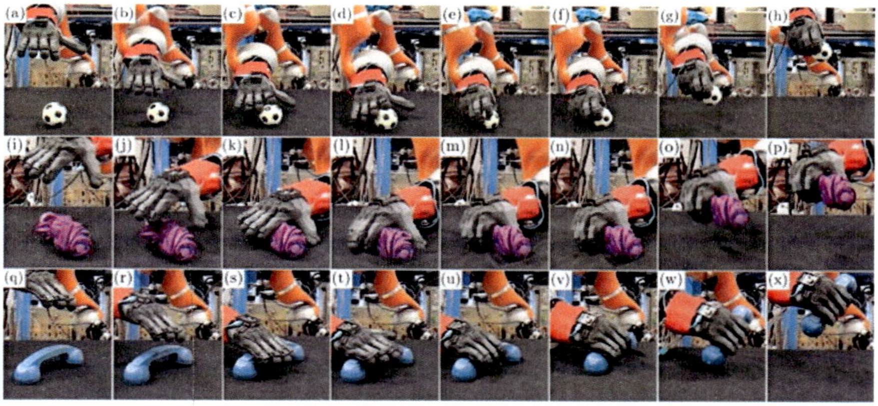

Fig. 11.7 Photosequences of grasps produced by the proposed architecture during validation: panels (**a–h**) present a top grasp of object 12, panels (**i–p**) a top-left grasp of object 5, and panels (q-x) a top-right grasp of object 16. Panels (**a–b**) depicts the approach phase. In (**c**) the contact is detected and classified using (11.2). In panels (**c–f**) the hand finely changes its relative position w.r.t. the object, as prescribed by the reactive routine, and grasps it. In (**g**) and (**h**) the item is firmly lifted

Fig. 11.8 Photosequence of a grasp produced by the proposed architecture during validation: Bottom grasp of object 14. The hand starts from the initial configuration of the primitive in panel (**a**). The contact happens in panel (**b**), triggering the reactive routine. In panel (**f**) the object is firmly lifted

sents - together with [5] - the first work that validates over a large set of objects a combination of deep learning techniques and soft hands. Any formal comparison between the two works is prevented by the fact that neither this nor the other paper used a standardized object set and protocol [4, 17]. With this as premise, it is worth noticing that our success rate is only fairly lower than the one in [5] (which reports 87% of successes, versus the 81% reported here). However, in our work, we considered a higher number of objects for the testing phase (20 versus 10), spanning a wider range of shapes, and with larger differences w.r.t. the learning set. Another interesting consideration arises from a more in-depth analysis of the results. If we remove from the statistics the three objects that would require a pinch grasps (i.e. 7, 8, 9) the success rate jumps over 88%. This can be explained by an intrinsic feature of the soft hand we used, which was designed to perform power grasp. Nonetheless, using the environment as an enabling constraint, the end-effector can still partially overcome this limitation. We are sure- and we will test it in the future—that using other versions of the SoftHand that can execute both pinch and power grasping see e.g. [9], the success rate will increase.

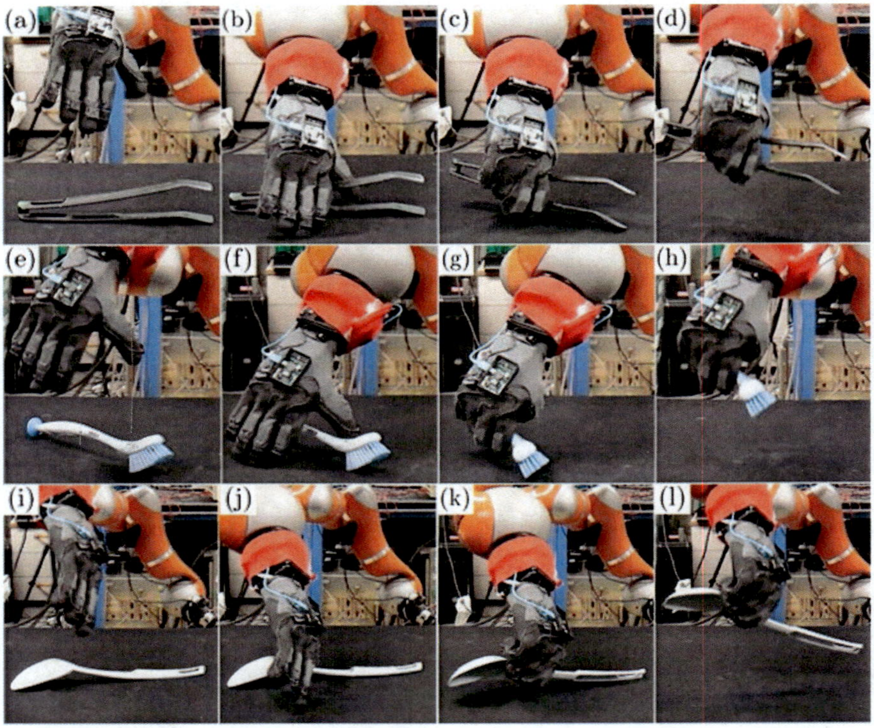

Fig. 11.9 Photosequences of grasps produced by the proposed architecture during validation: panels (**a–d**) present a pinch grasp of object 7, panels (**e–h**) a pinch-left grasp of object 8, and panels (**i–l**) a pinch-right grasp of object 9. Panel (**a**) shows the hand initial configuration. The contact is established in panel (**b**) through interaction with the environment, which also guides the hand towards the grasping achieved in panel (**c**). In (**d**) the object is firmly lifted

Fig. 11.10 Photosequence of a grasp produced by the proposed architecture during validation: slide grasp of object 3. Panels (**a–c**) depicts the approaching phase. In panels (**d–e**) the environment is exploited to guide the object to the table edge. In panels (**f–g**) the hand changes its relative position w.r.t. the object so to favor the grasp, which is established in panels (**h–i**). In panel (**j**) the item is firmly lifted

Fig. 11.11 Photosequence of a grasp produced by the proposed architecture during validation: lateral grasp of object 4. Panels (**a–c**) present the approaching phase. In panel (**c**) contact is detected, and in (**e**) the grasp is established. The object is lifted in panel (**f**)

11.7 Conclusions

In this chapter, we proposed and validated a data-driven human-inspired architecture for autonomous grasping with soft hands. We achieve this goal by: (i) introducing a novel deep neural network that processes the visual scene and predicts which action a human would perform to grasp the target object, (ii) formulating and implementing an artificial counterpart of the strategies that we observed in humans, (iii) combining them together in a integrated robotic platform, (iv) extensively testing the proposed architecture in the execution of 111 autonomous grasps, achieving an overall success rate of 81.1%. Future work will be devoted to testing the use of other anthropomorphic and soft hands within this framework, as e.g. SoftHand 2 [9], RBO hand [6].

References

1. Bianchi, M., Averta, G., Battaglia, E., Rosales, C., Tondo, A., Poggiani, M., Santaera, G., Ciotti, S., Catalano, M.G., Bicchi, A.: Tactile-based grasp primitives for soft hands: applications to human-to-robot handover tasks and beyond. In: 2018 IEEE International Conference on Robotics and Automation (ICRA). IEEE (2019)
2. Bicchi, A., Kumar, V.: Robotic grasping and contact: a review. In: ICRA, vol. 348, p. 353. Citeseer (2000)
3. Birglen, L., Laliberté, T., Gosselin, C.M.: Underactuated Robotic Hands, vol. 40. Springer (2007). https://doi.org/10.1007/978-3-540-77459-4
4. Calli, B., Singh, A., Bruce, J., Walsman, A., Konolige, K., Srinivasa, S., Abbeel, P., Dollar, A.M.: Yale-CMU-Berkeley dataset for robotic manipulation research. Int. J. Robot. Res. **36**(3), 261–268 (2017)
5. Choi, C., Schwarting, W., DelPreto, J., Rus, D.: Learning object grasping for soft robot hands. IEEE Robot. Autom. Lett. (2018)
6. Deimel, R., Brock, O.: A novel type of compliant and underactuated robotic hand for dexterous grasping. Int. J. Robot. Res. **35**(1–3), 161–185 (2016)
7. Della Santina, C., Arapi, V., Averta, G., Damiani, F., Fiore, G., Settimi, A., Catalano, M.G., Bacciu, D., Bicchi, A., Bianchi, M.: Learning from humans how to grasp: a data-driven architecture for autonomous grasping with anthropomorphic soft hands. IEEE Robot. Autom. Lett. **4**(2), 1533–1540 (2019)
8. Della Santina, C., Piazza, C., Gasparri, G.M., Bonilla, M., Catalano, M.G., Grioli, G., Garabini, M., Bicchi, A.: The quest for natural machine motion: an open platform to fast-prototyping articulated soft robots. IEEE Robot. Autom. Mag. **24**(1), 48–56 (2017)

9. Della Santina, C., Piazza, C., Grioli, G., Catalano, M.G., Bicchi, A.: Toward dexterous manipulation with augmented adaptive synergies: the Pisa/IIT softhand 2. IEEE Trans. Robot. **99**, 1–16 (2018)
10. Eppner, C., Deimel, R., Alvarez-Ruiz, J., Maertens, M., Brock, O.: Exploitation of environmental constraints in human and robotic grasping. Int. J. Robot. Res. **34**(7), 1021–1038 (2015)
11. Erdogan, C., Schröder, A., Brock, O.: Coordination of intrinsic and extrinsic degrees of freedom in soft robotic grasping. In: 2018 IEEE International Conference on Robotics and Automation (ICRA), pp. 1–6. IEEE (2018)
12. Feix, T., Romero, J., Schmiedmayer, H.B., Dollar, A.M., Kragic, D.: The grasp taxonomy of human grasp types. IEEE Trans. Hum.-Mach. Syst. **46**(1), 66–77 (2016)
13. Flash, T.: The control of hand equilibrium trajectories in multi-joint arm movements. Biol. Cybern. **57**(4–5), 257–274 (1987)
14. Gupta, A., Eppner, C., Levine, S., Abbeel, P.: Learning dexterous manipulation for a soft robotic hand from human demonstrations. In: 2016 IEEE/RSJ International Conference on Intelligent Robots and Systems (IROS), pp. 3786–3793. IEEE (2016)
15. Haas, M., Friedl, W., Stillfried, G., Höppner, H.: Human-robotic variable-stiffness grasps of small-fruit containers are successful even under severely impaired sensory feedback. Front. Neurorobot. **12**, 70 (2018)
16. Johansson, R.S., Edin, B.B.: Predictive feed-forward sensory control during grasping and manipulation in man. Biomed. Res.-Tokyo **14**, 95–95 (1993)
17. Leitner, J., Tow, A.W., Sünderhauf, N., Dean, J.E., Durham, J.W., Cooper, M., Eich, M., Lehnert, C., Mangels, R., McCool, C., et al.: The acrv picking benchmark: a robotic shelf picking benchmark to foster reproducible research. In: 2017 IEEE International Conference on Robotics and Automation (ICRA), pp. 4705–4712. IEEE (2017)
18. Miller, G.A., Galanter, E., Pribram, K.H.: Plans and the structure of behavior. Adams Bannister Cox (1986)
19. Nishimura, T., Mizushima, K., Suzuki, Y., Tsuji, T., Watanabe, T.: Thin plate manipulation by an under-actuated robotic soft gripper utilizing the environment. In: 2017 IEEE/RSJ International Conference on Intelligent Robots and Systems (IROS), pp. 1236–1243. IEEE (2017)
20. Piazza, C., Grioli, G., Catalano, M., Bicchi, A.: A century of robotic hands. Ann. Rev. Control Robot. Autonom. Syst. (2019) (In Press)
21. Redmon, J., Farhadi, A.: Yolo9000: Better, Faster, Stronger (2017)
22. Schmidt, P., Vahrenkamp, N., Wächter, M., Asfour, T.: Grasping of unknown objects using deep convolutional neural networks based on depth images. In: 2018 IEEE International Conference on Robotics and Automation (ICRA), pp. 6831–6838. IEEE (2018)
23. Szegedy, C., Vanhoucke, V., Ioffe, S., Shlens, J., Wojna, Z.: Rethinking the inception architecture for computer vision. In: Proceedings of the IEEE Conference on Computer Vision and Pattern Recognition, pp. 2818–2826 (2016)
24. Yan, X., Hsu, J., Khansari, M., Bai, Y., Pathak, A., Gupta, A., Davidson, J., Lee, H.: Learning 6-dof grasping interaction via deep geometry-aware 3d representations. In: 2018 IEEE International Conference on Robotics and Automation (ICRA), pp. 1–9. IEEE (2018)
25. Yang, Y., Liu, X.: A re-examination of text categorization methods. In: Proceedings of the 22nd Annual International ACM SIGIR Conference on Research and Development in Information Retrieval, pp. 42–49. ACM (1999)
26. Zeng, A., Song, S., Yu, K.T., Donlon, E., Hogan, F.R., Bauza, M., Ma, D., Taylor, O., Liu, M., Romo, E., et al.: Robotic pick-and-place of novel objects in clutter with multi-affordance grasping and cross-domain image matching. In: 2018 IEEE International Conference on Robotics and Automation (ICRA), pp. 1–8. IEEE (2018)

Chapter 12
Learning to Prevent Grasp Failure with Soft Hands: From On-Line Prediction to Dual-Arm Grasp Recovery

Soft-hands allow to simplify the grasp planning to achieve a successful grasp, thanks to their intrinsic adaptability. At the same time, their usage poses new challenges, related to the adoption of classical sensing techniques originally developed for rigid end-defectors, which provide fundamental information, e.g. to detect object slippage. Under this regard, model-based approaches for the processing of the gathered information are hard to use, due to the difficulties in modelling hand-object interaction when softness is involved. To overcome these limitations, in this chapter, we propose to combine distributed tactile sensing and machine learning (Recurrent Neural Network—RNN) to detect sliding conditions for a soft robotic hand mounted on a robotic manipulator, targeting the prediction of the grasp failure event and the direction of sliding. The outcomes of these predictions allow for an-on line triggering of a compensatory action performed with a second robotic arm-hand system, to prevent the failure. Despite the fact that the network was trained only with spherical and cylindrical objects, we demonstrate high generalization capabilities of our framework, achieving a correct prediction of the failure direction in 75% of cases, and a 85% of successful re-grasps, for a selection of twelve objects of common use.

12.1 Introduction

In recent years, the introduction of soft elements in robotic hands demonstrated to be an asset to easily provide capabilities never seen with rigid components [17, 21]. The intelligence, directly embedded into the mechanics, enables to fold the fingers around the object in a natural fashion, and to gently adapt the shape of the hand when interacting with the environment. This characteristic comes with the additional benefit that potential uncertainties in local relative placement between the

G. Averta, *Human-Aware Robotics: Modeling Human Motor Skills for the Design, Planning and Control of a New Generation of Robotic Devices*, Springer Tracts in Advanced Robotics 145, https://doi.org/10.1007/978-3-030-92521-5_12

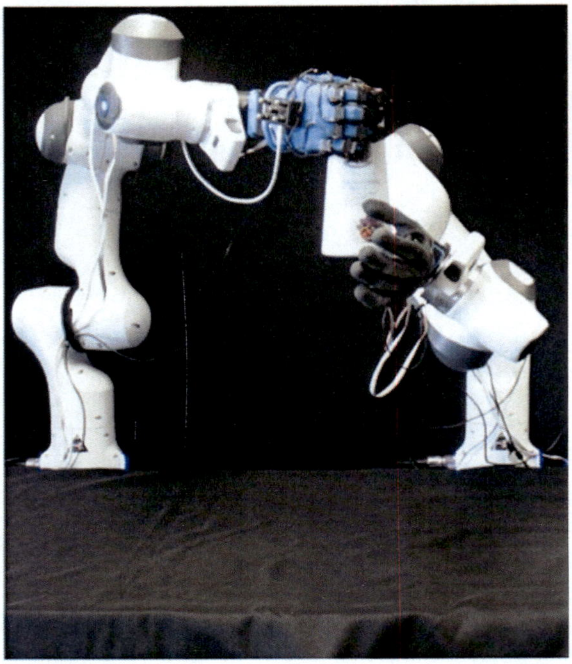

Fig. 12.1 A Franka Emika Panda robotic arm integrated with a Soft Hand equipped with IMU sensors is used to reach and grasp a generic object. A sliding event is detected by processing the IMU information with a Recurrent Neural Network, triggering a reactive re-grasping primitive with a second arm-hand system, to firmly hold the object

End-Effector and the object are compensated by the compliance of the hand, thus relaxing constraints in robot planning [4, 10, 11, 23].

However, this increased dexterity is also responsible for a reduced amount of information that the regulator may feed back to close a control loop. Indeed, because of the difficulties in defining accurate models of hands [25], of the hand-object interaction when softness is involved [14], and to the intrinsic uncertainties that elastic components produce in the measurements [16], it is in general not straightforward to use sensing techniques originally developed for rigid end effectors (e.g. rigid force sensors at the fingertips, encoder [24]), and to implement model-based feedback solutions that can react to unexpected situations [19]. Indeed, although the increased performances in terms of grasp success that characterize the usage of soft grippers, objects picking and grasping may still fail in many cases. In those events, it is important to have a system able to predict, within a reasonable time horizon, when a grasped object is going to slide w.r.t. the hand, and would probably fall, and eventually trigger a suitable corrective action (Fig. 12.1).

To solve this problem, one of the most common approaches relies on the direct measure of contact forces, usually relying on force/torque sensors at the joints or at the fingertip level [2, 18, 20, 22, 27]. However, these approaches are typically hardly feasible in practice, given the large cost of the hardware and the complexity of the sensing setup - which introduces significant computational effort—and are not

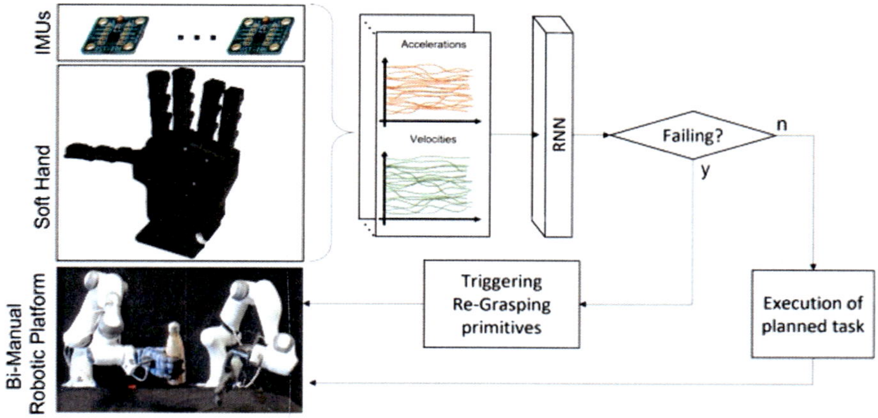

Fig. 12.2 General scheme of the proposed closed loop framework. A bi-manual robotic system, composed by two Franka Emika Panda Robots, endowed with two Pisa/IIT SoftHands, is used as test-bench. IMUs are fastened on the back of the fingers of the hand used to reach and grasp the object. A Recurrent Neural Network (RNN) architecture is used to continuously detect, on-line and with an inference rate of 5 Hz, the occurrence of a sliding event, and the direction of the relative motion. This information is then fed back to the controller to trigger a reactive re-grasping primitive, which depends on the output of the RNN and on the current status of the first manipulator

suitable in general for continuum-soft hands, where the shape and the mechanical response of the fingertip may be significantly different than rigid or articulated-soft hands.

As an alternative to force sensing, the community is recently exploring the usage of other sensory sources, such as audio signals [30], inertial sensing [7], video streams [26] and tactile sensors [13, 29], and infer contact forces through algorithms. However, little has been done so far to exploit such sensory information to predict when a grasp is going to fail, and to trigger reactive recovery primitives.

Recently, we proposed to exploit inertial sensing (accelerations and angular velocities) to feed a deep neural network which was able to accurately classify offline if the stream of data was associated to a grasp failure, and even predict online its occurrence [28]. More specifically, in [28] we demonstrated that Inertial Measurement Units (IMUs)—placed at the fingers level—are capable to record the vibrations caused by the sliding of grasped objects, and a deep architecture, trained to detect the occurrence of these conditions, can be used to predict when a grasp is going to fail.

In this chapter, we build upon our preliminary work and further extend our deep learning framework for grasp failure prediction. More specifically, while in previous experiments failures were caused by a rope which mechanically constrained the maximum distance between the grasped object and the table, resulting in an abrupt and non-ecological failure condition, in this work we completely re-designed the experimental part, generating failures as a consequence of a variable weight added to the object. Furthermore, a robotic arm was used to execute the reach-and-grasp task, in both success and failure cases, thus removing potential artifacts introduced by the manual handling of the robotic hand as done in [28]. Another significant contribution of this work with respect to [28] is that we now target not only the prediction of the failure event, but also the identification of the specific direction of slippage. This will enable the triggering of reactive re-grasp primitives performed by a second manipulator that can exploit the information of the direction of slippage to firmly secure the grasp.

In this work, we collected a grand total of 1800 independent trials. Of these, 56% was used to train the neural architecture, 24% for its validation and the remaining 20% for testing. Extensive research was carried out to identify the optimal recurrent neural architecture to use, aiming at maximizing the prediction accuracy over a dataset of testing trials while minimzing the footprint of the network. With respect to [28], where a Convolutional Neural Network CNN was combined with a Long Short Term Memory (two layers of 128 neurons) to perform the prediction, we removed the CNN for feature extraction and implemented a Recurrent Neural Network (RNN) architecture based on Gated Recurrent Units (GRU) [12] (one layer of 128 neurons). Finally, we also developed a completely new on-line feedback system which takes as input the inference of the proposed recurrent neural network (in terms of prediction of the sliding event as well as its direction, i.e. top and lateral) and selects a reactive re-grasping primitive, performed by a second robotic arm-hand system (see Fig. 12.2), that ultimately manages to firmly secure the grasp. Despite the fact that the network was trained only with spherical and cylindrical objects, we demonstrate high generalization capabilities of our framework, achieving a correct prediction of the failure direction in 75% of cases (approximately 2 s in advance), and a 85% of successful re-grasps, for a selection of twelve objects of common use. The content of this chapter is adapted from [5].

12.2 Methods

As introduced in the previous section, the goal of this work is to develop a closed-loop framework able to predict on-line if an object, grasped by a soft robotic hand, is sliding (and along which direction) and will likely drop. Such information is used to feed a reactive controller that triggers a re-grasping primitive which, in turn, firmly stabilizes the grasp. As a test-bench, we used two Franka Emika Panda manipulators [1], both endowed with two Pisa/IIT Softhands [9] as End-Effectors. To collect data for RNN training we used a 3D-printed object, composed by a interchangeable handle and a support where one or more masses were placed to modify the weight of the object. We considered two different shapes of handles, a sphere and a cylinder, which forced the shape of the hand in two different configurations. These were presented to the robot (i.e. one robotic arm and hand system) with two roughness level, one smooth and one covered with sandpaper (400 Grit). The handle was grasped following two main approaches: top grasp, i.e. with the palm parallel to the horizontal plane, and lateral grasp, i.e. with the palm parallel to a vertical plane. For each of these grasp approaches, we further considered two potential failures types: central slippage, i.e. when the object slips along the long fingers, and lateral slippage, i.e. when failure is caused by a relative motion perpendicular to the long fingers (examples are provided in Fig. 12.3a/b). Of note, to avoid that the network could identify failures only along the direction of the gravity, we included slippage data where the object was pulled off the SoftHand along a direction perpendicular to the gravity itself, by a second robotic arm-hand system (i.e. Franka Emika Panda equipped with a SoftHand) (see e.g. Figure 12.3a right; Fig. 12.3b left; Fig. 12.5b). The position and orientation of hand and object during experiments were continuously tracked through a 3D motion tracking system (Optitrack Flex 13, NaturalPoint Inc., Corvallis, Oregon (US), refresh 120 Hz). The robotic hand that performed the reach to grasp task was endowed with a soft glove, on which we mounted 17 Inertial Measurement Units (IMUs), one for each phalanx, fastened on the back of the hand as in [28]. Four IMUs were attached to the thumb, and three to each long finger. One additional sensor was placed on the hand dorsum, close to the wrist, for reference (see Fig. 12.3c).

Considering all the combinations discussed before, we performed a grand total of 1800 independent acquisitions, of which one third was composed of successful grasps, one third of central slippage and one third of lateral slippage. For each of these classes, we randomized the shape of the handle, the roughness level and the type of grasping approach (top vs. lateral), making sure that the different parameters were represented in a balanced manner. A random weight, ranging between 200 and 700 gr, was added to the object. For each trial, we recorded synchronously the stream of IMUs readings, the position of optical markers attached to hand and object, the encoder of the hand (which measures the degree of closure) and the robot joint positions, all with a refresh rate 70 Hz.

For each trial, then, we segmented the portion of data we intended to use as input for the neural architecture. More specifically, we identified as initial frame of the sequence the instant in which the arm start lifting the object. The final frame, instead, is identified as the one in which the distance between hand and object increases of 5 mm w.r.t. the previous values (i.e. the object is dropped). Finally, zero-padding was added at the beginning of each sample, to homogenize the trials length.

Once the dataset was built, to teach the network to recognise the event in advance, we removed from the dataset the final block of the signal, corresponding to a time slot immediately before the object drop. This has the twofold purpose of (i) removing high peaks in the signal stream caused by the drop of the object and (ii) learning to recognise small oscillations that are characteristics of failure events in the first frames of sliding, rather than larger oscillations evident in the final portion of the signal (see Fig. 12.4). We tested three different levels of anticipation, corresponding to one, two and three seconds before the actual drop, reported in Fig. 12.4 with a blue, green and red dashed line respectively. Hereinafter we will refer to the parameter quantifying this anticipation as Δ. Data were then randomized and splitted in three groups: 20% was devoted to testing and the remaining was further divided in 30% for validation and 70% for training.

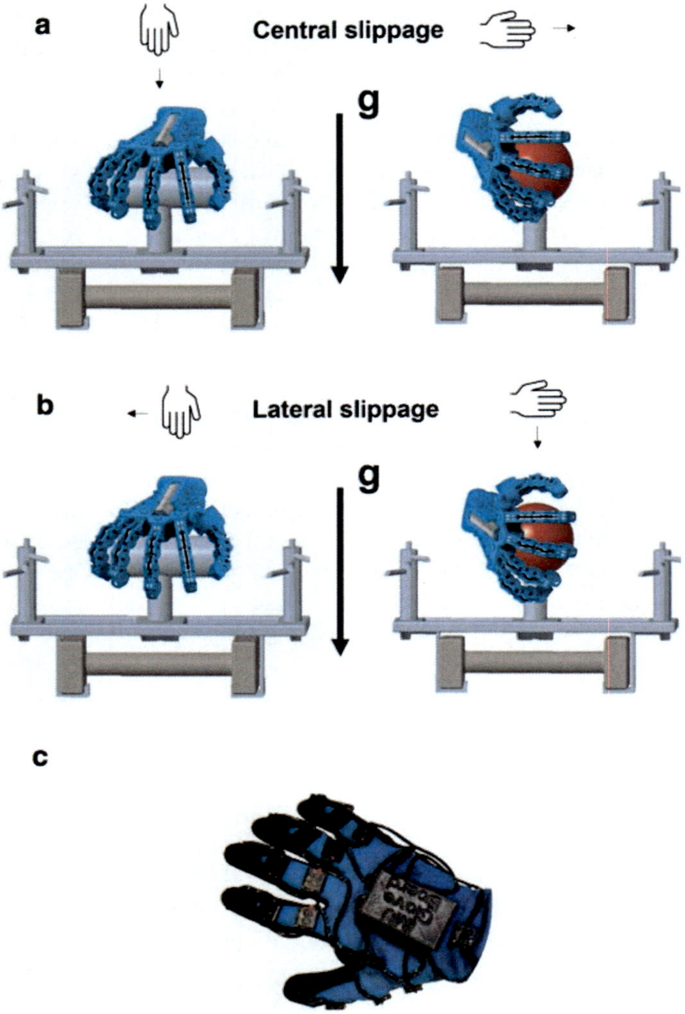

Fig. 12.3 Pictures of the experimental framework used for data collection. Two different handles (light gray in the figures) were used: a sphere (as in the right of panel a and b) and a cylinder (left of panel a and b). Handles were used with and without a sandpaper coverage to modify the roughness. Panel a shows the two hand configurations that we implemented to replicate central slippage. Panel b shows the two hand configurations for lateral slippage. In both panels, also the gravity vector is reported, to show that not all failures occur along the gravity direction. The object was endowed with two supports for the markers of the 3D motion tracking system (one on the right and one on the left) designed with the shape of a star to always ensure visibility of at least four markers as in [6]. The variable weight (in the range 200–700 g) was attached to the handle (dark gray in the figures). Panel c reports a picture of the IMU glove we mounted on the SoftHand to continuously collect inertial measurements from each hand phalanx

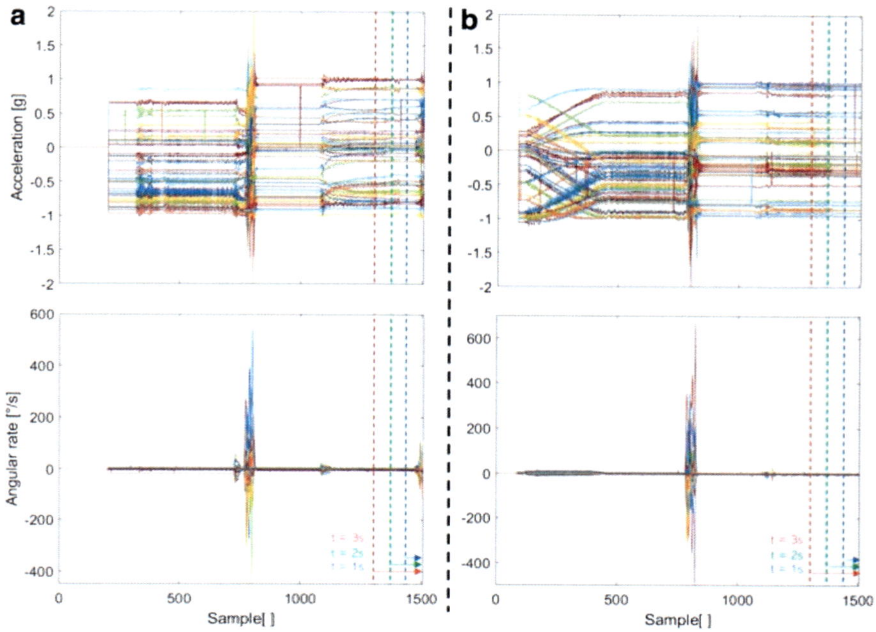

Fig. 12.4 Stream of raw accelerometers (top row) and gyroscopes (bottom row) during a failed grasp (subpolot A) and a successful grasp (B). Coloured dashed lines identify the initial frame of the block of signal that is removed from data before training, corresponding to one (blue), two (green) and three (red) seconds. We refer to this quantity as Δ

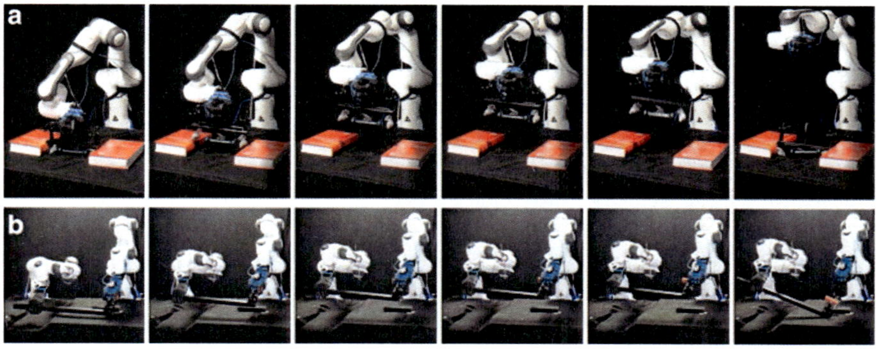

Fig. 12.5 Two examples of failures events executed during data collection. **a** Lateral slippage with a spherical smooth handle. **b** Lateral slippage with a rough cylindrical handle, caused by the second robot

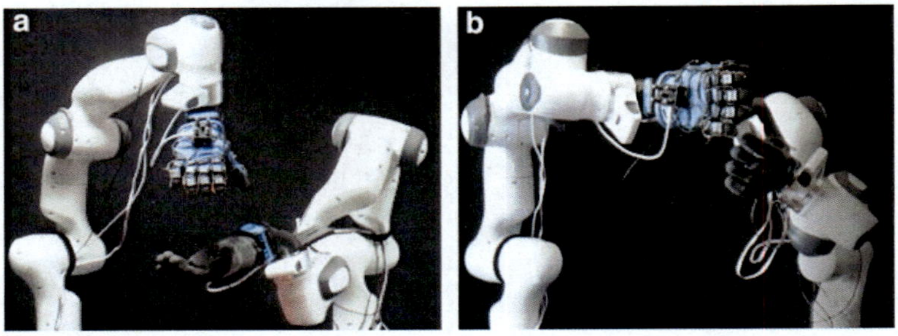

Fig. 12.6 Objects of common use selected for the experiments: a sauce bottle, an apple, a tennis ball, a squeeze tube, a mug, a box, a saucepan, a water bottle, a food box, a shampoo bottle, a rough l-shape, a smooth l-shape. We increased the weight of some of these objects by adding external weight (as done during experiments for dataset collection) to match the range between 200 and 700 g

The neural architecture we selected is based on Gated Recurrent Units (GRU) [12], which are neurons with a feedback channel, which enables to store, and learn from, the history of a time series. Training was performed using ADAM optimizer and Cross Entropy as loss function. Early stopping and dropout were also used to prevent overfitting. We tested different combination of hyperparameters, resulting in three different architectures that demonstrated the highest validation accuracy and the minimum footprint of the network (to minimize inference time), one for each Δ value considered. This is motivated by the fact that the larger is the model the larger is the time to perform inference. Among these, we selected for the implementation the network trained with $\Delta = 2\,\text{s}$, because this provided a time horizon sufficiently large to eventually plan a recovery action, while keeping high accuracy values over validation data.

To develop an on-line implementation of reactive primitives triggered by the output of our grasp failure predictor, we built a First-In-First-Our (FIFO) pile structure with a fixed size equal to the one used at training time, and containing fresh data coming from inertial sensors. During the execution of the on-line framework, the pile will always contain the last N readings coming from IMUs, where N is the number of time frames of acquisitions used during training (after pre-padding). At startup, the pile is initialized as a zero matrix. Because with the architecture we selected for implementation ($\Delta = 2\,\text{s}$) we observed an average inference time of $\approx 0.15\,\text{s}$, we implemented two ROS nodes, the first, running at 70 Hz, where data were read from the IMU glove and collected into a dynamic array, and a second one, running at 5 Hz, where the block of data collected by the first node was inserted in the FIFO pile, removing the exceeding samples from the top of the pile (i.e. the oldest samples). Data contained in the pile was then provided as input to the RNN. When the predicted value is constant for at least five consecutive inference rounds, and the classified entry is a lateral or a central slippage, then this signal is used to trigger the reactive behavior of the second manipulator. Note that five represents a trade-off

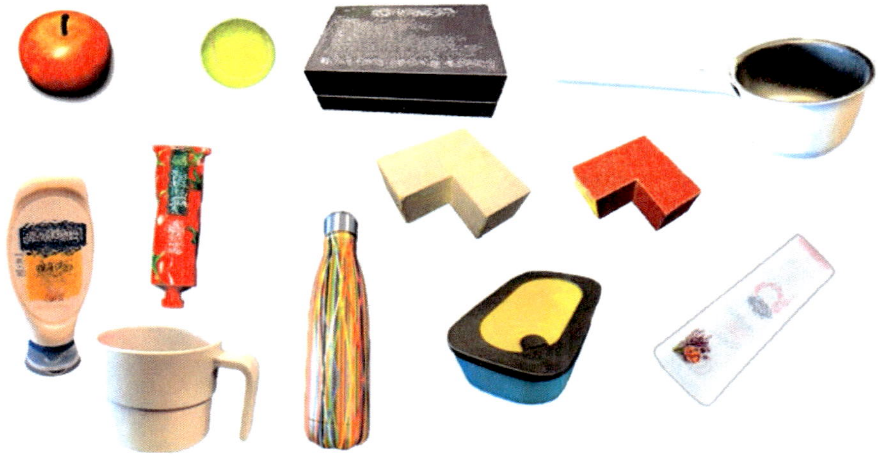

Fig. 12.7 Re-grasp primitives considered for top grasps (**a**) and lateral grasps (**b**)

between promptness of response for the controller and number of false positives and was manually and heuristically tuned.

To test our methods, we implemented two parametric reactive primitives for the secondary robot, one appropriate for the central slippage and one for the lateral slippage (i.e. the two failure classes considered in this work). More specifically, we programmed the first primitive (i.e. for central slippage) as a linear interpolation between the initial robot configuration and the Cartesian position of the first End-Effector. The orientation of the second hand is imposed to be with the palm upwards (see Fig. 12.6a). The estimation of Cartesian forces provided by the second manipulator is continuously read and fed back to the controller, in such a way that when the module of the readings overcome a certain threshold (2 N) we assume that the robot is in contact with the object and we stop the execution of the primitive. In case of lateral slippage, the second robot is programmed to reach via a linear interpolation the position of the first End-Effector. The orientation, instead, is rotated along the direction of the long fingers in such a way that the angle between the horizontal plane and the plane of the palm is 45 Degrees (see Fig. 12.6b). Also in this case, we command as reference the position of the first End-Effector, and exploit the estimation of contact forces to identify the contact.

We tested our on-line framework with two additional experiments. First, we replicated the failures with the same setup employed for data collection. Twenty trials were performed for each of the three classes, considering randomly one of the handles of Fig. 12.3. We then considered a selection of 12 objects of common use, of which ten are extracted from the YCB dataset [8] and two are l-shaped objects with smooth and rough surface (see Fig. 12.7). This selection was made with the purpose of forcing different types of power grasps, such as power circular, power prismatic, palm circular and palm prismatic (for terminology, we refer to [3]). We made sure that the objects' weight was in the range between 200 and 700 grams, by adding external weight when necessary. For each of these objects, we employed the grasping strategy afforded by the object. Indeed, as hypothesized in [15], the geometry of an object suggests one (or more) preferable grasping approaches, which we attempted to respect. For this reasons, tall objects, such as standing bottles, were grasped using a lateral grasp, while short ones were grasped using a top grasp. Of note, the bottle was presented in both the standing and lying down configuration. We repeated the grasp of each object ten times, forcing its ecological failure by regulating the strength of the hand closure [9], achieving a grand total of 130 samples.

12.3 Results

As already mentioned in the previous section, we tested our framework in two different ways. First, we validated the network by assessing the prediction accuracy over a pool of test data not used during training and validation, consisting of 360 independent samples of three classes: successful grasp, central slippage and lateral slippage. Then, we implemented our network in an on-line integrated framework of failure prediction and reactive re-grasp. We tested this implementation over a selection of 12 objects of common use extracted from the YCB dataset [8].

12.3.1 Validation of the Neural Architecture

Considering an anticipation time Δ of one, two, and three seconds, we converged to three optimized architectures, all based on GRU neurons. The optimal architecture with $\Delta = 1$ is composed by two layers of 64 neurons and was trained with a dropout of 0.3. This network demonstrated a validation accuracy of 0.93 ± 0.003 over ten different rounds of training (all starting from a random seed). For $\Delta = 2$, the optimal selection converged to a single layer of 128 neurons trained with a dropout of 0.5, achieving a validation accuracy of 0.91 ± 0.01 over ten different rounds of training (all starting from a random seed). Finally, with $\Delta = 3$s, the model consisted of two layers of 64 neurons, trained with a dropout of 0.3, yielding a validation accuracy of 0.87 ± 0.02 over ten different rounds of training (all starting from a random seed). After validation, we quantified the accuracy of prediction also over fresh data, not used during the training phase. This new dataset consisted of 360 samples, 120 for each class. Confusion matrices of the classification for different values of Δ are reported in Fig. 12.8. We obtained an overall test accuracy of 87, 84 and 76% for $\Delta = 1$s, $\Delta = 2$s and $\Delta = 3$s respectively.

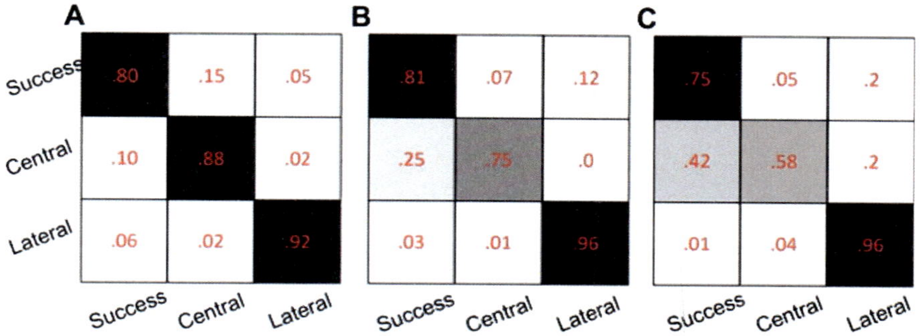

Fig. 12.8 Confusion matrices referred to failure classification over test data for $\Delta = 1$s (**a**), $\Delta = 2$s (**b**) and $\Delta = 3$s (**c**). Values are in percentage versus the total number of entries of each class. On the rows the real class, on the columns the predicted class. Cells are color-coded: black stands for 100%, white stands for 0%

12.3.2 *Validation of the On-Line Integrated Framework*

We decided to consider for the on-line implementation the architecture trained with $\Delta = 2$s, because this represents an appropriate trade-off between satisfactory prediction performances and capabilities of detecting small oscillations that are present in the early stages of sliding (minimizing the network footprint). As introduced in the Methods section, we performed two different experiments to test the capabilities of our framework in the on-line predict-and-regrasp task. The first one consisted in replicating the experimental setup already used to collect the training data. In this case, with a pile that continuously (with a refresh rate of 5 Hz) updates the stream of data given as input to the neural architecture, we obtained a correct classification in $\approx 78\%$ of cases, of which $\approx 87\%$ resulted in a successful robot regrasp. However, a correct prediction of the failure does not necessarily match with a successful regrasp, because after triggering the reactive primitive the robot could spend a certain amount of time to plan and execute the trajectory (approximately 1.5 s). For this reason, in certain cases, especially with very smooth objects, the total success rate of the re-grasp could be lower, and it is more appropriate to report on the number of successful regrasps over the ones correctly predicted. Indeed, in this first experiment we had that for occurrences of central slippage we were able to successfully prevent the failure in 80% of cases, while the performances increased to 94% for lateral slippage. This is caused by the fact that in the first case the time of sliding is shorter, on average, than the second one. We further validated our framework by performing a second experiment with 12 objects, of which 10 are extracted from the YCB dataset [8] (see Fig. 12.7). Also in this case, we verified the prediction and classification accuracy and then quantified the success rate of the failure prevention over the cases in which we were able to successfully predict the failure. Over a grand total of 130 experiments (see e.g. Fig. 12.9), we achieved a classification accuracy of 75% and, for the cases in which the type of failure was predicted correctly, we

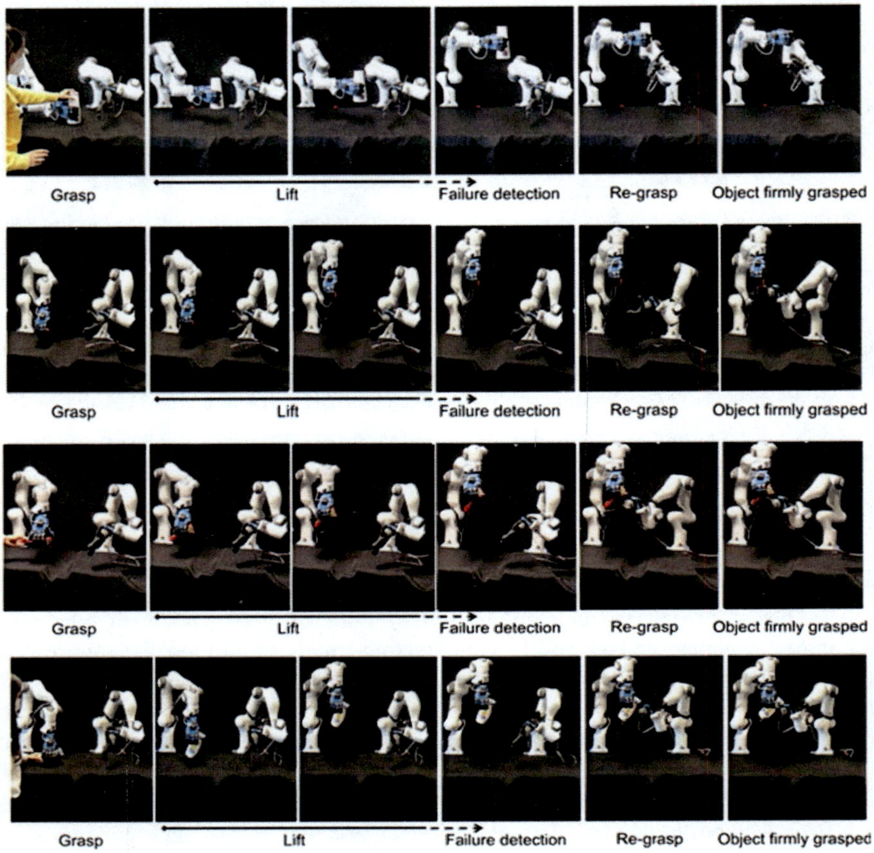

Fig. 12.9 Four examples of our framework while the occurrence and the direction of a failure was correctly predicted and a reactive primitive successfully prevented the drop of the object. From top to bottom, a shampoo bottle, an apple, a squeeze tube, a sauce bottle

successfully prevented the failure with our reactive primitive in 85% of cases. Of note, we observed marked differences across objects. More specifically, very smooth and spherical objects, such as the wooden apple in our pool of objects, although easily classified by our neural architecture (with a correct classification in 80% of cases), was successfully re-grasped in the 25% of cases of correct classification only. Others, instead, such as the mug, the saucepan, the tennis ball, the squeeze tube and the launch box, were successfully re-grasped in all the cases in which the neural architecture was able to correctly predict the failure.

12.4 Discussions and Conclusions

With this chapter, we demonstrated the feasibility, and provided an implementation of a neural architecture that can predict—up to 87% with test data—the occurrence and the direction of grasp failure, considering accelerations and angular velocities collected from a soft robotic hand (mounted on a robotic manipulator and equipped with an IMU glove) that autonomously grasp an object. We implemented our framework relying on a GRU architecture, a widely popular and consolidated recurrent neural network. Our implementation enables the triggering of reactive primitives, performed by a second robotic arm-hand system, achieving a correct prediction (approx. 2 s in advance) of the failure occurrence and its directions in 75% of cases, and—when correctly classified—firmly secure the grasp with a recovery action, with a success rate of 85%. These results refer to an experiment conducted with a pool of objects of common use, which were never used during the training phase, while the performances using the same experimental setup of the training phase were 78 and 87% respectively. Note that these results come from a combination of different factors, such as the smoothness of the object and the upper bound of the velocity of the manipulator. Of note, our implementation is completely on-line, from inference to motion execution, with an average inference time of 0.15 s and an average time required to complete the reactive behavior of 1.5 s. We noticed that certain objects, such as the wooden apple in our second experiments, were particularly harsh for our framework. Indeed, while the sliding was correctly predicted and classified in the 80% of cases, the failure was extremely quick and resulted in a successful re-grasp for the 25% of cases only. This is mainly related to the control of the hand itself and we believe that the extension of our reactive behavior to other re-grasping primitives may improve these performances.

To improve the prediction accuracy, on one side, and the re-grasp success on the other side, our future work will focus, on one side, on the investigation of the usage of other tactile sensors, such as the Tactip [29], to gather a larger amount of information during the grasp, which could improve the accuracy of our prediction. On the other side, we will consider the use of Neural Architecture Search (NAS) techniques to optimize the design of the neural network. Furthermore, we will also consider the usage of supplementary sensing sources, as for example cameras. In this way, sensor fusion could be exploited to feed with a more complete source of information the neural architecture and to improve the overall failure prediction accuracy. However, it is also worth mentioning that such an improvement would come with a significant increase in the dimensionality of raw data, resulting in higher complexity of the mechatronic system and in a larger footprint of the neural architecture. For this reason, we believe that a trade-off must be reached, depending on the resources available for a given application. For example, for fully autonomous robots, which should process the whole information with on-board electronics (possibly on the edge), one could use a minimalistic tactile sensing as the one employed in this chapter to minimize the footprint of the network, while for industrial scenarios, it may be feasible to have more complex systems. At the same time, we are planning to further expand the pool of reactive primitives considered, including also single-arm actions, such as End-Effector re-orientation and hand squeezing force regulation. To this aim, additional sources of information, as for example a vision layer, will be evaluated, which will help in discriminating which strategy could be the more appropriate for the specific case.

Ultimately, we believe that our work may represent a valuable contribution toward the development of intelligent manipulators, capable of identifying on-line whether a task is performed correctly, eventually triggering reactive behaviors to adapt the execution of the action to the expected goal [7]. This will help in developing grasp planning that minimize the force exerted during the object grasps, and demanding to the prediction-regrasping component the correction of a possible failure. This, together with the intrinsic adaptability of the softhands, could offer a viable solutions for the grasping of fragile and delicate objects.

References

1. Franka Emika. https://www.franka.de/
2. Ajoudani, A., Hocaoglu, E., Altobelli, A., Rossi, M., Battaglia, E., Tsagarakis, N., Bicchi, A.: Reflex control of the Pisa/IIT softhand during object slippage. In: 2016 IEEE International Conference on Robotics and Automation (ICRA), pp. 1972–1979. IEEE (2016)
3. Arapi, V., Della Santina, C., Averta, G., Bicchi, A., Bianchi, M.: Understanding human manipulation with the environment: a novel taxonomy for video labelling. IEEE Robot. Autom. Lett. (2021)
4. Averta, G., Angelini, F., Bonilla, M., Bianchi, M., Bicchi, A.: Incrementality and hierarchies in the enrollment of multiple synergies for grasp planning. IEEE Robot. Autom. Lett. **3**(3), 2686–2693 (2018)
5. Averta, G., Barontini, F., Valdambrini, I., Cheli, P., Bacciu, D., Bianchi, M.: Learning to prevent grasp failure with soft hands: from online prediction to dual-arm grasp recovery. Adv. Int. Syst. Wiley Online Library 2100146 (2021)
6. Averta, G., Della Santina, C., Battaglia, E., Felici, F., Bianchi, M., Bicchi, A.: Unveiling the principal modes of human upper limb movements through functional analysis. Front. Robot. AI **4**, 37 (2017)
7. Bianchi, M., Averta, G., Battaglia, E., Rosales, C., Bonilla, M., Tondo, A., Poggiani, M., Santaera, G., Ciotti, S., Catalano, M.G., et al.: Touch-based grasp primitives for soft hands: applications to human-to-robot handover tasks and beyond. In: 2018 IEEE International Conference on Robotics and Automation (ICRA), pp. 7794–7801. IEEE (2018)
8. Calli, B., Singh, A., Bruce, J., Walsman, A., Konolige, K., Srinivasa, S., Abbeel, P., Dollar, A.M.: Yale-CMU-Berkeley dataset for robotic manipulation research. Int. J. Robot. Res. **36**(3), 261–268 (2017)
9. Catalano, M.G., Grioli, G., Farnioli, E., Serio, A., Piazza, C., Bicchi, A.: Adaptive synergies for the design and control of the Pisa/IIT softhand. Int. J. Robot. Res. **33**(5), 768–782 (2014)
10. Choi, C., Schwarting, W., DelPreto, J., Rus, D.: Learning object grasping for soft robot hands. IEEE Robot. Autom. Lett. (2018)
11. Deimel, R., Brock, O.: Soft hands for reliable grasping strategies. In: Soft Robotics, pp. 211–221. Springer (2015)
12. Dey, R., Salem, F.M.: Gate-variants of gated recurrent unit (gru) neural networks. In: 2017 IEEE 60th International Midwest Symposium on Circuits and Systems (MWSCAS), pp. 1597–1600. IEEE (2017)

13. Donlon, E., Dong, S., Liu, M., Li, J., Adelson, E., Rodriguez, A.: Gelslim: a high-resolution, compact, robust, and calibrated tactile-sensing finger. In: 2018 IEEE/RSJ International Conference on Intelligent Robots and Systems (IROS), pp. 1927–1934. IEEE (2018)
14. Ficuciello, F., Migliozzi, A., Coevoet, E., Petit, A., Duriez, C.: Fem-based deformation control for dexterous manipulation of 3d soft objects. In: 2018 IEEE/RSJ International Conference on Intelligent Robots and Systems (IROS), pp. 4007–4013. IEEE (2018)
15. Gibson, J.J.: The theory of affordances. Hilldale, USA **1**(2), 67–82 (1977)
16. Haibin, Y., Cheng, K., Junfeng, L., Guilin, Y.: Modeling of grasping force for a soft robotic gripper with variable stiffness. Mech. Mach. Theory **128**, 254–274 (2018)
17. Hughes, J., Culha, U., Giardina, F., Guenther, F., Rosendo, A., Iida, F.: Soft manipulators and grippers: a review. Front. Robot. AI **3**, 69 (2016)
18. Kappassov, Z., Corrales, J.A., Perdereau, V.: Tactile sensing in dexterous robot hands. Robot. Autonom. Syst. **74**, 195–220 (2015)
19. Kitagawa, S., Wada, K., Okada, K., Inaba, M.: Learning-based task failure prediction for selective dual-arm manipulation in warehouse stowing. In: International Conference on Intelligent Autonomous Systems, pp. 428–439. Springer (2018)
20. Krug, R., Lilienthal, A.J., Kragic, D., Bekiroglu, Y.: Analytic grasp success prediction with tactile feedback. In: 2016 IEEE International Conference on Robotics and Automation (ICRA), pp. 165–171. IEEE (2016)
21. Piazza, C., Grioli, G., Catalano, M., Bicchi, A.: A century of robotic hands. Ann. Rev. Control Robot. Autonom. Syst. (2019) (In Press)
22. Pinto, L., Gupta, A.: Supersizing self-supervision: learning to grasp from 50 k tries and 700 robot hours. In: 2016 IEEE International Conference on Robotics and Automation (ICRA), pp. 3406–3413. IEEE (2016)
23. Pozzi, M., Marullo, S., Salvietti, G., Bimbo, J., Malvezzi, M., Prattichizzo, D.: Hand closure model for planning top grasps with soft robotic hands. Int. J. Robot. Res. **39**(14), 1706–1723 (2020)
24. Santaera, G., Luberto, E., Serio, A., Gabiccini, M., Bicchi, A.: Low-cost, fast and accurate reconstruction of robotic and human postures via imu measurements. In: 2015 IEEE International Conference on Robotics and Automation (ICRA), pp. 2728–2735. IEEE (2015)
25. Schramm, L., Sintov, A., Boularias, A.: Learning to transfer dynamic models of underactuated soft robotic hands. In: 2020 IEEE International Conference on Robotics and Automation (ICRA), pp. 4579–4585. IEEE (2020)
26. She, Y., Liu, S.Q., Yu, P., Adelson, E.: Exoskeleton-covered soft finger with vision-based proprioception and exteroception (2019). arXiv:1910.01287
27. Song, X., Liu, H., Althoefer, K., Nanayakkara, T., Seneviratne, L.D.: Efficient break-away friction ratio and slip prediction based on haptic surface exploration. IEEE Trans. Robot. **30**(1), 203–219 (2013)
28. V., A., Y., Z., G., A., Catalano, M.G., Rus, D., Della Santina, C., Bianchi, M.: To grasp or not to grasp: an end-to-end deep-learning approach for predicting grasping failures in soft hands. In: IEEE International Conference on Soft Robotics, Yale University, USA. IEEE (2020)
29. Ward-Cherrier, B., Pestell, N., Cramphorn, L., Winstone, B., Giannaccini, M.E., Rossiter, J., Lepora, N.F.: The tactip family: soft optical tactile sensors with 3d-printed biomimetic morphologies. Soft Robot. **5**(2), 216–227 (2018)
30. Zöller, G., Wall, V., Brock, O.: Acoustic sensing for soft pneumatic actuators. In: 2018 IEEE/RSJ International Conference on Intelligent Robots and Systems (IROS), pp. 6986–6991. IEEE (2018)

Chapter 13
Dexterity Augmentation of Robotic Hands: A Study on the Kinetic Domain

Postural hand synergies, or eigenpostures, are joint angle covariation patterns observed in common grasping tasks. A typical definition associates the geometry of synergy vectors and their hierarchy (relative statistical weight) with the principal component analysis of an experimental covariance matrix. In a reduced complexity representation, the accuracy of hand posture reconstruction is incrementally improved as the number of synergies is increased according to the hierarchy. In this work, we explore whether and how hierarchy and incrementality extend from posture description to grasp force distribution. To do so, we study the problem of optimizing grasps w.r.t. hand/object relative pose and force application, using hand models with an increasing number of synergies, ordered according to a widely used postural basis. The optimization is performed numerically, on a data set of simulated grasps of four objects with a 19-DoF anthropomorphic hand. Results show that the hand/object relative poses that minimize (possibly locally) the grasp optimality index remain roughly the same as more synergies are considered. This suggests that an incremental learning algorithm could be conceived, leveraging on the solution of lower dimensionality problems to progressively address more complex cases as more synergies are added. Second, we investigate whether the adopted hierarchy of postural synergies is indeed the best also for force distribution. Results show that this is not the case.

13.1 Introduction

The human hand is an extremely complex system, composed by many joints, muscles and sensory receptors, which constitute a highly sophisticated and dexterous apparatus of our body. Such an abundance, classically referred to as Bernstein's problem

G. Averta, *Human-Aware Robotics: Modeling Human Motor Skills for the Design, Planning and Control of a New Generation of Robotic Devices*, Springer Tracts in Advanced Robotics 145, https://doi.org/10.1007/978-3-030-92521-5_13

[3], would require a remarkable calculating capacity for its control. Nevertheless, several neuroscientific studies suggested that the human nervous system is able to cope with such complexity and organize it in a simple environment—as it happens for other natural phenomena according to the Simplexity concept [4]—leveraging on a control space of reduced dimensionality [17, 19, 22, 27], usually defined as synergistic control space.

This synergistic behavior seems to be used by the Central Nervous System to generate coordinated movements, simultaneously activating different Degrees of Freedom (DoFs), instead of acting separately on each joint or muscle. The existence of these patterns (aka synergies) was observed in different motor tasks and at different levels of the motor control architecture, i.e. neural [31], muscular [12, 30], kinematic [16, 18, 24]. For a review on these topics see e.g. [23].

Considering the kinematic level, such observations supported the idea that few combinations of the hand DoFs, e.g. described in terms of main principal components (PCs, i.e. postural synergies) of hand joint angles recorded in grasping tasks, can take into account large part of hand pose variability. Higher order PCs are likely involved to describe more complex tasks such as haptic exploration [32], or manipulation with the environment [13]. In geometrical terms, synergies can be regarded as a basis of hand principal movements: the more the elements are used (or “enrolled”), the more complex tasks can be executed [24].

While postural synergies provide a geometrical description of hand control in the kinematic space, such a model cannot be directly applied to explain force control, generation and distribution in grasping and manipulation tasks. To this goal, we need to consider also the mechanical compliance of the hand musculo-tendinous system, as introduced in the soft synergy model [6], inspired by the Equilibrium Point Hypotesis [14]. More specifically, according to this model, postural synergies are considered as a reference configuration and the hand is in equilibrium between the reference attraction and the repulsion forces exerted by the object. In [15], authors numerically demonstrated that for a paradigmatic human hand the same postural synergies that are important for pose generation are also involved in the optimal distribution of contact forces during the grasp. The study of optimal contact forces distribution for robotic hands has been tacked in licterature, as for example in [33], where the authors presented an algorithm for the evaluation of optimal contact forces distribution from tactile measurements, and [10], in which the authors faced the problem using dual theorem of non-linear programming. However, to the best of authors’ knowledge, there are no works that investigate the role that postural synergies play for the improvement of the grasp quality in terms of optimal contact force distribution.

Taking inspiration from the observation that the human development seems to progressively increase the hand control capabilities [1, 26]—which could be read as an increment of the available synergies—in this chapter, we pursue this investigation by analyzing through numerical simulations, the effect of an incremental enrollment of postural synergies for the execution of successful grasping strategies.

The proposed analysis takes also into account, for the first time, hand/object relative configuration. This investigation aims, firstly, at verifying that hand/object relative poses minimizing the force distribution are approximately invariant w.r.t. an incremental enrollment of synergies, and, secondly, at providing evidence on the existence of a correspondence between the hierarchy of hand posture synergies and the one of synergies controlling contact forces. An exhaustive inspection of these aspects requires an analysis that spans several hand grasp configurations. For this reason, to include a large part of the main graping hand configurations (e.g. referring to [11]), we simulated the grasp of four different objects with cubic and parallelepiped-like shape of variable dimensions. Two cubic objects are used to drive the hand shape for power and precision circular grasps, while two prismatic objects let the hand assume power and precision prismatic configurations.

The simulations we performed resulted in a large dataset, which allows a thorough analysis of the contact forces applied to the considered objects, varying the number and the types of the enrolled synergies. In particular, we focused on the grasp success rate and the value of contact forces to assess the quality and the effectiveness of the grasp w.r.t. the modification of the hand/object configuration.

For illustration of the general results of this study, we provide a worked-out example of a instantiation of a multifingered hand. The model used as example has the kinematic parameters and the CAD model of the anthropomorphic PISA/IIT SoftHand (SH) [9]. However, it differs in that all 19 articular joints are assumed to be independently controlled here (thus generalizing the results by ignoring the specific under-actuation scheme of the SH).

In this chapter we aim at evaluating if incrementality in the choice of the synergies to be enrolled—and eventually implemented in a robotic device—could be the right path toward a novel generation of robotic hands in which the implementation of the first postural synergy preserves the grasp simplicity, while other PCs are used to enhance the grasp quality and, potentially, the manipulation skills. Furthermore, we believe that this study could have an impact for the development of planning algorithms for highly actuated hands. Indeed, these robotic systems dramatically increase the grasping and manipulations capabilities but decrease the planning and control simplicity. In these cases, we believe that this analysis could be used to justify a first rough evaluation of a good grasping strategy in a lower dimensionality, which could be refined in a second step. The content of this chapter is adapted from [2].

13.2 Background

13.2.1 Modeling a Compliant Hand with Synergies

The problem of grasping with a compliant hand is broadly modelized through balance and congruence equations for hand and object. For a hand with n DoFs, which contacts an object in c_p points, we have that

$$\tau = J^T f, \qquad \xi_f = J\dot{q}, \tag{13.1}$$

where $J \in \mathbb{R}^{c \times n}$ is the hand Jacobian matrix, $c = 3c_p$ is the dimension of the contact force/torque vector, $\tau \in \mathbb{R}^n$ is the joint torque vector, $f \in \mathbb{R}^c$ is the contact force vector, $\xi_f \in \mathbb{R}^c$ is the twist of the contact points on the fingers and $q \in \mathbb{R}^n$ are the joint angles.

Similarly, the balance and congruence equations for the object are

$$w_e = Gf, \qquad \xi_o = G^T \xi_e, \tag{13.2}$$

where $G \in \mathbb{R}^{6 \times c}$ is the grasp matrix, $w_e \in \mathbb{R}^6$ is the object wrench, $\xi_o \in \mathbb{R}^c$ is the twist of the contact points on the object and $\xi_e \in \mathbb{R}^6$ is object twist.

The hand elasticity K is defined as

$$K = (C_s + JC_q J^T)^{-1}, \tag{13.3}$$

where $C_s = (1/k_{stru})I_{c \times c} \in \mathbb{R}^{c \times c}$ is the structural compliance matrix and $C_q = (1/k_{ss})I_{n \times n} \in \mathbb{R}^{n \times n}$ is the joint compliance matrix. K relates the contact forces with the displacement between the twist on the contact points on the fingers and on object as $\delta f = K(\delta\xi_o - \delta\xi_f)$ enabling to solve force indeterminacies in the rigid-body system and allowing to implement the soft synergy paradigm introduced in [6].

In neuroscience there are many studies suggesting that the control of the human hand can be simplified with a lower dimensionality description, usually implemented through Principal Component Analysis (PCA) [23]. Given a set of observations of a specific stochastic phenomenon, PCA calculates the orthonormal transformation that converts the original, possibly correlated, variables into a set of linearly uncorrelated and orthogonal variables called PCs or, in case of joint angles organized in a dataset of hand grasps, postural synergies. These components are usually obtained through Singular Value Decomposition (SVD) of the covariance matrix of the dataset. The eigenvectors (i.e. PCs) hence represent the uncorrelated variables that describe the considered dataset, while the eigenvalues associated to the PCs quantify the importance of each eigenvector in terms of explanation of dataset variability. The mapping between the synergistic displacement input $\delta\sigma \in \mathbb{R}^s$ and the joint reference positions δq_r can be formalized as

$$\delta q_r = S\delta\sigma, \tag{13.4}$$

where S arranges s PCs per-column, with $s \leq n$. Finally, the soft synergy paradigm relates the reference joint position δq_r with the actual joint position δq through

$$\delta q = \delta q_r - C_q \delta\tau. \tag{13.5}$$

Of note, numerical results depends on the dataset used to calculate S. There are several studies in literature focusing on different task-dependent datasets, [13, 28, 29, 34, 35]. One of the most popular also for robotic applications is [24], where the authors demonstrated that just two PCs account for 80% of the grasp dataset variance, while the first one alone explains more than 50% of pose variability. Findings reported in

[24] were also used for the design of robotic devices, as for the Pisa/IIT SH, which embeds the first synergy as single DoA. For the reader's convenience, we report in Fig. 13.1 the hand shapes generated using the first seven synergies.

13.2.2 Optimization of Grasping Force Distribution

Hand control through dimensionality reduction as suggested by hand synergies requires new approaches to analyze contact forces and grasp quality in terms of force-closure. A grasp has force closure if arbitrary large external wrenches applied to the grasped object can be compensated by the contact forces that the hand is able to apply on the object, see e.g. [20]. Adopting the notation introduced in Table 13.1, we briefly recall here the solution of the force closure problem, with no claim of being exhaustive and specifically targeting grasping through postural synergies and under-actuation. For a more detailed analysis refer to [5, 15, 20].

Given a general synergistic grasping problem of an under-actuated compiant hand, the solution of the force distribution problem is given by

$$f = \delta f_p + \delta f_{hr_s} + \delta f_{ho_s}, \tag{13.6}$$

where the three terms of the contact force vector $f \in \mathbb{R}^c$ are the particular solution δf_p, the active internal forces δf_{hr_s} and the passive (or preload) internal forces δf_{ho_s}. The latter are here omitted with no loss of generality since they are assumed null (the intersted reader could refer to [5] for a detailed discussion). The particular solution is given by

$$\delta f_p = G_K^R w_e, \tag{13.7}$$

where $G_K^R = KG^T(GKG^T)^{-1} \in \mathbb{R}^{c\times 6}$ is the K-weighted pseudoinverse of G, and $w_e \in \mathbb{R}^6$ is the wrench applied to the object. The active internal forces are

$$\delta f_{hr_s} = F_s\, \delta\sigma, \tag{13.8}$$

where $F_s := (I_{c\times c} - G_K^R G)KJS \in \mathbb{R}^{c\times s}$ and $\text{rank}(F_s) = e_s$. Given a basis $E_s \in \mathbb{R}^{c\times e_s}$ for the range space of the matrix F_s, (13.8) can be rephrased as

$$\delta f_{hr_s} = E_s y, \tag{13.9}$$

where $y \in \mathbb{R}^{e_s}$ is a vector parameterizing the active internal forces. Thus, the optimal force distribution can be evaluated by minimizing a cost function w.r.t. y. For a complete overview of this problem the reader could refer to [15]. I adopt here the definition of force-closure given in [5].

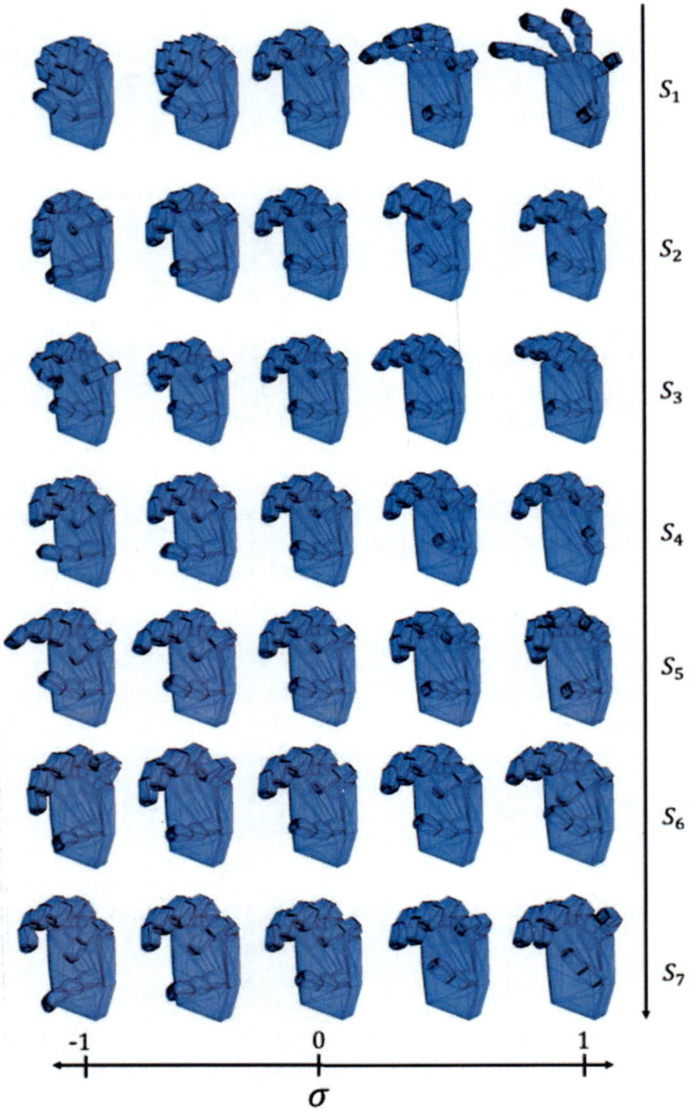

Fig. 13.1 Different hand shapes using the first 7 synergies introduced in [24] for pose generation. For each row, the central figure shows the mean pose of the hand, while the others report the effect of a specific synergy (from the first to the seventh) modulated by a coefficient $\sigma \in [-1, 1]$

Table 13.1 Notation for grasp analysis with postural synergies

Notation	Definition
$n \in \mathbb{R}$	Number of hand joints
$c_p \in \mathbb{R}$	Number of the contact points
$c \in \mathbb{R}$	Dimension of the contact force/torque vector
$\tau \in \mathbb{R}^n$	Joint torques
$q \in \mathbb{R}^n$	Joint angles
$q_r \in \mathbb{R}^n$	Joint reference angles
$f \in \mathbb{R}^c$	Contact force/torque vector
$w_e \in \mathbb{R}^6$	Object wrench
$\xi_e \in \mathbb{R}^6$	Object twist
$\xi_f \in \mathbb{R}^c$	Twists of the contact points on the fingers
$\xi_o \in \mathbb{R}^c$	Twists of the contact points on the object
$J \in \mathbb{R}^{c\times n}$	Hand Jacobian matrix
$G \in \mathbb{R}^{6\times c}$	Grasp matrix
$k_{stru} \in \mathbb{R}$	Structural stiffness
$k_{ss} \in \mathbb{R}$	Joint stiffness
$C_s \in \mathbb{R}^{c\times c}$	Structural compliance matrix
$C_q \in \mathbb{R}^{n\times n}$	Joint compliance matrix
$K \in \mathbb{R}^{c\times c}$	Stiffness matrix
$G_K^R \in \mathbb{R}^{c\times 6}$	$K-$weighted pseudo-inverse of G
$s \in \mathbb{R}$	Number of enrolled postural synergies
$\delta\sigma \in \mathbb{R}^s$	Synergistic displacement
$S \in \mathbb{R}^{n\times s}$	Synergy matrix
$Si \in \mathbb{R}^n$	Synergy vector ($i-$th column of S)
$\delta f_p \in \mathbb{R}^c$	Particular solution
$\delta f_{hr_s} \in \mathbb{R}^c$	Active internal forces
$\delta f_{ho_s} \in \mathbb{R}^c$	Passive (preload) internal forces
$F_s \in \mathbb{R}^{c\times s}$	Map of $\delta\sigma$ into active internal forces δf_{hr_s}
$e_s \in \mathbb{R}$	Rank of F_s
$E_s \in \mathbb{R}^{c\times e_s}$	Basis of the range space of F_s
$y \in \mathbb{R}^{e_s}$	Parameterizing vector of the active internal forces
$\mu \in \mathbb{R}$	Coulomb friction coefficient
$\mathcal{F}$	Composite friction cone
$P \in \mathbb{R}^{c\times c}$	Friction limit constraints matrix
$f_n \in \mathbb{R}^{c_p}$	Normal components of the contact force
$f_{min} \in \mathbb{R}$	Minimum value of f_n

Definition 13.1 (*Force-Closure*) A grasp is defined Force-Closure if and only if the following conditions are satisfied:

1 forces in arbitrary directions are resistible, i.e. $\text{rank}(G) = 6$;
2 the hand configuration is prehensile, i.e. $\exists\, y \in \mathbb{R}^{e_s}$ such that $f(y) \in \text{Int}(\mathcal{F})$.

$\text{Int}(\mathcal{F})$ denotes the internal part of the composite friction cone $\mathcal{F}$. The satisfaction of this friction limit constraint is equivalent to the positive definiteness of the matrix $P \in \mathbb{R}^{c \times c}$, defined as in [8], which contains contact wrenches and friction coefficients.

The problem of finding the optimal distribution of contact forces f in the grasp of an object subject to the external load w_e with regard to the minimization of a suitable cost function $\Psi(y)$ can be formalized as

Definition 13.2 *(Grasping Force Optimization)* Given a grasp characterized by G_K^R, E_s, and P, and an object wrench $w_e \in \mathbb{R}^6$, find y^* in Eq. (13.6), such that $f(y^*) \in \text{Int}(\mathcal{F})$, and the cost function $\Psi(f(y^*))$ is minimized.

I assume here zero preload at the beginning of the grasp and an auxiliary constraint on the minimum value f_{min} for all the normal components f_n of the contact force. Under these hypotheses, the grasping force optimization problem is set up as

$$\begin{aligned} y^* &= \text{argmin}\Psi(y) \\ \text{s.t. } f &= G_K^R w_e + E_s\, y, \quad P(f) \succ 0, \quad f_n \succeq f_{min} \end{aligned} \tag{13.10}$$

where $f \in \mathbb{R}^c$ are the contact forces and $\Psi(y) = \|f(y)\|_2$.

In this chapter, we study such a force optimization problem varying the number of enrolled postural synergies and for different hand-object relative poses.

13.3 Materials and Methods

Given the hand kinematics and the synergy dataset, the analysis described in the previous section does not depend on the particular shape of the object. In fact, the optimization procedure depends only on the joint angles, the position of the contact points, i.e. the 3D points in which the fingers are in contact with the object, and the direction of the contact normals. In this chapter, we assume that the directions of the contact normals are perpendicular to the object local tangent plane. We obtain these grasp-related quantities through the simulator described in [21], which allows simulating the whole kinematics and dynamics of a generic fully-actuated compliant robotic hand.

In this chapter, we used the kinematic model of the SH [9], which has 19 joints, four for each long finger, and three for the thumb. The synergy set used is the one reported in [24]. The kinematic model of the SH has a good mapping with the human hand model considered in [24] allowing a direct mapping of the postural grasp synergies.

I simulated the grasp of four objects: two cubes (obj #1: $30 \times 30 \times 30$mm and obj #3: $60 \times 60 \times 60$mm) and two rectangular bars (obj #2: $30 \times 60 \times 60$mm and

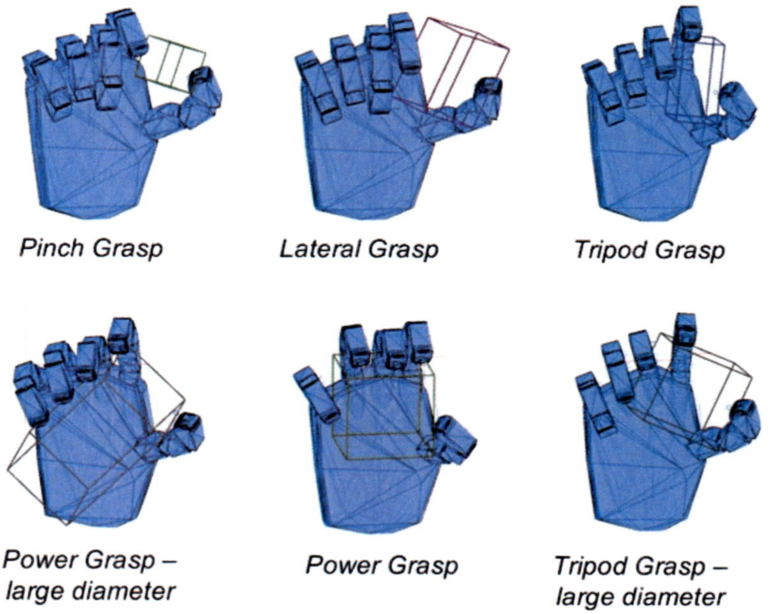

Fig. 13.2 The particular selection of objects dimensions, in conjunction with the large variation of hand/object relative configuration, allow to simulate a wide set of grasp typologies

obj #4: 100 × 60 × 100mm). The shapes and the dimensions of the objects are suitably chosen to include in the analysis several grasp configurations, in accordance with [11]. Indeed, the objects #1 and #2 suggest a precision grip while the objects #3 and #4 a power configuration. The size of the objects allows to generate both circular shaped grasps and prismatic configurations. Furthermore, objects #1 and #2 in conjunction with hand/object relative pose variations allow to generate other grasp approaches, such as lateral pinch. Figure 13.2 shows some meaningful grasp strategies simulated. The weight is equal to 1Kg for all the objects. Finally, we assume a modeling of the contacts as Point Contact With Friction (PCWF) with a friction coefficient $\mu = 1.5$, minimal normal force $f_{min} = 0.1$N, structural stiffness $k_{stru} = 1$N/mm and admissible joint stiffness $k_{ss} = 200$Nmm/rad, in accordance to [15]. The simulation is intended as a free exploration of the space: the hand shape follows the closure path imposed by the first synergy and adapts around the object according to the soft synergies paradigm. The hand completes its movement when all the fingers are in contact with the object or fully closed. The information about joint angles and the position of the contact points are then stored and passed to the optimization tool for the evaluation of the optimal force distribution (Fig. 13.3).

Finding a valuable strategy for grasp success and quality in terms of contact force by changing the hand/object relative configuration represents a key topic for planning and control of robotic hands. This is particularly true considering soft adaptable robotic hands where the importance of hand/object relative pose is crucial to fully

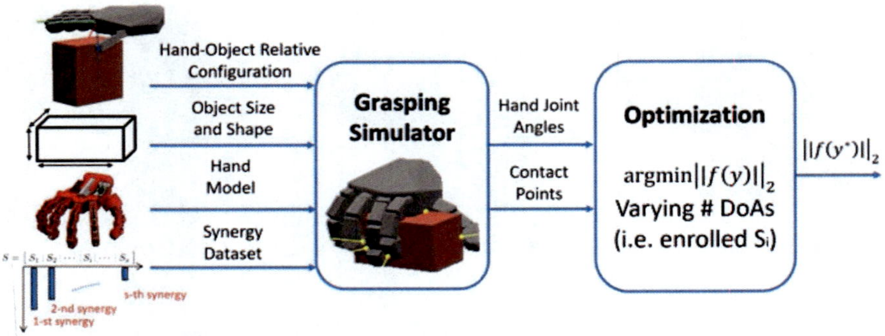

Fig. 13.3 Block scheme of the simulation and optimization procedure. Simulations take as input the synergy dataset, the hand kinematic model, geometrical properties of the object and the hand/object relative pose. The simulator outputs, i.e. joint angles and contact points, are used as argument of the optimization problem which—given a set of synergies—calculate the optimal distribution of contact forces

take advantage from end-effector adaptability in shaping around different items [7]. This motivates my investigation for different hand poses.

However, the problem of evaluating the best hand configuration is not trivial at all and computationally expensive. The principal parameters that make this issue time-consuming are: (i) the range of the evaluated relative poses, and (ii) the number of DoA enrolled for the analysis. While for (i) the best choice could be a trade-off between a reduced number of samples and an adequate span of the whole workspace, for the number of synergies considered the solution is not obvious. Indeed, the quality of the grasp generally varies w.r.t. the number of enrolled synergies. Relying on that, simulations were performed varying hand/object relative pose, in particular we considered rotations around the axis normal to the palm (abduction-adduction) and rotations around the axis along the long fingers (pronation-supination). The first DoF is spanned in the range $A = [-180, 180]^\circ$ with resolution of 5°, while the second DoF is spanned in the range $B = [-60, 60]^\circ$ with resolution of 5°. The total number of simulated hand/object poses is 1800 for each object.

In order to clarify the whole optimization procedure, we summarize it in Algorithm 3. First, an initialization procedure is required to load the object geometries, a hand kinematic model, a synergy dataset $\bar{S} \in \mathbb{R}^{n\times n}$ mapped on that model and the hand/object relative configurations, i.e. vectors $A \in \mathbb{R}^{a}$ (abduction-adduction) and $B \in \mathbb{R}^{b}$ (pronation-supination). Then, if the focus is the incrementality analysis we run Algorithm 4. This is composed of two iterative loops. In the external one we iterate on the hand/object relative configurations, simulating each pose $(\alpha, \beta)_i$ to obtain the contact points and hand joint angle values. In the internal loop we incrementally consider the columns of the synergy dataset $\tilde{S} \in \mathbb{R}^{n\times n}$. Indeed, at each step we compute the synergy enrollment set $S \in \mathbb{R}^{n\times s}$ as $S = \tilde{S}(:, 1 : s)$ (Eq. (13.4)). We then solve the optimization problem (13.10) for the considered hand pose $(\alpha, \beta)_i$ and the considered synergy enrollment set S. Otherwise, if we focus on the hierarchy analysis, we run Algorithm 5. This cycles over several permutations of the columns of the synergy dataset $\bar{S}$, running Algorithm 4 at each step. It is worth noting that the whole procedure is completely independent on the particular hand model, synergy dataset and object geometry.

Algorithm 3 Grasp Analysis

procedure INITIALIZATION
  Load Object Geometries
  Load Hand Kinematic Model
  Load Synergy Dataset Matrix $\bar{S} \in \mathbb{R}^{n \times n}$
  Define $A \in \mathbb{R}^a$ and $B \in \mathbb{R}^b$ ▷ Grasp Poses Sets
  Choose *Analysis Type*
procedure GRASP ANALYSIS
  switch *Analysis Type* **do**
    case *Incrementality*
      Compute $\tilde{S} \in \mathbb{R}^{n \times n} = \bar{S}$
      Run *Incrementality Procedure* ▷ Algorithm 4
    case *Hierarchy*
      $Z = perms([1:n])$
      Run *Hierarchy Procedure* ▷ Algorithm 5

Algorithm 4 Incrementality Procedure

procedure INCREMENTALITY PROCEDURE
  $i = 0$
  do
    Set $(\alpha, \beta)_i \in \{A, B\}$ ▷ Sample of Hand Pose
    Simulate the Grasp, i.e. Compute Contacts
    $s = 0$
    do
      $s = s + 1$
      $S = \tilde{S}(:, 1:s) \in \mathbb{R}^{n \times s}$ ▷ Eq. (13.4)
      Solve Optimization Problem ▷ Eq. (13.10)
    while $s \neq n$
    $i = i + 1$
  while $i \neq a \cdot b$

Algorithm 5 Hierarchy Procedure

procedure HIERARCHY PROCEDURE
  $j = 0$
  do
    $j = j + 1$
    Compute $\tilde{S} \in \mathbb{R}^{n \times n} = \bar{S}(:, Z(j,:))$
    Run *Incrementality Procedure* ▷ Algorithm 4
  while $j \neq size(Z, 1)$

13.4 Results and Discussion

In this section, we evaluate the grasp performances in terms of contact forces. In particular, we analyze the norm of contact forces w.r.t. the hand/object relative poses incrementally enrolling the grasp synergies and focusing on the inspection of local minima (*Analysis Type* = *Incrementality* in Algorithm 3). Secondly, we question the role that each synergy plays in success rate and grasping performances identifying the synergies that give the more dominant contribution (*Analysis Type* = *Hierarchy* in Algorithm 3).

13.4.1 Synergy Incrementality and Optimal Hand Configuration

In this first analysis, we evaluate the optimal contact forces w.r.t. the hand/object relative configurations in the case of the enrollment of one or more synergistic DoAs. In particular, we sequentially enroll the synergies from S_1 to S_{15} and evaluate the variations of the optimal grasping forces distribution. Figure 13.4 shows the results for the case of object #4 (for the sake of space only cases $\{S_1\}$, $\{S_1, S_2\}$, $\{S_1, S_2, S_3\}$ and $\{S_1, S_2, S_3, \dots S_{15}\}$ are reported). On the left, we report the force values w.r.t. the relative hand/object pose as colormap. As expected, reading the figure from top to bottom (i.e. increasing the number of enrolled synergies), the value of the contact forces decreases and the grasping success rate increases. Furthermore, results reveal that the local minima are preserved increasing the number of enrolled synergies. To clarify the latter point, we reported on the right—as an example—an exploded view of a local minimum neighborhood. This could suggest that the poses which represent local minima for the case of a reduced set of enrolled synergies seem to maintain their role increasing the number of DoA, hence remarking a local minimum also for the full actuated hand case. In order to generalize this analysis and to evaluate if this behavior is maintained for all the local minima and for all the objects, we evaluated—for all the objects—the local minima obtained with the set $\{S_1\}$, then we incremented the number of enrolled synergies, i.e. the set $\{S_1, S_2\}$, and evaluated if a local minimum of this second case exists in a neighborhood of $[-10, +10]^\circ$ both for abduction and pronation axis. This procedure can be iterated sequentially increasing the cardinality of the synergy set and evaluating the permanences between $\{S_1, \dots S_k\}$ and $\{S_1, \dots, S_k, S_{k+1}\}$, until the case $\{S_1, \dots, S_{15}\}$. Considering that the data analyzed present missing values—which correspond to no-force closure conditions—we decided to evaluate the local minima using the following policy. Given a synergy set and the correspondent contact force values w.r.t. the relative hand/object pose, we consider as local minimum candidates all the values lower than the median value. Then, for each local minimum candidate, we consider different cases: (i) if none of the poses in the considered neighborhood produces force closure, then it is assumed as a local minimum; (ii) if in the considered neighborhood there

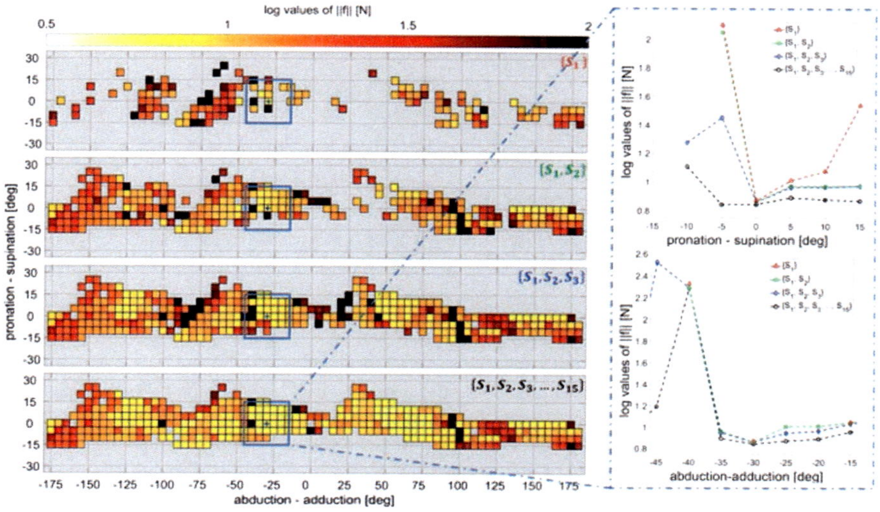

Fig. 13.4 Norm of the contact forces w.r.t. the hand/object relative abduction and pronation for the case of object #4. On the left, a colormap is used to report the log of contact force values for each 2D hand/object pose. From top to bottom, the analysis reports on an incremental enrollment of synergies. Empty cells refer to poses, which do not provide force closure with the specific synergy set. On the right, we report the exploded view of a local minimum neighborhood. Top figure shows the contact force values w.r.t. variations of hand pronation for a fixed exemplary value of abduction equal to $-30°$ (corresponding to a local minimum). Bottom figure shows the contact force values w.r.t. variations of hand abduction for a fixed value of pronation equal to $0°$. Conclusions that can be drawn are: (i) increasing the number of enrolled synergies decreases the contact force values and increases the grasping success rate; (ii) the exploded view shows that the local minimum is preserved in a neighborhood for different sets of enrolled synergies

are other relative configurations for which there is force closure but the value of the candidate is lower than the other, then it is assumed as a local minimum; (iii) if none of the previous conditions is satisfied, then the candidate is discarded. Numerical results show that the minimum preservation condition is verified in the 96% of the cases, with a minimum of 92% for the object #1 and a maximum of 99% for object #4.

13.4.2 *The Role of Synergy Hierarchy*

The conclusions drawn in the previous section assume an incremental enrollment of synergies following the order suggested by their importance for the control of hand shape described in [24]. Notwithstanding, this particular selection results in an arbitrary choice, considering that the control of the hand posture is regulated independently from the control of the contact forces. To evaluate, hence, if there is

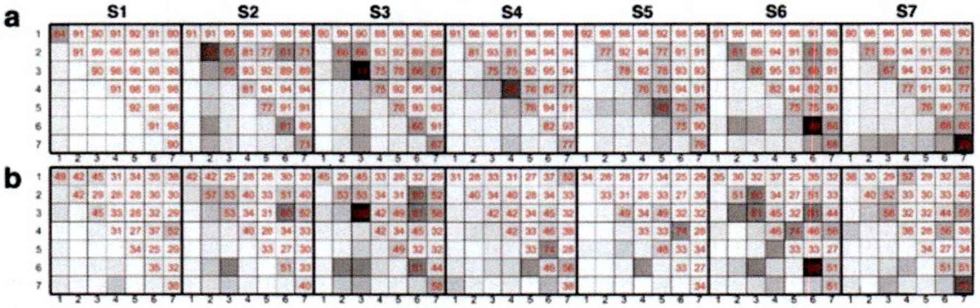

Fig. 13.5 **a** Grasping percentage, averaged for the four objects, while enrolling a 3−permutation of 7 synergies w.r.t. the hand/object relative configurations. Each 7 × 7 matrix labeled as S_k represents the set of 3−permutations starting with the synergy S_k. The element (i, j) of the S_k matrix contains the round percentage of the successful grasps for the synergies triplet $\{S_k, S_i, S_j\}$. These values are also depicted with a gray scale representation, where black represents 0% and white 100%. Note that, in the 7 × 7 matrix associated to S_k, the element (k, k) represents the percentage of successful grasps for the case of k as single DoA. The elements on the k−row, on the k−column and on the diagonal represent the percentage of successful grasps considering two synergies, i.e. S_k and one of the other six respectively. Finally, all the other elements contain the percentage of successful grasps enrolling three synergies (including S_k). **b** Optimal contact forces averaged w.r.t. all the hand/object relative configurations and the object considered. The data presentation follows the same paradigm of Fig. 5-a. Note that the matrices are symmetrical

a one-to-one mapping between the importance of synergies for hand shape and for the control of contact forces, we repeated the optimizations while randomizing the order and the number of the enrolled synergies (Fig. 13.5).

13.4.2.1 A Reduced Set of Synergies

For each hand/object relative configuration the number of combinations to be tested is equal to the sum of the the binomial coefficients $\sum_{k=1}^{N}\binom{N}{k}$. For high values of N, analysis workload would be too high, then to reduce computational complexity, we evaluated the minimum value of N that enables to approximate the grasp capacity of the full synergy set case. For the objects considered in this chapter, we obtained that with the first seven PCs the hand is able to produce force closure in all the configurations in which also the full set of synergies does (see Table 13.2). Leveraging on this result, from now on $N = 7$ will be the upper extremity of the dimension of the synergy set considered.

13.4.2.2 The Importance of Synergies in Optimal Contact Forces Distribution

To analyze the actual role that each grasp synergy plays in the minimization of contact forces we evaluated their optimal distribution while randomizing the number and the

Table 13.2 Normalized percentage of successful grasps per-object with the first k synergies

	1	2	3	4	5	6	7	8
#1	87.16	99.80	100	100	100	100	100	100
#2	62.88	93.51	99.82	100	100	100	100	100
#3	55.67	87.42	97.52	99.38	99.79	99.79	100	100
#4	50.29	83.05	97.41	100	100	100	100	100

order of synergies that are incrementally enrolled. To face this problem, leveraging on the result of the previous section, we take into consideration the first 7 synergies, which are selected in randomized sub-sets of one, two and three elements, for a total number of combination equal to 63 and, hence, a maximum number of optimization per-object equal to 113400. For each sub-set and for each object, we calculate the number of successful grasps performed and the optimal distribution of contact forces w.r.t. the hand/object relative configuration. I report the results of this analysis—in a compact form—in Fig. 13.5. In particular, in Fig. 13.5a we show the percentage of successful grasps (normalized w.r.t. the case of the full synergy set), averaged w.r.t. the four object considered. These values are depicted as a confusion matrix in which the white color is associated to 100% and the black to 0%. Each cell of the matrix contains the percentage of successful grasps of a specific triplet. The first index (k) of the triplet is identified by the index reported on the top of the matrix (S_1, S_2, ..., S_k, ..., S_7), which is in common for all the cells under the label S_k . Fixed the matrix associated to the first index of the triplet, the element (i, j) of the matrix contains the round percentage of the successful grasps for the synergies triplet $\{S_k, S_i, S_j\}$. Note that, in the 7×7 matrix associated to S_k, the element (k, k) represents the percentage of successful grasps for the case of S_k as single DoA. The elements on the $k-$row, on the $k-$column and on the diagonal represent the percentage of successful grasps considering two synergies, S_k and one of the other six respectively. Finally, all the other elements contain the percentage of successful grasps enrolling three synergies (including S_k). In Fig. 13.5b, the same representation of Fig. 13.5a is used to report the optimal contact forces distribution averaged w.r.t. all the possible hand/object relative configurations and all the object considered. Note that the matrices are symmetrical.

These results show that, in the case of single DoA, S_1 is the PC which produces force closure with the highest probability, i.e. in the 65% of the cases. Besides S_1, the PCs with the highest percentages of successful grasps are S_4 (34%) and S_5 (48%). What is also noticeable is that, for the case of two DoAs, the highest percentage of successful grasps is achieved enrolling S_1 and S_5. These values show that S_1, S_4 and S_5 are the components more related to the increasing of the percentage of grasp successful. This observation—as expected—is still true when these synergies are enrolled as second DoA of a pair or triplet. For example, for the triplets starting with S_3, it can be shown that, even if the sub-set containing only S_3 produce force closure only in the 13% of the cases, the pairs $\{S_3, S_1\}$, $\{S_3, S_4\}$ and $\{S_3, S_5\}$ produce force closure in the 90, 75 and 78% of the cases.

Besides the fundamental role that S_1 plays for the control of contact forces distribution, which was an expected result, we demonstrated that also S_4 and S_5 have a relevant role in increasing the probability to perform force closure. These roles are confirmed by the analysis of the actual optimal contact forces that the hand exert when actuated following the randomized triplets. In fact—with reference to Fig. 13.5b—it can be shown that in the case of single DoA the synergies that exert a mean lower value of contact force are, in order, S_4, S_5 and S_1. This confirms the role of S_4 and S_5 in the control of contact forces also under the point of view of forces minimization. Moreover, what can be also shown is that S_4 and S_5 are the components which minimize the averaged value of the contact forces of the couple of synergies in conjunction with S_1.

13.5 Conclusions

In this chapter, we have investigated whether an incrementality or a hierarchy for synergy enrollment exist for optimal force distribution at different hand/object relative poses. To achieve this goal we performed simulations where the model of a synergy inspired compliant hand was used to grasp different objects, and we evaluated successful grasp rate and force distribution. This analysis has been performed while testing several hand/object relative poses and synergistic pattern combinations.

Main findings of this chapter are: (i) the minima of contact forces are invariant w.r.t. an incremental enrollment of synergies within the variation of hand/object relative configurations, and, (ii) a hierarchy exists on eigenposture selection. Indeed, the first synergy seems to play the most important role to achieve successful grasps and optimized force distribution and, together with the fourth and fifth, are those mainly devoted to producing compression movements, which are coherent with force closure requirements. This suggests that hierarchy of pre-grasp related synergies can not be directly mapped to contact force distribution. This could be a springboard towards the implementation of novel analysis techniques which can take into consideration different grasp-related quantities.

Note that an experimental validation of these results could be performed employing a highly actuated robotic hand controlled using postural synergies. The control strategy should also implement the hand compliance. Moreover, tactile sensors may be used to estimate the contact forces.

I believe that, taking inspiration from the Uncontrolled Manifold Hypothesis [25], these results could pave the path towards a new generation of soft hands, embedding synergy inspiration, compliance and a minimalistic usage of resources, for more effective robotic grasp strategies. Future works will be devoted to (i) extending this analysis to more complex manipulation tasks, and (ii) involving a larger set of objects, e.g. taking inspiration from datasets available in literature.

References

1. Asada, M., Hosoda, K., Kuniyoshi, Y., Ishiguro, H., Inui, T., Yoshikawa, Y., Ogino, M., Yoshida, C.: Cognitive developmental robotics: a survey. IEEE Trans. Autonom. Mental Dev. **1**(1), 12–34 (2009)
2. Averta, G., Angelini, F., Bonilla, M., Bianchi, M., Bicchi, A.: Incrementality and hierarchies in the enrollment of multiple synergies for grasp planning. IEEE Robot. Autom. Lett. **3**(3), 2686–2693 (2018)
3. Bernstein, N.A.: The Control and Regulation of Movements (1967)
4. Berthoz, A.: Simplexity: Simplifying Principles for a Complex World (G. Weiss, Trans.). Yale University Press (2012)
5. Bicchi, A.: On the closure properties of robotic grasping. Int. J. Robot. Res. **14**(4), 319–334 (1995)
6. Bicchi, A., Gabiccini, M., Santello, M.: Modelling natural and artificial hands with synergies. Philos. Trans. R. Soc. B: Biol. Sci. **366**(1581), 3153–3161 (2011)
7. Bonilla, M., Farnioli, E., Piazza, C., Catalano, M., Grioli, G., Garabini, M., Gabiccini, M., Bicchi, A.: Grasping with soft hands. In: 2014 14th IEEE-RAS International Conference on Humanoid Robots (Humanoids), pp. 581–587. IEEE (2014). https://doi.org/10.1109/HUMANOIDS.2014.7041421
8. Buss, M., Hashimoto, H., Moore, J.B.: Dextrous hand grasping force optimization. IEEE Trans. Robot. Autom. **12**(3), 406–418 (1996)
9. Catalano, M.G., Grioli, G., Farnioli, E., Serio, A., Piazza, C., Bicchi, A.: Adaptive synergies for the design and control of the Pisa/IIT softhand. Int. J. Robot. Res. **33**(5), 768–782 (2014)
10. Cornellà, J., Suárez, R., Carloni, R., Melchiorri, C.: Dual programming based approach for optimal grasping force distribution. Mechatronics **18**(7), 348–356 (2008)
11. Cutkosky, M.R.: On grasp choice, grasp models, and the design of hands for manufacturing tasks. IEEE Trans. Robot. Autom. **5**(3), 269–279 (1989)
12. D'avella, A., Lacquaniti, F.: Control of reaching movements by muscle synergy combinations. Front. Comput. Neurosci. **7**, 42 (2013)
13. Della Santina, C., Bianchi, M., Averta, G., Ciotti, S., Arapi, V., Fani, S., Battaglia, E., Catalano, M.G., Santello, M., Bicchi, A.: Postural hand synergies during environmental constraint exploitation. Front. Neurorobot. **11**, 41 (2017)
14. Feldman, A.G., Levin, M.F.: The equilibrium-point hypothesis–past, present and future. In: Progress in Motor Control, pp. 699–726. Springer (2009)
15. Gabiccini, M., Bicchi, A., Prattichizzo, D., Malvezzi, M.: On the role of hand synergies in the optimal choice of grasping forces. Autonom. Robot. **31**(2–3), 235–252 (2011)
16. Grinyagin, I.V., Biryukova, E.V., Maier, M.A.: Kinematic and dynamic synergies of human precision-grip movements. J. Neurophysiol. **94**(4), 2284–2294 (2005)
17. Latash, M.L.: Synergy. Oxford University Press (2008). https://doi.org/10.1093/acprof:oso/9780195333169.001.0001
18. Mason, C.R., Gomez, J.E., Ebner, T.J.: Hand synergies during reach-to-grasp. J. Neurophysiol. **86**(6), 2896–2910 (2001)
19. Mussa-Ivaldi, F.A.: Modular features of motor control and learning. Curr. Opin. Neurobiol. **9**(6), 713–717 (1999). https://doi.org/10.1016/s0959-4388(99)00029-x
20. Prattichizzo, D., Trinkle, J.C.: Grasping. In: Springer Handbook of Robotics, pp. 955–988. Springer (2016)
21. Rocchi, A., Hauser, K.: A generic simulator for underactuated compliant hands. In: IEEE International Conference on Simulation, Modeling, and Programming for Autonomous Robots (SIMPAR), pp. 37–42. IEEE (2016)
22. Saltiel, P., Wyler-Duda, K., D'Avella, A., Tresch, M.C., Bizzi, E.: Muscle synergies encoded within the spinal cord: evidence from focal intraspinal NMDA iontophoresis in the frog. J. Neurophysiol. **85**(2), 605–619 (2001)

23. Santello, M., Bianchi, M., Gabiccini, M., Ricciardi, E., Salvietti, G., Prattichizzo, D., Ernst, M., Moscatelli, A., Jörntell, H., Kappers, A.M., et al.: Hand synergies: integration of robotics and neuroscience for understanding the control of biological and artificial hands. Phys. Life Rev. (2016)
24. Santello, M., Flanders, M., Soechting, J.F.: Postural hand synergies for tool use. J. Neurosci. **18**(23), 10105–10115 (1998)
25. Scholz, J.P., Schöner, G.: The uncontrolled manifold concept: identifying control variables for a functional task. Exp. Brain Res. **126**(3), 289–306 (1999). https://doi.org/10.1007/s002210050738
26. Stanton, B.F., Geme III, J.W.S., Schor, N.F., Behrman, R.E., Kliegman, R.M.: Nelson Textbook of Pediatrics, 19/e. Elsevier, India (2007)
27. Stratmann, P., Lakatos, D., Albu-Schäffer, A.: Neuromodulation and synaptic plasticity for the control of fast periodic movement: Energy efficiency in coupled compliant joints via PCA. Front. Neurorobot. **10** (2016). https://doi.org/10.3389/fnbot.2016.00002
28. Thakur, P.H., Bastian, A.J., Hsiao, S.S.: Multidigit movement synergies of the human hand in an unconstrained haptic exploration task. J. Neurosci. **28**(6), 1271–1281 (2008)
29. Todorov, E., Ghahramani, Z.: Analysis of the synergies underlying complex hand manipulation. In: 26th Annual International Conference of the IEEE, Engineering in Medicine and Biology Society, 2004. IEMBS'04, vol. 2, pp. 4637–4640. IEEE (2004)
30. Tresch, M.C., Jarc, A.: The case for and against muscle synergies. Curr. Opin. Neurobiol. **19**(6), 601–607 (2009)
31. Tresch, M.C., Saltiel, P., Bizzi, E.: The construction of movement by the spinal cord. Nat. Neurosci. **2**(2), 162 (1999)
32. Turvey, M.T.: Action and perception at the level of synergies. Hum. Mov. Sci. **26**(4), 657–697 (2007)
33. Villani, L., Ficuciello, F., Lippiello, V., Palli, G., Ruggiero, F., Siciliano, B.: Grasping and control of multi-fingered hands. In: Advanced Bimanual Manipulation, pp. 219–266. Springer (2012). https://doi.org/10.1007/978-3-642-29041-1_5
34. Vinjamuri, R., Sun, M., Chang, C.C., Lee, H.N., Sclabassi, R.J., Mao, Z.H.: Temporal postural synergies of the hand in rapid grasping tasks. IEEE Trans. Inf. Technol. Biomed. **14**(4), 986–994 (2010)
35. Zatsiorsky, V.M., Latash, M.L.: Prehension synergies. Exerc. Sport Sci. Rev. **32**(2), 75 (2004)

Chapter 14
Exploiting Principal Components for Robots Walking: An Approach for Sub-Optimal Locomotion

Recently, the method of choice to exploit robot dynamics for efficient walking is numerical optimization (NO). The main drawback in NO is the computational complexity, which strongly affects the time demand of the solution. Several strategies can be used to make the optimization more treatable and to efficiently describe the solution set. In this letter, we present an algorithm to encode effective walking references, generated offline via numerical optimization, extracting a limited number of principal components and using them as a basis of optimal motions. By combining these components, a good approximation of the optimal gaits can be generated at run time. The advantages of the presented approach are discussed, and an extensive experimental validation is carried out on a planar legged robot with elastic joints. The biped thus controlled is able to start and stop walking on a treadmill, and to control its speed dynamically as the treadmill speed changes.

14.1 Introduction

The energy efficiency of robotic locomotion is strongly related to the careful exploitation of the natural dynamics of the robots. For example, looking at passive and heavily under-actuated bipeds, a fine tuned design and control is crucial to get [4, 7, 25] highly efficient systems. However, these robots still exhibit limited flexibility if compared to fully actuated humanoids, which, on the other hand, present a high Cost of Transport (CoT) [2, 7]. Main reasons for this poor energy performance of humanoids can be ascribed to the mismatch between the complexity of robot dynamics and the over-simplified modeling (e.g. cart-table, spring loaded inverted pendulum, etc.), which is usually employed for robotic locomotion planning. References [15, 16, 23].

G. Averta, *Human-Aware Robotics: Modeling Human Motor Skills for the Design, Planning and Control of a New Generation of Robotic Devices*, Springer Tracts in Advanced Robotics 145, https://doi.org/10.1007/978-3-030-92521-5_14

Such a mismatch is even more evident in the emerging generation of elastic joint robots (e.g. [18, 29]) since the actuation compliance, purposefully introduced in the mechanical design to guarantee robustness [22], efficiency [30], peak performance [12], is usually neglected or not properly taken into account.

NO could represent a viable approach to tackle this issue and hence take advantage of the full robot dynamics for motion planning. Recently, in [14] NO was applied to plan the walking motion of a 23 Degree-of-Freedom (DoF) humanoid robot. The results show a cost of transport slightly above 1, which currently represents the best performance achievable with fully actuated humanoids. However, NO comes with a cost, which is related to time demand: e.g. the footstep optimization presented in [14] requires from 910 to 380 s. For these reasons, to successfully use NO for online applications, a suitable combination of off-line and on-line computations is required. This usually implies the off-line evaluation of a large set of trajectories and their encoding. For example, in [17] relevant features of optimal walking trajectories (e.g. cost of transport and trajectory of the center of mass) were encoded via quadratic functions of the initial state and footsteps.

In [8] the authors moved from an off-line dataset of optimal gaits, and applied supervised learning to identify state-variable feedback control policies to manage speed transition and disturbance rejection.

In [6] walking gaits were synthesized via motion primitives [27]. A similar approach is commonly used also in computer graphics [20, 26]. In [26] Principal Component Analysis (PCA) was employed to encode a large set of behaviors (from an existing motion capture database) in motion primitives to animate a 60 DoF character. In [20] reinforcement learning was adopted to synthesize locomotion controllers that correspond to specific motion skills (e.g. jumping movements extracted from a motion capture database).

In [6] parametric motion primitives were learned from a dataset of optimized trajectories. Once the motion primitives were defined, their parameters were used as optimization variables to synthesize trajectories not belonging to the original dataset.

In this chapter, we present an approach that allows to drastically reduce the space complexity of the original data set. Furthermore, it also enables the robot to update its stride on-line and to synthesize gaits not belonging to the original data set (avoiding further optimizations that are required e.g. in [6]).

More specifically, we propose a technique that leverages upon NO (see e.g. [14]), and dimensionality reduction of PCA for on-line planning of efficient locomotion of robots with elastic actuation. We built a set of optimized gaits for different speeds and swinging foot clearances. The gaits we obtained minimize robot energy consumption fulfilling both stability and dynamic feasibility requirements, while a reduced representation of the set of optimal gaits was identified through PCA. Similarly to [6], the relationship between the principal component (PC) weights and task parameters (namely, forward speed and step clearance) was approximated by a parametric representation, named Component Map (CM), while a mostly feed-forward strategy was used to preserve the mechanical compliance of the system introduced by design as in [9] – which is not generally guaranteed by learning a feedback policy (which was described e.g. in [8]).

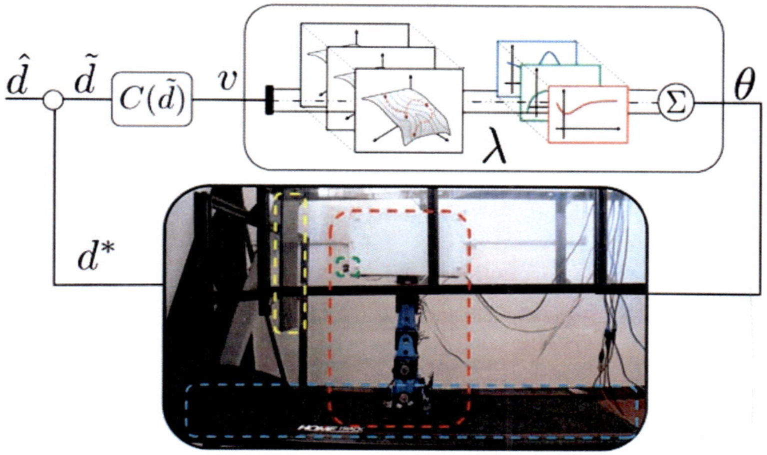

Fig. 14.1 A six DoF biped robot (in red), named *SoftLegs*, walks on a treadmill (in blue) and a target (in yellow) is placed in front of it. An infrared sensor (in green) retrieves the current distance d^* of the robot from the target. Given the desired distance $\hat{d}$, the task goal is to regulate the error $\tilde{d} = \hat{d} - d^*$ while the treadmill speed changes from 0.3 to 0.7 km/h. The proposed control approach includes a PI controller $C(\tilde{d})$ that provides as output the forward speed of the robot v. Then the principal component combination generates the desired motor trajectories θ that approximate the optimal (i.e. efficient) gait at the given speed v

One of the main contributions of this chapter is the extensive numerical and experimental validation of the method on a planar biped robot powered by compliant actuators (see Fig. 14.1) named *SoftLegs*. Numerical results show that: (i) only two PCs are needed to satisfactorily reconstruct a walk (explained variance 99%) and, (ii) CMs allow to reduce the complexity of the original data set by two orders of magnitude.

Finally, experiments demonstrated that with our techniques the robot was able to successfully start and stop walking, perform transitions between different forward speeds, and walk on a treadmill while the velocity of the belt changed (see Fig. 14.1). The content of this chapter is adapted from [11].

14.2 Problem Definition

In this section we compare the CoT given by optimized walking gaits with the CoT obtained by using a classical ZMP based method. The comparison is carried out on a six DoF planar biped (a scheme is reported in Fig. 14.2a, b)

The biped is composed of two planar three-DoF legs, connected to a pelvis on which a one-DoF trunk is mounted. Each joint of the robot is powered by a Series

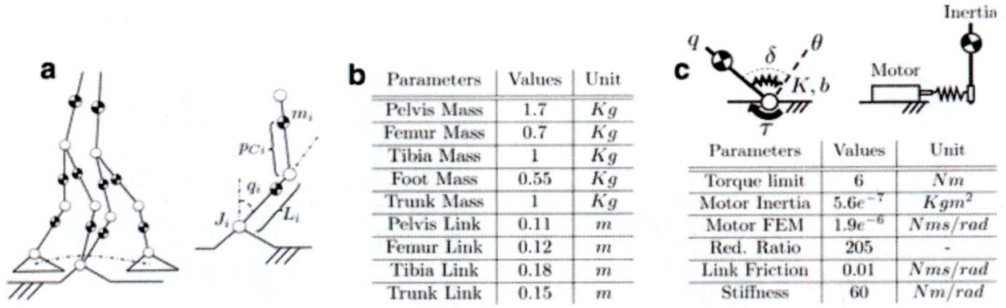

Parameters	Values	Unit
Pelvis Mass	1.7	Kg
Femur Mass	0.7	Kg
Tibia Mass	1	Kg
Foot Mass	0.55	Kg
Trunk Mass	1	Kg
Pelvis Link	0.11	m
Femur Link	0.12	m
Tibia Link	0.18	m
Trunk Link	0.15	m

Parameters	Values	Unit
Torque limit	6	Nm
Motor Inertia	$5.6e^{-7}$	Kgm^2
Motor FEM	$1.9e^{-6}$	Nms/rad
Red. Ratio	205	-
Link Friction	0.01	Nms/rad
Stiffness	60	Nm/rad

Fig. 14.2 Model of the biped: biped kinematics in (**a**), kinematic and dynamic parameters in (**b**), scheme and parameters of the SEA in (**c**). The link and the motor displacement are represented by q and θrespectively, k is the elastic transmission stiffness and b the motor damping (more details in the appendix). Quantities in table are named in analogy with their biological counterparts

Elastic Actuator (SEA) [24]. The parameters of the motor dynamics are showed in Fig. 14.2c (details in appendix).

Given desired forward speed v and swing foot clearance f_h, we derive, in accordance to [19], the robot step length:

$$s = \alpha \nu^{\beta} , \tag{14.1}$$

where $s = S/l$ is the normalized step length, l is the leg length and $\nu = v/\sqrt{lg}$ is the normalized speed. The stride parameter sets used in the study are: $f_h = [0.01, 0.03]\,\mathrm{m}$ with a granularity of 5 mm, and $v = [0.04, 0.5]\,\mathrm{m/s}$ with a granularity of 1 cm/s. These choices resulted in 485 trajectories. The upper bound of the velocity range has been chosen to include the walk-to-run transition speed[1] in order to guarantee to span dynamic gaits, while the lower bound was selected to allow a smooth transition to the standing posture. The swing-foot clearance range has been chosen to avoid compass gaits (discovered by the optimizer for $f_h = 0$) by preferring more practical gaits up to a step height equal to ten percent of the leg length.

For the sake of clarity, a detailed formulation of the optimal control problem to derive a walking gait is provided in appendix. As examples we show in Fig. 14.3 the photo sequence of two optimal walks.

The ZMP-based trajectories were determined according to [16]. Combining step parameters and centre-of-mass evolution, it is possible to obtain the link trajectories, and, through inverse dynamics, motor positions and torques to finally evaluate the CoT.

Figure 14.4a presents the values of the CoT of the two approaches over a range of forward speeds. The simulations show that the NO gaits perform at least four times better if compared to the ZMP based gaits. Figure 14.4c shows the joint torques during the walking cycle for the case with $\nu = 0.1\,\mathrm{m/s}$ and $f_h = 0.03\,\mathrm{cm}$. It is possible to

[1] §The walk-run transition speed, defined as the speed at which running becomes more convenient than walking [28], has been determined with another optimization campaign whose results is not included due to space limits.

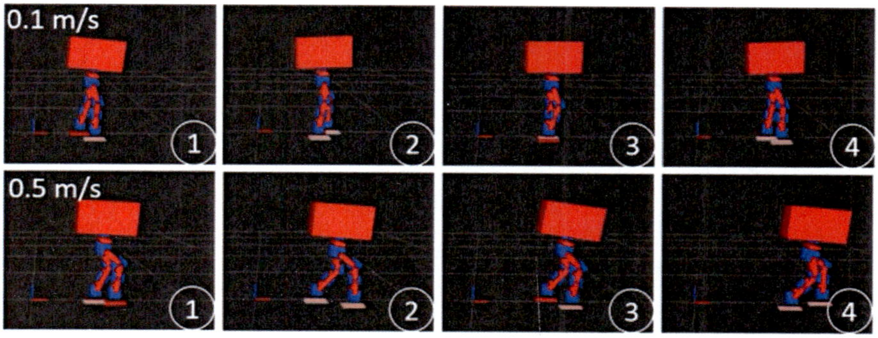

Fig. 14.3 Photo sequence of two optimal gaits for two different forward speeds: on the top $v = 0.1\,\mathrm{m/s}$ and on the bottom $v = 0.5\,m/s$. The feet are colored in white when in contact with the ground

notice that there are minor differences in the swinging leg squared torques of the two methods. Conversely the stance leg squared torques of NO are substantially lower than ZMP ones. This because NO fully exploits the leg singularity in the central part of the motion (see Fig. 14.4b). The application of NO presents one major limitation: for the majority of cases of practical interest the optimization problems are too complex to be solved at run time.

14.3 Optimal Gait Encoding via Principal Components

In this section we apply PCA to encode the data set of optimal trajectories into a small set of PCs and their weights. Then we interpolate the dependence of the weights on the walk parameters through a CM.

The starting dataset is determined as follows. Assume that: (i) the robot is composed of n_j joints, (ii) the trajectories are described by n_t sample times, (iii) the walking task is described by n_p parameters (e.g. forward speed, step height, lateral speed, etc.) and (iv) $n_s(i)$ parameter samples are needed to explore the range of the i-th parameter. Hence, the starting data set can be stored in a matrix $\Sigma \in \mathbb{R}^{m_{jt} \times m_{sp}}$, where $m_{jt} = n_j \times n_t$ and $m_{sp} = \Pi_{i=1}^{n_p} n_s(i)$. Σ has been evaluated accordingly to Sect. 14.3 ($n_p = 2$, $n_s = \{5, 95\}$). In Fig. 14.5 we report the complete set of trajectories for Hip, Knee, and Ankle used in this study. Figure 14.5 suggests a strong correlation between motor trajectories, motivating the application of motion primitives.

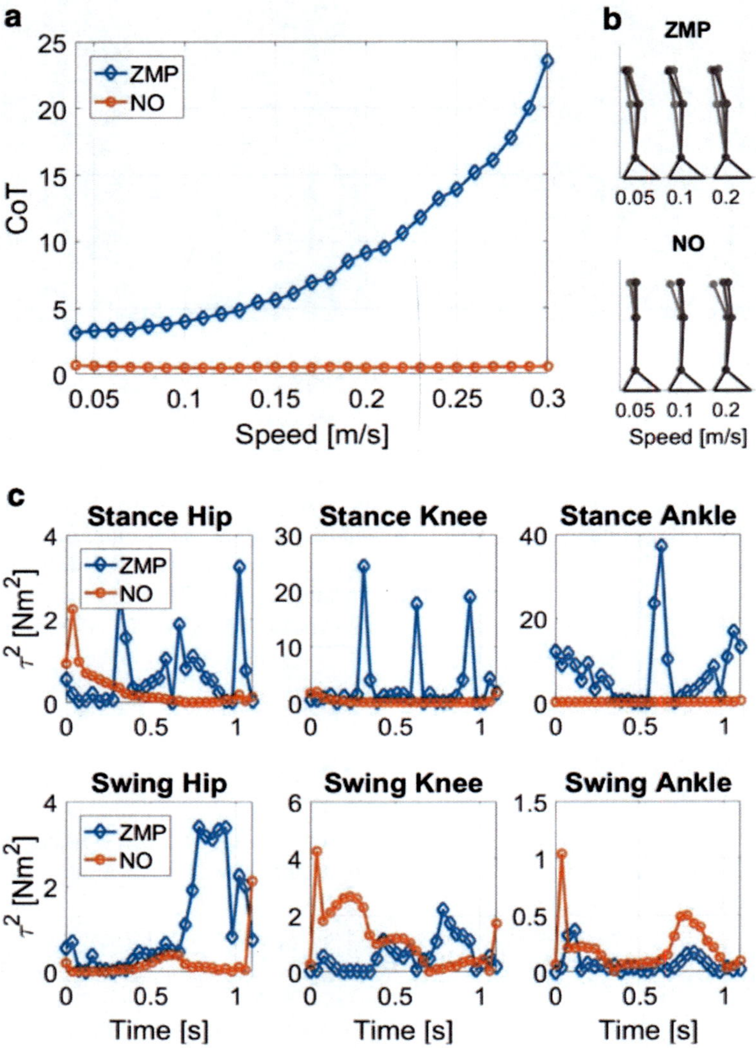

Fig. 14.4 **a** Comparison between ZMP-based and NO-based CoT at different speed averaged over the swing/foot step height. **b** Stance leg configuration at different speed. **c** Squared torque evolution of Stance (top) and Swing (bottom) leg for a ZMP-based and NO-based step for two different forward speed $v = 0.1$ m/s and $f_h = 0.03$ m

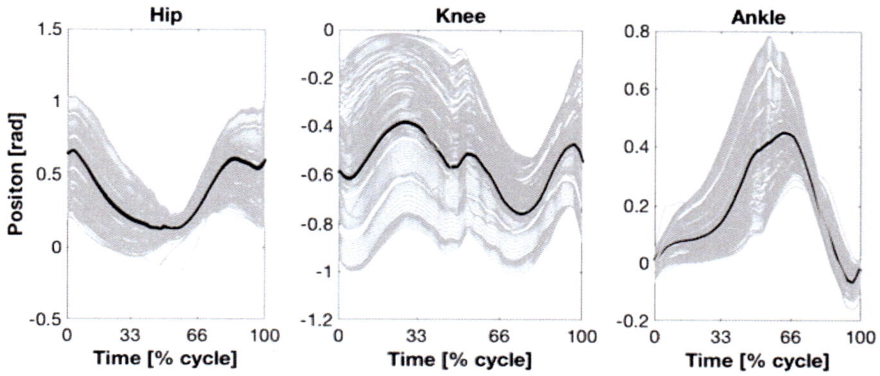

Fig. 14.5 Motor trajectories of Σ (in gray) and a representative trajectory (in black).eps

14.3.1 Principal Components Analysis

Among the many methods proposed to identify motion primitives (e.g. radial basis function, gaussian process, etc.), we decide to rely on PCA which represents an effective compromise between complexity and computational cost. PCA uses an orthogonal transformation to convert a set of possibly correlated variables, to which the mean value of each observation is subtracted, into a set of values of linearly uncorrelated components, called PCs. In order to apply the PCA to the robot trajectories we proceed as follows. Each motor trajectory $\theta_j \in \mathbf{R}^{n_t}$ of the starting set is split into two components: its mean value $\bar{\theta}_j \in \mathbf{R}$ and a null mean vector $\tilde{\theta}_j = \theta_j - \bar{\theta}_j$ for $j \in \{1, ..., n_j\}$. Hence Σ can be described by two matrices: $\Sigma_m \in \mathbf{R}^{m_{jt} \times m_{sp}}$ and $\Sigma_p \in \mathbf{R}^{n_j \times m_{sp}}$. Each column of Σ_m is a vector composed of $[\tilde{\theta}_1, ..., \tilde{\theta}_{n_j}]^T$. Each column of Σ_p is a vector composed of $[\bar{\theta}_1, ..., \bar{\theta}_{n_j}]^T$ that represents the robot mean pose (hereinafter Pose) of the corresponding movement (column) in Σ. We applied PCA to Σ_m and Σ_p for the case study of the planar biped robot with elastic joints ($n_j = 6$, $n_t = 30$). The variance explained by the PCs, reported in Table 14.1, shows that few principal components account for a large amount of variation in both Σ_m and Σ_p. This leads to hypothesize that all the trajectories of Σ can be reconstructed with high precision using only two PCs from a set of 180 PCs.

Table 14.1 Explained variance as a function of the combination of an increasing number of principal components

N PC	Σ_m	Σ_p	RMSE
1	87.04%	98.81%	0.06
2	98.75%	≃100%	0.017
3	99.43%	≃100%	0.005

Table 14.2 Space Complexity Comparison

	Σ	PCs	PCs + CMs
$O(N)$	$O(m_{jt}m_{ps})$	$O(p_{k_m}(m_{jt}+m_{ps})+ p_{k_p}(n_j+m_{ps}))$	$O(p_{k_m}(m_{jt}+d^{n_p})+ p_{k_p}(n_j+d^{n_p}))$
N	87300	2312	408

In the following we evaluate how much the space complexity of Σ can be reduced via the application of PCs. Let $O(m_{jt}m_{sp})$ and $O(n_j m_{sp})$ be the space complexities of Σ_m and Σ_p respectively. Let p_{k_m} be the chosen number of PCs to represent the robot movements, Σ_m can be encoded in the eigenvector matrix $\Sigma_{m,e} \in \mathbb{R}^{m_{jt}\times p_{k_m}}$ and in the weight matrix $\Sigma_{m,w} \in \mathbb{R}^{p_{k_m}\times m_{sp}}$. Let p_{k_p} be the chosen number of PCs to represent the robot Poses, the same procedure can be adopted for Σ_p. This results in a total complexity of $O(p_{k_m}(m_{jt}+m_{sp})+p_{k_p}(n_j+m_{ps}))$. Table 14.2 reports the data related to the case study presented in this chapter.

14.3.2 Component Maps

Since in practical cases $m_{sp} \gg m_{jt}$ it is worth to further encode the weight dependence on the task parameters. This can be achieved by exploiting second order polynomial functions. In Fig. 14.6 we present the results of the approximation of the weight dependence on v and f_h for both trajectories and Pose.

The very low values of RMSE suggest that the CMs can approximate $\Sigma_{m,w}$ and $\Sigma_{p,w}$. To quantify the level of approximation introduced by applying the CMs we evaluated the RMSE in the reconstruction of the trajectories of Σ. The results of PCs only and the combination of PCs and CMs are compared in Fig. 14.7a. Moreover, in Fig. 14.7b, we report the performance of the two approaches in the reconstruction of a novel set of optimal trajectories not included in Σ. Both the validation tests show that the reconstruction errors in case of combination of PCs and CMs are comparable to the ones in case of PCs only.

The reduction in the space complexity of $\Sigma_{m,w}$ and $\Sigma_{p,w}$ allowed by the encoding via CMs is evaluated in the following. Assuming that CMs are polynomial functions of order d in each parameter, an upper bound of the space complexity of each CM is $O(d^{n_p})$. Hence the total complexity of $\Sigma_{m,w}$ after the encoding becomes $O(p_{k_m}m_{jt}+p_{k_m}d^{n_p})$. The same holds for $\Sigma_{p,w}$. The complete complexity analysis is reported in Table 14.2.

In conclusion, both PCs and PCs plus CMs show a low RMSE when reconstructing the trajectories of Σ. Moreover, the space complexity of Σ is reduced by one order of magnitude via PCs and two orders of magnitude via PCs plus CMs.

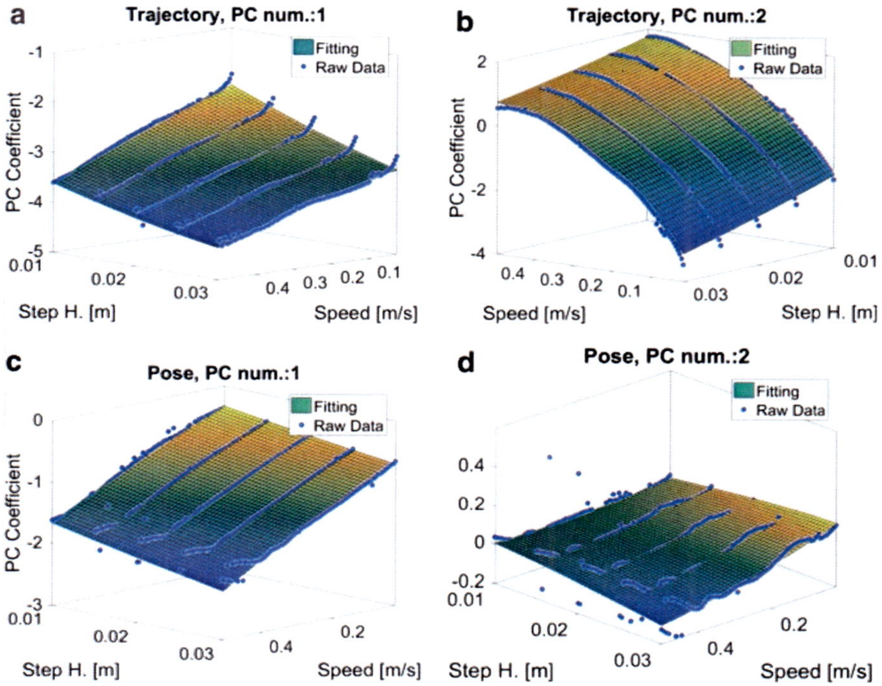

Fig. 14.6 Principal Component Mapping functions. Surfaces represent the CMs, while the points represents the actual PC weights (Raw Data). In (**a**), first PC weight of $\Sigma_{m,w}$—RMSE 0.038, in (**b**), second PC weight of $\Sigma_{m,w}$—RMSE 0.023, in (**c**), first PC weight of $\Sigma_{p,w}$—RMSE 0.043 , in (**d**), second PC weight of $\Sigma_{p,w}$—RMSE 0.035. Each CM is a second order polynomial on each stride variable. The CMs approximate worse both the pose and trajectory PCs for higher speed levels. This explains the RMSE trend in Fig. 14.7a. However, this phenomenon occurs just at the boundaries of the explored region of parameters and produces a low RMSE on the whole set

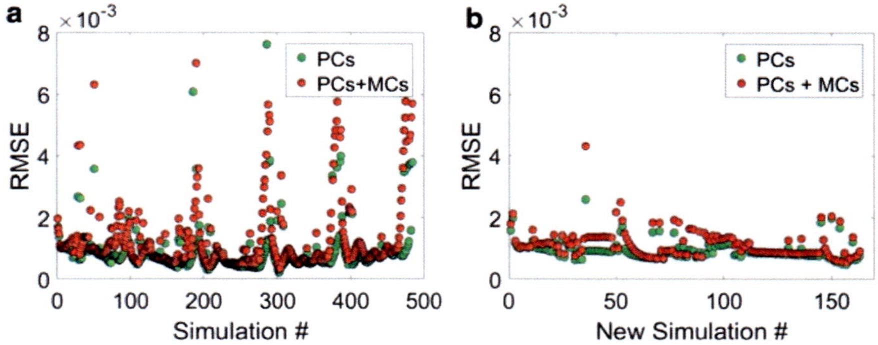

Fig. 14.7 RMSE of the reconstruction of the optimal trajectories via two PCs (green) and via the combination of PCs and CMs (red). Panel (**a**) shows the error obtained approximating Σ, while panel (**b**) shows the same for a new validation set

14.4 Experimental Validation

The proposed approach has been experimentally validated in the following cases: (i) constant speed, (ii) variable speed with off-line computation of controls via PC combination, (iii) variable speed with on-line computation of controls via PCs and CMs.

The experimental setup is shown in Fig. 14.8. Each joint of the robot is actuated by a *qbmove* Advanced (an improved version of [5]). Eight spherical bearings, rolling on two parallel walls, constrain the robot trunk in the sagittal plane. A four bar mechanism drives the ankle joints. Above the centre of rotation of each of them are located the ankle actuators. To retrieve contact forces, four Opto Force sensors [1] are mounted in the sole of each foot.

The *qbmove* actuator dynamics is modeled here as:

$$I\ddot{\theta} + b\dot{\theta} = -h(\delta, \sigma) + u\,, \tag{14.2}$$

where $\theta, \dot{\theta}, \ddot{\theta}$ are the motor position, speed and acceleration respectively, σ is the stiffness adjuster, I is the motor inertia and b is the damping coefficient. $h(\delta, \sigma)$ represents the nonlinear elastic characteristic of the actuator (see [5, 10] for more details). In the following, we consider the linearized actuator stiffness at $\bar{\delta} = 0$, i.e.

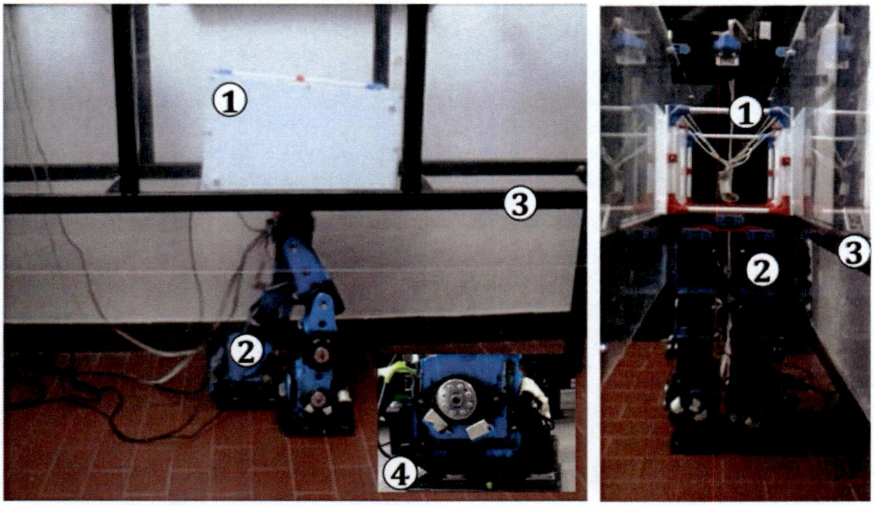

Fig. 14.8 *SoftLegs* experimental setup: side view (left), and front view (right). The system is composed of two planar legs (2) with three degrees of freedom each and a trunk mounted on the robot pelvis (1). The flat feet are equipped with four force sensors each (4). An external structure (3) constraints the robot to move in its sagittal plane

$$k = \left.\frac{\partial h(\delta, \sigma)}{\partial \delta}\right|_{\bar{\delta}=0} . \quad (14.3)$$

Given the stiffness adjuster which reproduces the stiffness used in the optimization, the motor position is set to replicate the optimal behavior returned by the numerical optimization. Reference tracking is guaranteed by a PID control on motor position, i.e. there is no direct feedback of the link position. The motor dynamic parameters are reported in Fig. 14.2c. The optimization results are concatenated coherently with the robot phase (single or double) to accomplish the locomotion task for several steps. The transition between the initial steady state configuration, and the regime is obtained with few steps generated by low pass filtering the optimal trajectories. The same procedure is used, in reverse, to stop the robot.

14.4.1 Walking with PCs—Constant Speed

In the first experiment we evaluate the walking capability of the robot driven by (i) the nominal optimal trajectories, (ii) the ones related to the first PC (for both trajectory and pose), and (iii) the ones obtained through the combination of the first and second PCs. The nominal walking conditions are generated for a forward speed of $v = 0.1$ m/s with foot clearance $f_h = 0.015$ m. Figure 14.9a presents the measured joint values of the left leg. In the case of two PCs the RMSE of the left leg joints per sample, hereinafter JRMSE, is comparable to the one obtained for the nominal case (JRMSE $= 0.0028$) which is almost the half of the one obtained in the case of application of just the first PC, (JRMSE$= 0.0056$). Moreover it is possible to argue that the obtained trajectories are dynamically consistent and that a small set of PCs can be successfully applied to reproduce a walk. Figure 14.9b shows the photo sequence of the experimental result obtained by applying two PCs.

14.4.2 Walking with PCs and with PCs and CMs—Speed Variation

In the second experiment we evaluate the applicability of the PC combination in case of speed variation. Firstly, PCs are used to compute off-line motor references for the cases $v_1 = 0.1$ and $v_2 = 0.04\,\mathrm{m/s}$ with $f_h = 0.015$ m. (Other case studies have been tested, e.g. $v_1 = 0.1,\ v_2 = 0.2\,\mathrm{m/s}$, but are not reported here due to space limitations). Then, the transition from v_1 to v_2 and viceversa is obtained simply switching PC weights and mean poses when the error between the motor references is minimum in norm. The desired and measured link positions are presented in Fig. 14.10a, b. These results show that it is possible to modify the robot speed by a proper switch of the references obtained by PC combination.

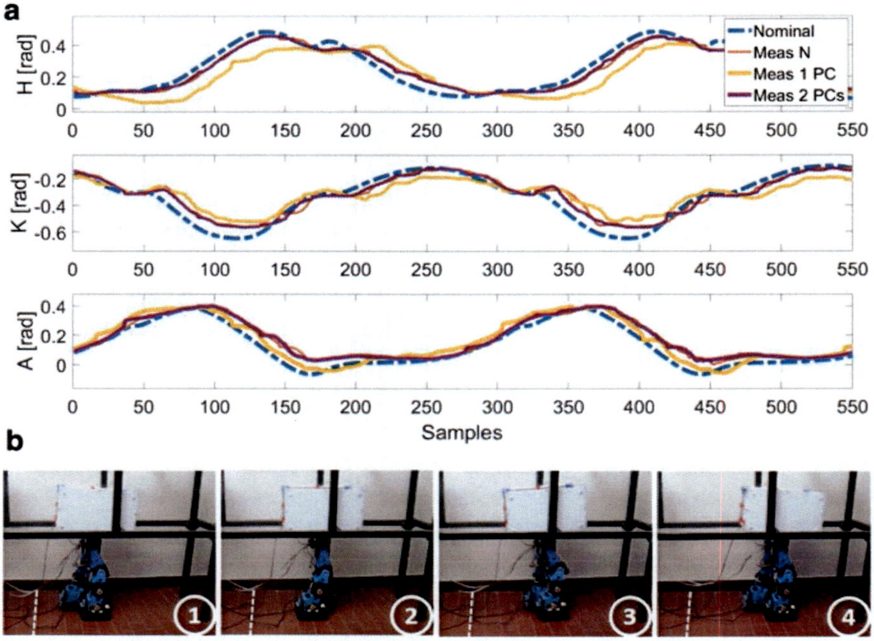

Fig. 14.9 **a** Link measurements of the left leg joint (H,K,A for Hip, Knee and Ankle) obtained by applying: optimal trajectories (Meas N), the first PC (Meas 1 PC) and the combination of the first two PCs (Meas 2 PCs). The signals are compared to the link measurements returned by the optimization (Nominal). **b** Photo sequence of the experiment for the speed $v = 0.1\,\mathrm{m/s}$ and foot clearance $f_h = 0.015\,\mathrm{m}$

In the final experiment the biped is placed on a treadmill (*Home Track HT2000* min/max speed: {0.03, 16} km/h) and an on-line control policy adjust the biped speed v in order to maintain a desired position $\hat{d} = 0.4\,\mathrm{m}$, while the treadmill velocity is varied (see the attached video). The external structure, which constrains the robot motion, is adapted to the treadmill. An infrared distance sensor, *SHARP 2y0a21*, is mounted on the biped trunk and retrieves the robot distance d^* from a target placed in front of the robot (see Fig. 14.1). When the treadmill is activated the robot speed reference $v = C(\tilde{d})$ is provided by a PI controller which uses as input the distance error $\tilde{d} = \hat{d} - d^*$. Obtained v and given f_h the CMs return the PC weight vector λ which is then used to combine the PCs and to obtain the motor references θ. These signals are interpolated and resampled at $T_s = 0.004\,\mathrm{s}$, i.e. the motor position control runs at 250 Hz while the speed control loop runs at a sample rate of 100 Hz. Control strategy, PCs and CMs, and signals management are implemented in C++/ROS and run on an Intel Core i7 ThinkPad Notebook. Figure 14.1 shows the experimental setup and control scheme while Fig. 14.11 presents the experimental results. In Fig. 14.11a d^* (*Distance* in figure) is always within a neighborhood of $\hat{d}$ throughout all the experiment, i.e. the control action successfully adapts the robot gait in response

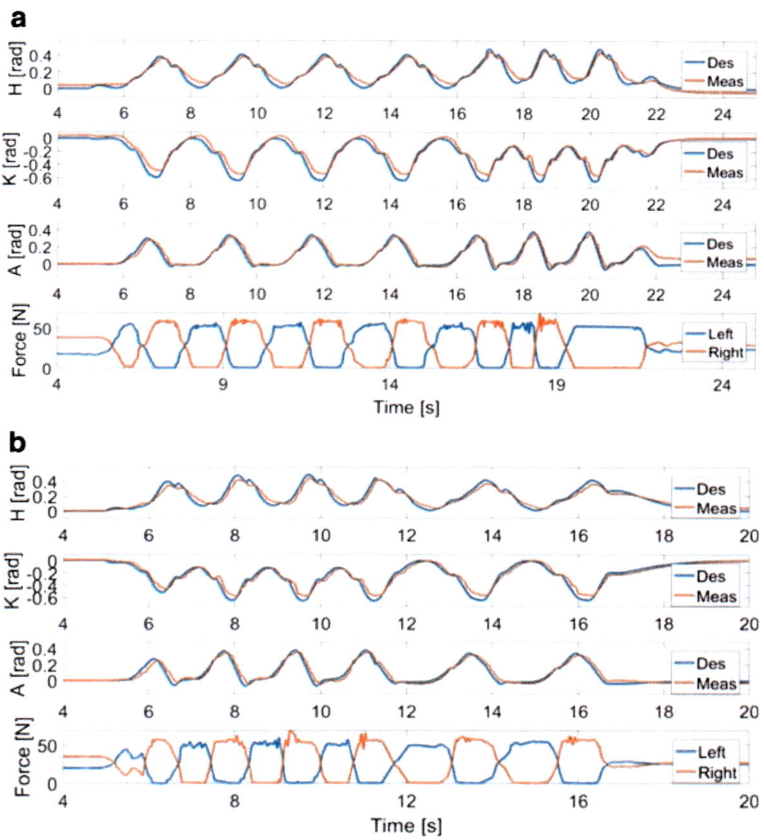

Fig. 14.10 Experimental robot behaviour during the speed transients: accelerating from 0.04 to 0.1 m/s (**a**) and decelerating from 0.1 to 0.04 m/s (**b**). Top three plots show the desired (Des) and measured (Meas) values of the Link positions of Hip (H), Knee (K), Ankle (A). Bottom plot shows the contact forces of the two feet

to the varying treadmill speed (mean distance 0.38 m, min distance 0.26 m, max distance 0.48 m). We also notice that v (*Speed* in figure) is larger than the treadmill speed (*T.Speed* in figure). This is probably due to the not negligible friction on the supporting structure and to the elasticity of the treadmill belt both not considered in the model.

Two time intervals are colored in red and in green which refer to the slowest and fastest experimental conditions respectively, i.e. [0.3,0.7] km/h. Figure 14.11b shows the desired and current motor position of the left leg joints and contact forces for the slowest speed case. Figure 14.11c presents analogous signals for the fastest speed case. The experimental result suggests that the considered transition policy and the application of PCs and CMs is indeed admissible. In this chapter we privileged the presentation of experimental validation of the trajectories at different forward

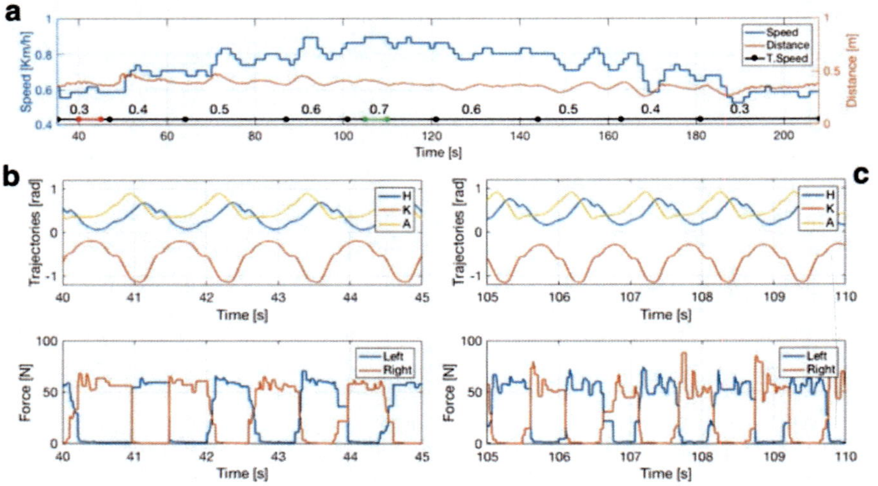

Fig. 14.11 Experimental results of the treadmill experiments. In (**a**) the robot *Distance* (in orange) is showed while the treadmill speed *T.Speed* is modified. The time line of the varying T.Speed is colored in black. The speed reference provided to the robot, *Speed*, is reported in blue in figure. On treadmill time line two intervals are colored in red and in green respectively, which refer to the slowest and fastest experimental case, i.e. 0.3km/h in (**b**) and 0.7km/h in (**c**). Of these intervals, on the top of the figure the motor positions (H,K,A for Hip, Knee and Ankle) are showed while on the bottom the contact forces (Left and Right foot) are presented

speeds over step height since this is the most important parameter toward the distance regulation test.

14.4.3 Robot Performance Evaluation

By applying PCs plus CMs the robot successfully adapted its stride to match the treadmill speed by using a reduced parameter set and approximating solutions for unknown case studies. It is worth noting that explored tasks cover both static and dynamic gaits. The sum of the norm of the inertial, Coriolis and centrifugal joint torques $||\tau_{MC}||_2 = [3.52, 7.62, 12.01, 14.37, 19.06]^T$ Nm, and the sum of the norm of the gravitational joint torques $||\tau_G||_2 = [9.99, 12.50, 14.74, 16.68, 18.55]^T$ Nm, have been evaluated for the following forward speeds $v = [0.05, 0.1, 0.15, 0.20, 0.25]$ m/s. The results show that at low speeds the gravitational torques play the most important role, while at higher speeds the dynamic torques become larger.

The stability evaluation of the approximated trajectories has not been considered in this chapter. A strategy to guarantee the fulfillment of the first-order approximation of the constraints in case of application of the approximated trajectories has been investigated in [21]. Future works will merge this idea with the approach presented in this chapter.

14.5 Conclusions

In this chapter numerical optimization has been used to obtain a data set of optimal and efficient locomotion patterns for a six-DoF biped with compliant actuation.

Then PCA and polynomial regression were applied to the trajectories belonging to the data set. This allowed to synthesize gaits that approximate the ones of the data set with a low number of parameters, facilitating the on-line generation of new trajectories.

Finally, this approach has been extensively experimentally validated on a planar biped robot with elastic joints.

Future works will be devoted to make a comparison of our approach with other state-of-art works as [6] and to exploit the reduced representation to optimize trajectories on-line, so as to react to possible external disturbances. The validation of optimal 3D efficient walk for [29] is an ongoing work.

Appendix

Let $n_a = 6$ be the number of the robot actuated joints, $q \in \mathbb{R}^{n_a}$ and $\theta \in \mathbb{R}^{n_a}$ be the link and motor position vectors respectively and $\dot{q}, \ddot{q}, \dot{\theta}, \ddot{\theta}$ their time derivatives. The output shaft of each j-th motor is connected to the j-th joint link via a viscoelastic transmission whose parameters are the stiffness k_j and the damping b_j with $j \in \{1, ..., n_a\}$.

Let $\delta \in \mathbb{R}^{n_a}$ be the deflection vector of the elastic elements and $u \in \mathbb{R}^{n_a}$ be the motor torques. Defined the stiffness, damping and motor inertia matrices $K = \text{diag}(k_j)$, $B = \text{diag}(b_j)$, $I = \text{diag}(I_j)$ with I_j the j-th motor inertia, the actuator dynamics is reported in (14.5).

Let $q_p \in \mathbb{R}^{n_{na}}$ with $n_{na} = 3$ be the vector which uniquely describes the robot position in the sagittal plane, i.e. $q_p = [p_x, p_y, \phi]$ where p_x and p_y are the translation and ϕ the rotation terms. The dynamics of the system reads:

$$M(\xi)\ddot{\xi} + n(\xi, \dot{\xi}) + \sum_{i=1}^{n_{p_i}} J_i^\top w_i = Q(k, \delta, \dot{q}) , \tag{14.4}$$

$$I\ddot{\theta} + B\dot{\theta} + K\delta = u \tag{14.5}$$

$$J_i\ddot{\xi} = \gamma_i \, , \; i = \{1, ..., n_{p_i}\} \tag{14.6}$$

where $\xi = [q_p, q]^T$ is the robot configuration vector and $\dot{\xi}$, $\ddot{\xi}$ are its time derivatives. $M(\xi)$ is the robot inertia matrix, $n(\xi, \dot{\xi})$ is the vector of nonlinear contributions due to gravity, centrifugal and Coriolis effects and the remaining terms of (14.4) describe the effects of the exerted forces and motor couplings. In particular the j-th component of the vector of generalized forces $Q(K, \delta, \dot{q})$ can be expressed as $Q_j = k_j\delta_j - f_j\dot{q}_j$ with f_j the j-th friction term in case the j-th DoF is actuated, $Q_j = 0$ otherwise.

$w_i \in \mathbb{R}^3$ is the planar external wrench acting on the i-th support foot at its contact point p_i, whose Jacobian is J_i. Finally, n_{p_i} is the number of feet that are in contact with the ground.

During the contact, neither sliding nor interpenetration between the foot and the ground are allowed, these result in holonomic constraints on the position of the foot $c_i(\xi)$. These constraints must be taken into account in the dynamics, through the additional equations in (14.6) where $J_i = \partial c_i/\partial \xi$ and $\gamma_i = -\left((\partial J_i/\partial \xi)\,\dot{\xi}\right)\dot{\xi}$.

A multiphase formulation is employed to describe the dynamics of the robot through the different contact phases.

The Optimal Control Problem (OCP) that has to be solved in order to obtain the walking trajectories can be stated as follows:

$$\min_{x(\cdot),u(\cdot),w(\cdot),t_s} \quad \int_0^{T_h} \frac{\|u(t)\|^2}{mgS} dt \tag{14.7}$$

$$\text{s.t.} \quad F_c(x(t), \dot{x}(t), w(t), u(t), K) = 0, \;\; c \in C \tag{14.8}$$

$$\Delta(x(t_s^+), x(t_s^-)) = 0 \tag{14.9}$$

$$x_{\text{red}}(0) = \Pi x_{\text{red}}(T_h) \tag{14.10}$$

$$h(x(t), u(t), w(t), f_h) \geq 0 \tag{14.11}$$

where $x = (\xi, \theta, \dot{\xi}, \dot{\theta})^\top$ indicates the state of the system.

We assume the mean speed in the locomotion direction is given, as we prescribe both the step length S and the gait period T. In (14.7), we minimize the CoT. A cost index that realistically represents the energy consumption of an electrical powered geared motor is complex as this would require the modeling of both electrical and mechanical dynamics, see e.g. [31]. To enhance the tractability of the problem, we relate the actuator energy consumption to the squared torque which, with the squared power, is one of the two cost indices most commonly used in literature (e.g. [14]). This choice appears particularly appropriate due to the characteristics of *SoftLegs* actuators, that are electrically powered and backdrivable. This makes the energy efficiency of the system strictly related to the intensity of the torques. In (14.8), the function $F_c(\cdot)$ represents the dynamics of the system during the c-th contact phase. Because of periodicity and symmetry, the set of contact phases is $C = \{1, 2\}$, where 1 and 2 represent the single and the double support phase respectively. In (14.9), the function $\Delta(\cdot)$ represents the state discontinuity at the phase change occurring at time $t_s \in [0, T_h]$ where $T_h = T/2$, caused by the impact of the floating foot with the ground. In (14.10) the gait is constrained to be periodic, by imposing the relation between $x_{\text{red}}(0)$ and $x_{\text{red}}(T_h)$. The additional inequality constraints, stated in (14.11), guarantee that the external contact wrench w belongs to the static friction cone, that the contact is unilateral and that the Center of Pressure (CoP) is inside the support area. Moreover, limits on joint positions and velocities, as well as on motor torques, are imposed. Finally, during the single support phase, a minimum foot clearance f_h is ensured between the floating foot and the ground. The OCP was solved numerically, by employing a direct collocation method [13]. Specifically, this method consists in:

(i) discretizing the time horizon into a finite sequence of time intervals, consistently parameterizing the state and control trajectories by finite sequences of optimization variables, (ii) solving the resulting finite-dimensional nonlinear program (NLP) through a Newton-based solver.

Specifically, the problem was formulated in the CasADi framework [3] and solved numerically, using the interior-point solver IPOPT [32]. 485 OCPs have been solved successfully and, despite the problem complexity, each optimization lasted on average about 56 s. This method is not free from local minima and the global optimal solution search is out of the purpose of this chapter. Future works will include strategies to avoid local minima, such as convex relaxation of the nonlinear programming problem.

References

1. Optoforce sensors. http://optoforce.com/
2. The w prize, for unprecedented speed, efficiency, and dexterity in terrestrail locomotion. http://www.wprize.org/aboutus.html
3. Andersson, J., Åkesson, J., Diehl, M.: Casadi: A symbolic package for automatic differentiation and optimal control. In: Recent Advances in Algorithmic Differentiation, pp. 297–307. Springer (2012)
4. Bhounsule, P.A., Cortell, J., Grewal, A., Hendriksen, B., Karssen, J.D., Paul, C., Ruina, A.: Low-bandwidth reflex-based control for lower power walking: 65 km on a single battery charge. Int. J. Robot. Res. **33**(10), 1305–1321 (2014)
5. Catalano, et al.: Vsa-cubebot: a modular variable stiffness platform for multiple degrees of freedom robots. In: IEEE International Conference on Robotics and Automation (ICRA), pp. 5090–5095. IEEE (2011)
6. Clever, D., Harant, M., Koch, H., Mombaur, K., Endres, D.: A novel approach for the generation of complex humanoid walking sequences based on a combination of optimal control and learning of movement primitives. Robot. Autonom. Syst. **83**, 287–298 (2016)
7. Collins, S., Ruina, A., Tedrake, R., Wisse, M.: Efficient bipedal robots based on passive-dynamic walkers. Science **307**(5712), 1082–1085 (2005)
8. Da, X., Hartley, R., Grizzle, J.W.: Supervised learning for stabilizing underactuated bipedal robot locomotion, with outdoor experiments on the wave field. In: 2017 IEEE International Conference on Robotics and Automation (ICRA), pp. 3476–3483. IEEE (2017)
9. Della Santina, C., Bianchi, M., Grioli, G., Angelini, F., Catalano, M.G., Garabini, M., Bicchi, A.: Feedback and feedforward control for soft robots. IEEE Robot. Autom. Mag.
10. Della Santina, C., Piazza, C., Gasparri, G.M., Bonilla, M., Catalano, M.G., Grioli, G., Garabini, M., Bicchi, A.: The quest for natural machine motion: an open platform to fast-prototyping articulated soft robots. IEEE Robot. Autom. Mag. **24**(1), 48–56 (2017)
11. Gasparri, G.M., Manara, S., Caporale, D., Averta, G., Bonilla, M., Marino, H., Catalano, M., Grioli, G., Bianchi, M., Bicchi, A., et al.: Efficient walking gait generation via principal component representation of optimal trajectories: application to a planar biped robot with elastic joints. IEEE Robot. Autom. Lett. **3**(3), 2299–2306 (2018)
12. Haddadin, S., Weis, M., Wolf, S., Albu-Schäffer, A.: Optimal control for maximizing link velocity of robotic variable stiffness joints. In: IFAC World Congress, pp. 6863–6871 (2011)

13. Hargraves, C.R., Paris, S.W.: Direct trajectory optimization using nonlinear programming and collocation. J. Guid. Control Dyn. **10**(4), 338–342 (1987)
14. Hereid, A., Cousineau, E.A., Hubicki, C.M., Ames, A.D.: 3d dynamic walking with underactuated humanoid robots: a direct collocation framework for optimizing hybrid zero dynamics. In: 2016 IEEE International Conference on Robotics and Automation (ICRA), pp. 1447–1454. IEEE (2016)
15. Hopkins, M.A., Hong, D.W., Leonessa, A.: Compliant locomotion using whole-body control and divergent component of motion tracking. In: 2015 IEEE International Conference on Robotics and Automation (ICRA), pp. 5726–5733. IEEE (2015)
16. Kajita, S., Kanehiro, F., Kaneko, K., Fujiwara, K., Harada, K., Yokoi, K., Hirukawa, H.: Biped walking pattern generation by using preview control of zero-moment point. In: Proceedings of the IEEE International Conference on Robotics and Automation, 2003, ICRA'03, vol. 2, pp. 1620–1626. IEEE
17. Kim, J., Pollard, N.S., Atkeson, C.G.: Quadratic encoding of optimized humanoid walking. In: 2013 IEEE-RAS International Conference on Humanoid Robots (Humanoids), pp. 300–306. IEEE
18. Knabe, C., Griffin, R., Burton, J., Cantor-Cooke, G., Dantanarayana, L., Day, G., Ebeling-Koning, O., Hahn, E., Hopkins, M., Neal, J., et al.: Team valor's escher: a novel electromechanical biped for the darpa robotics challenge. J. Field Robot. (2017)
19. Kuo, A.D.: A simple model of bipedal walking predicts the preferred speed-step length relationship. J. Biomech. Eng. **123**(3), 264–269 (2001)
20. Levine, S., Lee, Y., Koltun, V., Popović, Z.: Space-time planning with parameterized locomotion controllers. ACM Trans. Graph. (TOG) **30**(3), 23 (2011)
21. Marcucci, T., Garabini, M., Gasparri, G.M., Artoni, A., Gabiccini, M., Bicchi, A.: Parametric trajectory libraries for online motion planning with application to softrobots. In: 2017 IEEE International Conference on International Symposium on Robotics Research (ISRR)
22. Negrello, F., M., G., G., C.M., Malzahn, J., Caldwell, D.G., Bicchi, A., Tsagarakis, N.G.: A modular compliant actuator for emerging high performance and fall-resilient humanoids. In: HUMANOIDS2015, in 15th IEEE RAS Humanoids Conference (2015)
23. Pierre-Brice Wieber, R.T., Kuindersma, S.: Modeling and control of legged robots. In: Springer Handbook of Robotics (2015)
24. Pratt, G., Williamson, M.M., et al.: Series elastic actuators. In: Proceedings of the 1995 IEEE/RSJ International Conference on Intelligent Robots and Systems 95.'Human Robot Interaction and Cooperative Robots', vol. 1, pp. 399–406. IEEE (1995)
25. Ramezani, A., Hurst, J.W., Hamed, K.A., Grizzle, J.W.: Performance analysis and feedback control of atrias, a three-dimensional bipedal robot. J. Dyn. Syst. Meas. Control **136**(2), 021012 (2014)
26. Safonova, A., Hodgins, J.K., Pollard, N.S.: Synthesizing physically realistic human motion in low-dimensional, behavior-specific spaces. ACM Trans. Graph. (TOG) **23**(3), 514–521 (2004)
27. Schaal, S., Peters, J., Nakanishi, J., Ijspeert, A.: Learning movement primitives. Robot. Res. 561–572 (2005)
28. Smit-Anseeuw, N., Gleason, R., Vasudevan, R., Remy, C.D.: The energetic benefit of robotic gait selection–a case study on the robot Ramone. IEEE Robot. Autom. Lett. **2**(2), 1124–1131 (2017)
29. Tsagarakis, N.G., Caldwell, D.G., Negrello, F., Choi, W., Baccelliere, L., Loc, V., Noorden, J., Muratore, L., Margan, A., Cardellino, A., et al.: Walk-man: a high-performance humanoid platform for realistic environments. J. Field Robot. **34**(7), 1225–1259 (2017)

30. Velasco, A., Gasparri, G.M., Garabini, M., Malagia, L., Salaris, P., Bicchi, A.: Soft-actuators in cyclic motion: analytical optimization of stiffness and pre-load. In: 2013 13th IEEE-RAS International Conference on Humanoid Robots (Humanoids), pp. 354–361. IEEE (2013)
31. Verstraten, T., Beckerle, P., Furnémont, R., Mathijssen, G., Vanderborght, B., Lefeber, D.: Series and parallel elastic actuation: impact of natural dynamics on power and energy consumption. Mech. Mach. Theory **102**, 232–246 (2016)
32. Wächter, A., Biegler, L.T.: On the implementation of an interior-point filter line-search algorithm for large-scale nonlinear programming. Math. Program. **106**(1), 25–57 (2006)

Conclusions and Lessons Learned

As extensively discussed so far, all the work I carried out to produce this work is driven by a common idea, i.e. the observation and the study of humans as an extraordinary source of inspiration to produce significant advancements in multiple research fields. I have demonstrated that the study of human motion coordination is still largely unexplored, especially considering the incredible opportunities that this can provide to the development of technologies and methodologies in Robotics, Prosthetics and Rehabilitation.

The characterization of kinematic synergies at the hand and upper limb level has provided key design ideas for novel concepts of wrist prosthesis development, enabling the design of a prototype that is lighter, simpler, and more dexterous than most of the state-of-the-art solutions, still preserving very low potential market costs and simplifying the components to minimize the probabilities of system failure. At the same time, a clever exploitation of the functional characterization of human upper limb movements has provided a new way to quantitatively and effectively assess the level of impairment for stroke subjects. This may represent a major step change in Rehabilitation and impairment evaluation.

Finally, the work presented in this work provides a strong evidence regarding the significant opportunities that the study of human skills may offer for robot control and planning. Indeed, I demonstrated how functional-based motor primitives can be profitably exploited to simplify the planning problem of generic anthropomorphic manipulators. The solution I propose represents an effective approach to automatically generate complex movements for redundant robotic arms, strongly simplifying the standard problem of trajectory planning. This approach can be complemented by the usage of Machine Learning techniques to harness the astonishing and abundant information nested in complex human behaviour, and translate it for the implementation of complex manipulation tasks.

G. Averta, *Human-Aware Robotics: Modeling Human Motor Skills for the Design, Planning and Control of a New Generation of Robotic Devices*, Springer Tracts in Advanced Robotics 145, https://doi.org/10.1007/978-3-030-92521-5

To conclude, I believe that this work can be regarded as a small building block for pushing further the perspective of technologies made for humans, able to closely and actively interact with humans, because built “on” humans. And, I can guarantee, the view is marvelous and exciting.